Classification, evolution, and the nature of biology

Historically, naturalists who propose theories of evolution, including Darwin and Wallace, have always done so to explain their perception of natural classification. This book begins by exploring the intimate historical relationship between patterns of classification and patterns of phylogeny. But if the pattern of classification is the *explanandum* for phylogeny, classification and homology cannot be "evidence for evolution".

The author presents the historical development of a rigorous logical argument for evolution from other evidence. He then gives a history of methods of classification and their relationship to the reconstruction of phylogeny. The author makes the important claim that if the hierarchical pattern of classification is a real phenomenon, then biology is unique as a science in making taxonomic statements. This conclusion is reached by way of historical reviews of theories of evolutionary mechanism and the philosophy of science as applied to biology.

The book is addressed to biologists, particularly taxonomists, concerned with the history and philosophy of their subject, and to philosphers of science concerned with biology. It is also an important source on methods of classification and the logic of evolutionary theory for students and for professional biologists and palaeontologists.

Classification, evolution, and the nature of biology

ALEC L. PANCHEN
Reader in Vertebrate Zoology
University of Newcastle upon Tyne

Published by the Press Syndicate of the University of Cambridge
The Pitt Building, Trumpington Street, Cambridge CB2 1RP
40 West 20th Street, New York, NY 10011-4211, USA
10 Stamford Road, Oakleigh, Victoria 3166, Australia

First published 1992

Printed in the United States of America

Library of Congress Cataloging-in-Publication Data

Panchen, Alec L.

Classification, evolution, and the nature of biology / Alec L. Panchen

p. cm.
Includes bibliographical references and index.
ISBN 0-521-30582-9 (hardback). – ISBN 0-521-31578-6 (pbk.)
1. Biology – Classsification. 2. Evolution (Biology) 3. Biology – Philosophy. I. Title.
QH83.P35 1992
574'.012–dc20 91–26274
CIP

A catalog record for this book is available from the British Library.

ISBN 0-521-30582-9 hardback
ISBN 0-521-31578-6 paperback

To those of my former students
who learned to be sceptical

CONTENTS

ACKNOWLEDGEMENTS

The orginal stimulus to write this book came in the rather unexpected form of an invitation from the Cambridge University Press to present a book proposal. During the lengthy period of its gestation, the editorial staff at the press have accepted my excuses for delay with grace and good humour. Dr Robin Smith was the eventual friendly midwife. Most of the writing was done in the Department of Zoology (subsequently Biology) at the University of Newcastle, but during the summer of 1987 I was privileged to be a Senior Visiting Scholar at St John's College, Oxford, and enjoyed the facilities of the College, the University Museum (thanks to the good offices of Dr Tom Kemp) and the Radcliffe Science Library, while writing several chapters. Professor Wallace Arthur has read and commented on early drafts of Chapters 2, 4, 5, 6, 11, and 12; Dr Adrian Friday on Chapters 9 and 10; and Dr Henk Littlewood on Chapters 1, 13, and 14. While I have benefitted from the criticisms of these friends, it goes without saying that all errors of fact and interpretation, and all the ideas and opinions expressed, are my own.

The staff of the Audio-Visual Centre at Newcastle University helped with the preparation of the figures. Earlier drafts of a number of chapters were typed by Mrs Liz Kinghorn and at Oxford, but the onerous task of typing the whole manuscript and subsequent revisions fell to Mrs Carol Weiss. The indexes were compiled by Harriet and Nicholas Luft. I am grateful to all.

1

Introduction

> There have been many authorities who have asserted that the basis of science lies in counting or measuring, i.e. in the use of mathematics. Neither counting nor measuring can however be the most fundamental processes in our study of the material universe – before you can do either to any purpose you must first select what you propose to count or measure, which presupposes a classification.
>
> (*Crowson 1970, p. 2*)

It is perhaps an analytic statement – that is, self-evidently true – that the only way in which members of our species, *Homo sapiens,* can order their perceptions of the world and the ideas to which they give rise is to produce a classification. There is every reason to believe that this need also applies to other animals and is by no means confined to those animals closely related to man. It is literally vital to any animal that it must have a series of metaphorical compartments in which it places perceived phenomena – food, drink, shelter, danger, own species (sub-divided as parent, sib, rival, sexual partner, etc.). We also know that with greater intelligence, learning ability, and general mental flexibility, the classifications employed by higher animals, although articulated only in our own species, are more complex and subtle than those of lower animals. In humankind our knowledge is ordered in many precise, ranked classifications, although the individual person is unlikely to be aware of the structure of the classification he or she is using.

In the simple example above, two ranks are specified. "Own species" is of a higher rank than those items included within it ("parent, sib, rival, sexual partner . . . "). But it is also easy to see that "own species" could be included with other items to compose a *taxon* of higher rank. That higher taxon could be "all living things", or, more limited but precise, "all living things of about my size/shape/colour/pattern/behaviour". Furthermore, any item in the lower rank could be subdivided into two or more taxa at a lower rank still – for example, "rival whom I can intimidate" *versus* "rival who can intimidate me". One can now see that it is possible to draw a diagram of the pattern of such a classification, a *dendrogram,* with a number of ranks each at a different level and with the taxa at each rank grouped together to compose a smaller number of taxa at the next higher rank. The pattern of the dendrogram would look like that of a human family tree of the sort that shows all the descendants of a single patriarch, but whereas the latter is a pedigree representing a sequence of generations in time, the classification is a static *inclusive hierarchy*.

In my discussion of the classificatory hierarchy, I used the word "taxon" (plural "taxa"). Strictly speaking, this term should be confined to a particular type of classification, the systematic classification of organisms – animals, plants, fungi, and so on. Organisms can of course be classified in a number of ways – by size, by their ecology, or by their use or danger to man: the word "fish", as in food sold by a fishmonger, means something very different in the English language from "fish" (the obsolete class Pisces) as used by a professional taxonomist. The job of the taxonomist is, amongst other things, to produce systematic classifications of groups of organisms, or, in other words, to elucidate their relationships. If that taxonomist believes that he or she is literally producing a pattern of relationships rather than simply a taxonomic hierarchy, then this signals an acceptance of the theory that evolution has occurred. A theory of evolution, in the broadest sense, is a theory that the apparent relationships of classification are real, indicating community of descent.

I have been giving zoology tutorials in the University of Newcastle upon Tyne for nearly thirty-five years. Until recently, I often began one of the first meetings of the year by asking students the question, What is the theory of evolution? Now I tend to *tell* them rather than *ask* them, for in all those years I have never had an immediate and satisfactory reply. My experience was the same when I was giving undergraduate courses as a visiting professor in both the United States and Australia. In every case I was dealing with intelligent students who had been given a grounding in biology at school and, in most cases, for at least a year at university. It is only fair to add that generally we came up with a satisfactory formulation after a short discussion, but the initial lack of response, apart from being dispiriting, strikes me as remarkable – remarkable because evolution is supposed to be the foundation on which all biological science is built, and most academic biologists at least pay lip-service to its importance. But the truth is that little if anything of evolutionary theory is taught at school, at least in Britain and for different reasons in North America; and in many British universities, particularly those with schools of biology dominated by the biochemical–molecular end of the spectrum of levels of analysis, the treatment is so perfunctory as to be contemptible.

So students cannot characterise the theory of evolution. Admittedly some will give a (usually erroneous) account of the theory of Natural Selection, which they will associate (correctly) with the name of Charles Darwin and, much less often, also with that of Alfred Russel Wallace. Natural selection is one component of evolutionary theory as proposed by Darwin and Wallace, but the other, for which selection is merely a hypothesis of mechanism, is the theory that evolution has occurred. But that theory must have been proposed to explain some body of data and/or lower-level theories. What I wanted to hear from my students was what that corpus of knowledge was. The answer, as we have seen, is that the theory of evolution states that the apparent relationships of organisms in a systematic classification are *real* relationships, because "relationship" in such a classification is not a metaphor but is actually to be ascribed to community of descent. Thus the theory of evolution was proposed by Darwin and Wallace to explain the pattern of relationships in what we may now term "Natural Classification".

The purpose of this book is, firstly, to deal with the complex relationship between the concepts of natural classification and evolution, or "transmutation" as Darwin and Wallace and their forerunners would originally have called it. In a broader sense I shall be dealing with what scientists and philosophers saw (and see) as the natural order of living things and the way in which they explain that order. Darwin and Wallace saw the order as an irregular hierarchical pattern, but they were not the first evolutionists. Other patterns were proposed and sometimes explained as caused by transmutation. Nor was it always taken *a priori* that the correct pattern of classification should be the same as the natural order of organisms. We must therefore explore both.

But if the theory that evolution has occurred was proposed to explain the phenomenon of natural classification, the latter, in logic, cannot be "evidence for evolution"; so we must ask whether there is any evidence that is not merely composed of taxonomic data.

The enormous diversity of organisms on Earth is unique beyond all reasonable doubt. Furthermore, the differences among individual organisms, among species of organisms, and among the taxa at every rank into which those species can be grouped must, if we accept that evolution has occurred, be based on *contingent* properties of the organisms that form the basis of the hierarchy. One must therefore ask whether there are any general statements that can be made about living things. One view of the proper business of scientists is that it should be to produce statements of ever increasing generality until a series of statements that are asserted to be universally true is produced. This view of science is claimed to be correct by many physical scientists, by some biologists who suffer from "physics envy", and by those philosophers of science for whom physics is the paradigm science. In the latter part of this book, I shall be asking if there are, or possibly can be, any universal statements in biology.

Tackling such a fundamental question probably seems like *hubris* of a high order. Because of this I have felt constrained to present the evidence on which my opinions are based in some detail, but I hope this detail may be of use to the reader in other ways. Thus I give an account of methods of classification and phylogeny reconstruction which I hope will be accessible to the layperson but still sufficiently up-to-date, detailed, and rigorous to be useful to students of biology and professional colleagues. I have also tried to give each subject I discuss a historical basis, and, while I would not claim that any part of the book is a work of historical scholarship, I have attempted to play fair by making it clear, but without overt statement, when I have used primary and when secondary historical sources.

But mostly the book is meant to be one long logical argument. Thus in the spirit of those scholars, particularly professional philosophers, who summarise their work in their introduction in terms of "I will first show . . . I will then go on to demonstrate . . . ", I now present an abstract of that argument, itemised as the content of individual chapters.

Thus in Chapter 2: Until early in the nineteenth century, the dominant idea of the natural order was that of a *scala naturae,* an unbroken sequence from the most primitive to the most advanced organisms. Various other patterns were suggested, but the natural order was not accepted as an irregular hierarchy until the time of Darwin and Wallace, despite the fact that the pattern of classification was accepted as hierarchical long before. Subsequent classifications usually claimed to be phylogenetic and emphasised either grade, as in the *scala naturae,* or the pattern of branching. Two recent techniques, phenetics and cladistics, yield dichotomous dendrograms, but the latter is based on a hierarchy of characters, the former on aggregate similarity.

In Chapter 3: Theories of evolution were proposed to explain the authors' perceptions of the natural order. At the beginning of the nineteenth century, Lamarck's original theory, subsequently modified, was formulated to explain the *scala naturae.* Darwin and Wallace emphasised the contingent nature of the hierarchy, but later authors revived the *scala* by accepting an evolutionary pattern of classification similar to the ancient "Tree of Porphyry" pattern of classifications. The *scala* was also present in the search of evolutionists for ancestor-descendant sequences, but cladistics took the hierarchical pattern *a priori,* originally as representing phylogeny but latterly as a pattern to be explained by phylogeny.

In Chapter 4: If the pattern of classification is logically prior to phylogeny, the characters on which it is based should have logical priority over the pattern. There should therefore be a Natural Hierarchy of characters, whose similarity in all the members of a taxon is recognised as *homology.* Diagnostic homologies are *taxic homologies,* whereas characters incongruent with the pattern are *homoplastic.* These concepts pre-date proposals of phylogeny and are based on comparative anatomy, as is that of transformational homology. Homology is not evidence for evolution, but the existence of vestigial organs is.

In Chapter 5: Both geology and biogeography yield evidence for evolution that is independent of the pattern of classification. Stratigraphic palaeontology corroborates a picture of "progression" through geological time which cannot be explained by successive catastrophic extinction and creation: evolution is the only rational answer. This is reinforced by chronoclines of transformational homology and of species within major taxa. The patterns of geographical distribution of animals cannot be explained by adaptation to environmental conditions alone; the biogeographical history of organisms must be invoked together with evolution. The cladistic technique of vicariance biogeography reconstructs patterns and sequences of geological events, not of phylogeny. Thus it is not logically prior to phylogeny, but is powerful evidence for evolution.

In Chapter 6: Techniques of classification date back to the logical division of Plato and Aristotle, later to appear as the tree of Porphyry. The same methods were used by Linnaeus but were regarded as a means of summarising knowledge rather than representing the natural order. The two aims were reconciled by the time of Darwin's *Origin of Species,* but post-Darwinian taxonomy remained largely *ad hoc,* with contradictions between the pattern of classification and that of phylogeny arising from the different emphases given to phyletic evolution (and thus grade), to cladogenesis (the pattern of splitting in evolution), and also to pre-Darwinian tradition.

In Chapter 7: Phenetics and cladistics were rival techniques developed after the middle of this century; their methods are described in some detail. Phenetics claims objectivity and freedom from any theoretical presumptions. Aggregate differences among taxa are represented as distances in character hyperspace, and a hierarchy is constructed from the clusters so formed. There is a null hyothesis of no hierarchical structure in the data. Cladistics (originally "Phylogenetic Systematics") was introduced by Hennig as a technique based on the pattern of speciation, division in evolution of one species into two or more, but with the *a priori* assumption of a hierarchical pattern of taxic homologies. Like phenetics it aims to produce dichotomous dendrograms, but these are based on unique characters at every rank ("shared derived characters"). Thus characters have to be "polarised" as unique to a given taxon or characterising a taxon of higher rank – originally by "out-group comparison". Cladograms are "Steiner trees" with real "terminal taxa" (those to be classified) only at the tips of the branches; as phylogeny the internal nodes can only represent hypothetical ancestors. Thus real fossils cannot be recognised as ancestors by cladists.

In Chapter 8: "Transformed cladistics" was developed from phylogenetics by dropping the assumption that speciation yields the branch points in the cladogram, thus restoring the priority of classification to evolution: the pattern of natural classification is the *explanandum* of which evolution is the *explanans*. But this leaves the *a priori* assumption of a hierarchy of characters with no justification. Transformed cladists polarise characters using patterns of ontogeny (individual development) where these are available; this is compared with out-group comparison. Otherwise their methods are similar to those of phylogenetics, but with no assumption of a phylogeny, the distinction between homology and homoplasy is vital to the *a priori* hierarchy. The distinction is made by the use of parsimony. Of three taxa, those two with a majority of shared unique characters are "sister-groups"; characters suggesting any other pairing are "mistakes".

In Chapter 9: "Numerical cladistics" and techniques of phylogeny reconstruction based on biochemical and molecular data developed concurrently with transformed cladistics but emphasised rather than rejected the presumption of phylogeny. Numerical cladistics, which diverged from phenetics and converged on cladistics, is described in detail. Like cladistics it polarises characters and uses parsimony, but to minimise branch lengths in the dendrogram representing total character difference among taxa. But early molecular techniques, notably immunology and DNA hybridisation, produced only *distance data* comparable to the calculated distances in phenetics. Comparing sequences of amino acids between homologous proteins of different organisms or bases in DNA does, however, yield unit character differences, but, given the limited possible number of acids or bases, they are not unique characters. Techniques of aggregate similarity (as in phenetics), parsimony, and "likelihood" (based on models of random change) are rivals for resolving them.

In Chapter 10: The merits of these latter techniques and that of compatibility analysis, described previously, are discussed with respect to both molecular taxonomy and taxonomy in general. It is concluded that *if* the hierarchical pattern of the natural order of organisms, and the hierarchy of characters on which it is based, are accepted *a priori, then* cladistics is the only valid technique for discovering that order. This re-establishes the logical priority of classification to phylogeny. But there is no extrinsic method of establishing that the natural order of organisms is a divergent hierarchy, and empirical evidence from hybrid species and the difficulty of classifying animal taxa of high rank suggest rejection of the universal hierarchy. Without it evolutionary theory loses its *explanandum*. There is no solution to this paradox. Furthermore, if all the features of organisms result from their individual history, one can ask whether there are any other universal statements in biology.

In Chapter 11: Perhaps a true hypothesis of the mechanism of evolutionary change might be such a statement. A history of such theories of mechanism is presented. They comprise internal factors, producing bodily change in organisms, and external factors to which organisms respond so as to produce adaptation to the environment. Lamarck's theory, and that of many post-Darwinian naturalists, proposed directional (orthogenetic) internal factors and the inheritance of responses to the environment. Darwin and Wallace proposed "random" individual changes (of unknown causality) and the differential inheritance of those better adapted (*Natural Selection*). It is shown that the theory of natural selection is not "tautologous" but has empirical content. The development of genetics in the twentieth century added the "missing ingredient" to Darwin's theory but at first suggested a rival theory of evolution by "saltation".

In Chapter 12: The reconciliation came with the development of population genetics, but by concentrating on changes of gene frequency in populations, the "Synthetic Theory" is both tautologous and can make no predictions about emergent properties in evolution beyond the species level. Other criticisms are that adaptation by selection is taken as *a priori* for all characters of an organism, and that insufficient attention is paid to systematic constraints in development of the characters on which selection is supposed to act.

In Chapter 13: Two problems have dogged epistemology (theory of knowledge) since the time of the ancient Greeks; they are causality and induction. The problem of induction is that, despite the claims of Logical Positivism, there is no logical way of proceeding from individual scientific data to universal laws. But the alternative of testing such laws, proposed *ad hoc,* by falsification is also flawed. Laws (or theories) may succeed one another sequentially, or replace one another for sociological reasons. It is even doubtful whether universal laws play much part in the progress of science.

In Chapter 14: A distinction can be made in the philosophy of science between Natural History and Natural Philosophy, in which the former uses general principles to explain particular phenomena. Laws, theories, and other empirical generalisations occur in all sciences, but comparison of the theories of plate tectonics and evolution, both theories in Natural History, suggest a special status for the taxonomic hierarchy, the *explanandum* for evolutionary theory. The special status of *taxonomic statements* arises from this but poses questions about the philosophical status of the entities classified: species properly defined are individuals, as are higher taxa to the phylogenetic cladist; thus their taxonomic hierarchy must be a unique and contingent entity. But all taxa in transformed cladistics are classes and logically prior to phylogeny. Cladistics should use *methodological* essentialism to construct cladograms, which may subsequently be interpreted as phylogeny and tested as classifications. Laws in biology concern classes of entities such as taxonomic categories, but predictive generalisations about individuals are taxonomic statements.

2

Patterns of classification

It is taken for granted today, at least by zoologists, that systematic classifications of organisms can be represented by branching diagrams (dendrograms: Mayr, Linsley, and Usinger 1953) that represent hierarchical arrangements – Darwin's (1859) "groups within groups". The nested groups are taxa, each of which belongs to a category that represents its level in the hierarchy (Simpson 1961a). In the tenth edition of Linnaeus's *Systema Naturae* (1758), he proposed the following categories: *Regnum* (Kingdom), *Classis, Ordo, Genus, Species,* to which the categories Phylum and Family were added later. All the taxa at the same level in the hierarchy occupy the same *rank* and are given the same category. Thus "the rank of a taxon is that of the category of which it is a member" (Simpson 1961a). Modern biological classification is therefore a process of "ordinally stratified hierarchical clustering" (Jardine and Sibson 1971, p. 127), and the result is an *aggregational hierarchy* (Mayr 1982, pp. 64–6) in which the units, usually species, which constitute its lowest rank, are aggregated in successively higher ranks. The hierarchy is also an *inclusive* one (Mayr 1982, pp. 205–8) as opposed to an *exclusive* one:

> Military ranks from private, corporal, sergeant, lieutenant, captain up to general are a typical example of an exclusive hierarchy. A lower rank is not a subdivision of a higher rank; thus lieutenants are not a subdivision of captains. The *scala naturae* . . . is another good example of an exclusive hierarchy. Each level of perfection was considered an advance (or degradation) from the

next lower (or higher) level in the hierarchy, but did not include it. [See below and Rieppel 1988.]

Two other features of the taxonomic hierarchy are important: first, the hierarchy is *divergent,* so that a taxon of specific rank belongs only to one taxon of higher rank (genus or above); second, the hierarchy is usually *irregular:* it is not expected that the whole will necessarily have a fixed symmetrical pattern. Thus when represented by a dendrogram, the treelike pattern will not have branches that rejoin after separating, and the tree will also be irregular.

Most taxonomists today regard the "correct" pattern of classification as in some way "natural", although they differ in the meaning of the phrase "natural classification". But it has not always been the case that those classifying animals and plants have aspired to a natural *pattern.* Acceptance, however, that there was such a thing was necessary before a satisfactory theory of evolution could be proposed. In this chapter we explore the development of patterns of classification and their authors' attitude to them.

I. The *scala naturae*

One of the most pervasive ideas in Western civilisation is that of the *scala naturae* or "Great Chain of Being" as it is known in Lovejoy's (1936) famous discursive essay on the subject. Two ideas are inherent in the concept of the Great Chain of Being. The first, termed by Lovejoy the principle of plenitude, is usually attributed to Plato; the second, that of a linear series, to Aristotle and the Neoplatonists, notably Plotinus. The principle of plenitude arises out of Plato's philosophical idealism in which all material phenomena are imperfect representations of an unchanging world of "Ideas". One of the attributes that contributes to the perfection of that world is that it contains no gaps. Correspondingly, there are no gaps in the phenomenal world which is its imperfect representation: everything that could *possibly* exist does exist – somewhere. Both concepts persisted into the era of the rise of modern science. That of plenitude appears in *The Ethics* of Spinoza (1677), Locke's *Essay Concerning Human Understanding* (1690), and most notably, in the philosophy of Liebniz, in each case associated with the concept of the *scala naturae*. However, the *scala naturae* was not just a plaything of philosophers. Probably its best known advocate was the Swiss naturalist Charles Bonnet. His *scala* (as an *Échelle des Êtres*) is a long list of organisms "degenerating" to inanimate entities at the bottom and ascending to man as the topmost organism, but continuing upwards through angels and archangels, and so on, into the spiritual realm. The natural part of the *Échelle,* spelt out only in words by Bonnet in his *Contemplation de la Nature* (1764), is presented as a diagram by Ritterbush (1964) (see Fig. 2.1).

G. L. Leclerc, Comte de Buffon, in his massive *Histoire Naturelle* (1749–1804), was also an advocate of the principle of plenitude, but students differ as to his attitude toward the *scala naturae* (Winsor 1976). His views are set out in the first of the forty-four volumes, and his advocacy of plenitude forms a background to all his work. He was like advocates of the *scala* in asserting that man is a part of the system rather than a thing apart from other organisms. Buffon resolutely declined to attempt any classification. He gave grudging recognition to the reality of species but would not even essay an artificial classification beyond the species level. However, advocates of the *scala naturae* did in fact construct hierarchical classifications. We shall see this in the next section in the case of Linnaeus; for its justification we can turn to the first great scientific (or at least semi-scientific) evolutionist, Lamarck.

Lamarck distinguished two distinct operations in his methodology of animal classification: first, *distribution générale* and second, *classification* (*sensu* Lamarck). The first is the allocation of animal species to their correct place on the *scala naturae;* the second is the drawing of divisions of the length of the *scala* to yield a convenient clustering of taxa for a hierarchical classification. The second was regarded as a purely artificial operation. The anatomical gaps between different classes were demonstrations of our ignorance of connecting forms and similarly for orders, families (used by Lamarck but not Linnaeus), genera, and species (Lamarck 1809, Part 1). As we shall see, Linnaeus's view was in essence similar, at least in his earlier work.

Fig. 2.1. *Idée d'une échelle des êtres naturelles*. (Diagram from Ritterbush 1964, after the written list of Bonnet 1764.)

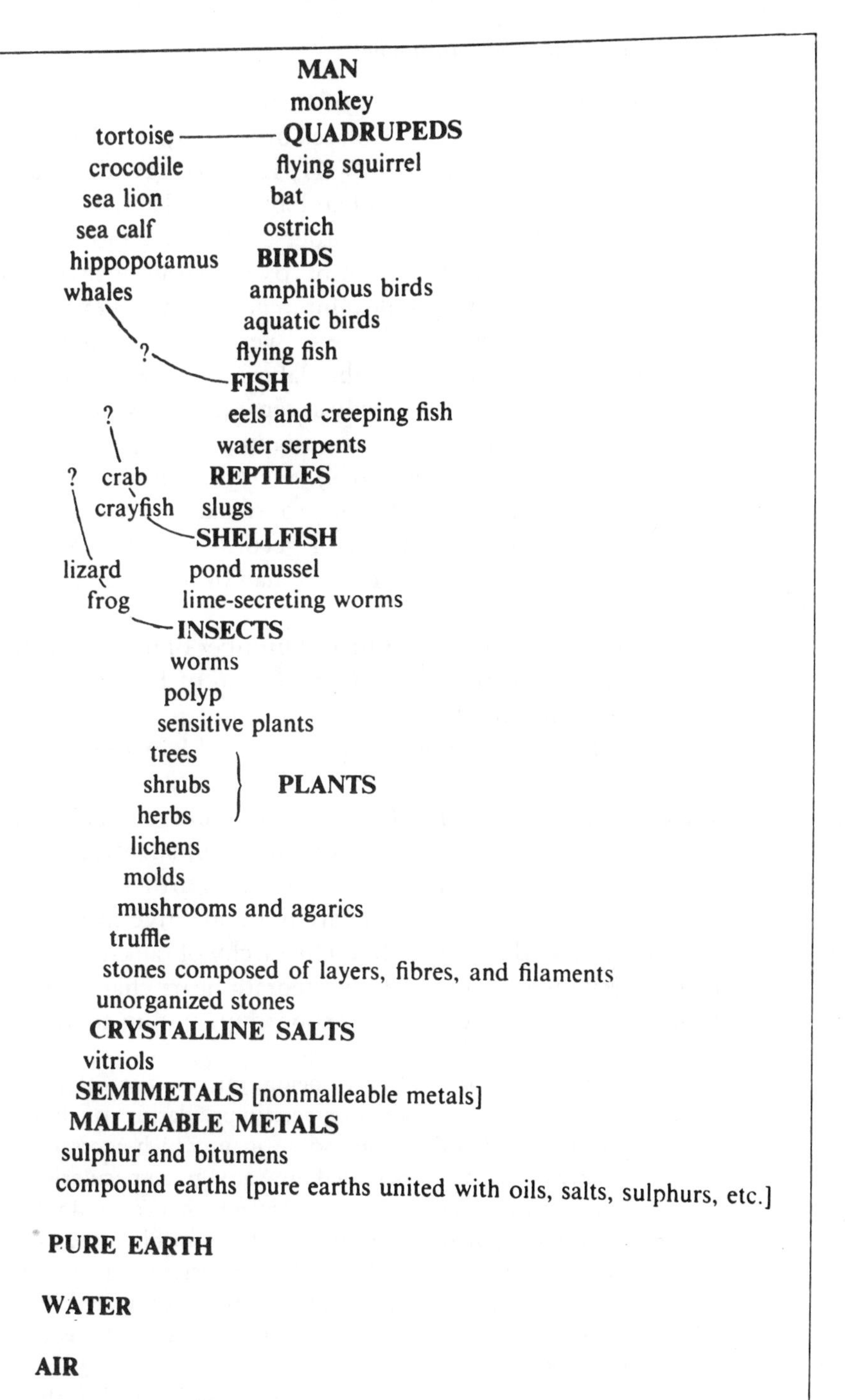

MAN
monkey
tortoise
QUADRUPEDS
crocodile
flying squirrel
sea lion
bat
sea calf
ostrich
hippopotamus
BIRDS
whales
amphibious birds
aquatic birds
?
flying fish
FISH
?
eels and creeping fish
water serpents
?
crab
REPTILES
crayfish
slugs
SHELLFISH
lizard
pond mussel
frog
lime-secreting worms
INSECTS
worms
polyp
sensitive plants
trees
shrubs
herbs
PLANTS
lichens
molds
mushrooms and agarics
truffle
stones composed of layers, fibres, and filaments
unorganized stones
CRYSTALLINE SALTS
vitriols
SEMIMETALS [nonmalleable metals]
MALLEABLE METALS
sulphur and bitumens
compound earths [pure earths united with oils, salts, sulphurs, etc.]
PURE EARTH
WATER
AIR
ETHEREAL MATTER

The idea of the Great Chain of Being was carried into the nineteenth century not only by Lamarck but also by the German school of *Naturphilosophie.* Oken (1809–11, 1839–41) saw the chain as the pattern of cosmic development and introduced a new element, the parallelism between the linear order of the *scala* and the mode of development of (vertebrate) embryos which he insisted climbed the *scala* in their individual development, an early version of Haeckel's *Biogenetic Law.* Oken's views were turned into a more specific theory of evolution, with the elaboration of a threefold parallel – *scala,* ontogeny, and phylogeny as represented by the fossil record – by Robert Chambers in his anonymous work *Vestiges of the Natural History of Creation* (1844). Here Chambers was building on the non-evolutionary idea of progression (particularly of successive vertebrates) in the fossil record, which was common amongst palaeontologists in the first half of the nineteenth century (Bowler 1976).

Early in the nineteenth century, a number of naturalists revolted against the dominance of the Great Chain of Being to assert that organisms, particularly animals, could not possibly be arranged naturally in a linear series. This was true of Lamarck's later work (Chapter 3, Section I), but also of Lamarck's younger colleague and dominant rival, Baron Cuvier. According to Cuvier, the characters by which animals are classified are all adaptive; there is no underlying organising principle of a *scala.* Furthermore, the characters can be arranged in a hierarchy of adaptive importance and thus stability so that they define a hierarchy of ranks. The structure of the nervous system or the vertebrate heart characterises large groups; subordinate characters, subgroups within them:

> The separate parts of every being must . . . possess a mutual adaptation; there are therefore, certain pecularities of conformation which exclude others, and some again which necessitate the existence of others. When we know any given pecularities to exist in a particular being we may calculate what can and what cannot exist in conjunction with them. The most obvious, marked and predominant of these, those which exercise the greatest influence over the totality of such a being, are denominated its *important* or *leading characters;* others of minor considerations are termed *subordinate.*
>
> (*Cuvier 1827, vol. I, p. 9: quoted by Oldroyd 1980, p. 38*)

The classification of the Animal Kingdom that arose from Cuvier's principles was characterised by four *embranchements* (vertebrates, articulates, molluscs, and radiates), diagnosed by the nature of the nervous system (*the* leading character). Within each *embranchement* four classes were diagnosed, which were then further subdivided. Each *embranchement* was a separate adapted creation within the Kingdom Animalia, and they were not to be ranged as a *scala*. Cuvier's system was adopted with enthusiasm by another famous nineteenth-century palaeontologist, Louis Agassiz, in his *Essay on Classification* (1857), as well as being independently arrived at by von Baer (see below). Agassiz also supported the idea of the threefold parallel of classification, embryology, and fossil succession, but he was a "progressionist" and emphatically rejected evolution: the fossil record was one of a succession of creations at the beginning of each geological period. As a result, not only the species but also every taxonomic category was a real entity:

> *Branches* or *types* are characterised by the plan of their structure,
> *Classes,* by the manner in which that plan is executed, as far as ways and means are concerned,
> *Orders,* by the degrees of complication of that structure,
> *Families,* by their form, as far as determined by structure,
> *Genera,* by the details of the execution in special parts, and
> *Species,* by the relations of individuals to one another and to the world in which they live, as well as by the proportions of their parts, their ornamentation, etc.
>
> (*Agassiz 1857, p. 170*)

An extreme example of a "natural" arrangement represented by a series of parallel *scala naturae* is the classification of mammals proposed by Thomas Henry Huxley (1880). Despite Huxley's fame as "Darwin's bulldog", his views on evolution were very different (Desmond 1982, and Chapter 3). Huxley's assumption was that the divergence of the various clades of mammals represented by the natural orders had taken place at some distant unrecorded period in geological time and that all the orders had then evolved in parallel through a series of grades, all then unrecorded, to their definitive grade. Thus a natural arrangement of mammals would be represented by a series of parallel *scalae,* one for each order. I suggest, in Section III, that in this arrangement he is showing the early influence of William Sharp MacLeay.

II. The Linnean hierarchy and the Tree of Porphyry

I have already noted that Linnaeus appears, at least at first, to have accepted a *scala naturae* arrangement of organisms, or at least of animals, as the natural order, and yet he is often regarded as the "inventor" of the natural hierarchical arrangement. Although the classification of animals set out by Linnaeus in the *Systema Naturae* can be represented by a dendrogram, the relationship of Linnaeus's classification to his idea of the natural order was at first not unlike that of the early Lamarck, with the natural order a *scala naturae* but the classification at least in part artifice. This was more strongly the case for Linnaeus's classification of plants (Ritterbush 1964), in which he looked for a *scala naturae* of all his classes of plants. Later, in a letter to Johann Gesner, Linnaeus suggested an analogy of a map to represent the interrelationship of all the plant orders. This was realised as a diagram by his former student, P. D. Giseke, editor of a posthumous work by Linnaeus published in 1792 (Fig. 2.2). It seems probable that both regarded the arrangement of large and small circles, each representing a taxon, in the same way that modern numerical taxonomists envisage taxa in hyperspace (Chapter 7).

Whatever his views on the natural arrangement of organisms, Linnaeus's system of classification followed an ancient Greek model, that of logical division (Chapter 6). Associated with this method from the time of Plato onwards was a particular pattern of classification later known as the "tree of Porphyry". Its first known appearance is in Plato's *Sophist,* in which the human activity of angling is classified into successively higher ranks (Fig. 2.3). Porphyry, born in the third century A.D. was a populariser of the works of Plotinus and the founder of the school of Neoplatonism, with roots going back to both Plato and Aristotle.

The essence of the tree (Fig. 2.4) is that at any rank, from the most inclusive one (*Summum genus*) to the penultimate one (*Infima species*), each taxon is divided to give two taxa at the rank below, using character(s) (*Differentia*) that separate them unambiguously. It is important to notice, however, that in the original tree of Porphyry, only one taxon of the pair at any rank is divided further, and that the other, while implicit, is not speci-

Fig. 2.2. A two-dimensional clustering of the Vegetable Kingdom by P. D. Giseke. (In Linnaeus 1792.)

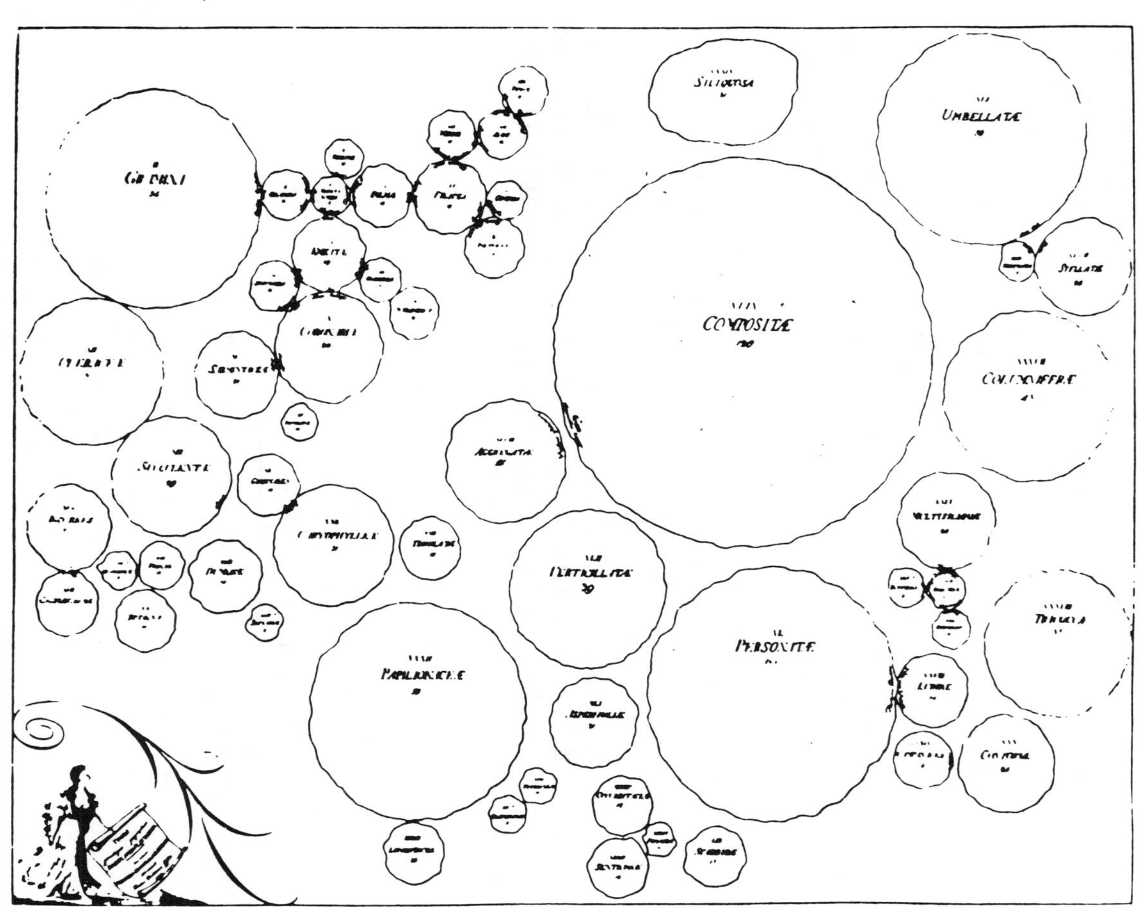

not specified. Another way of looking at this is to say that the tree illustrates the classification of an individual (*individuum*) – Socrates in this case – in successively higher categories. The tree of Porphyry has the same structure as a type of cladogram, dubbed by me (Panchen 1982, 1991) a "Hennigian comb" after the founder of cladistic classification. In the eighteenth century, its pattern would have been regarded as simply a representation of the method of logical division, so that Linnaeus would have accepted the tree as representing the pattern of classification; whereas a *scala,* or later the map, would represent his views on

Fig. 2.3. The earliest recorded "cladogram" from Plato's *Sophist*, with placement of "angling" in a dichotomous classification of the Arts. (From A. L. Panchen, "The Early Tetrapods: Classification and the Shapes of Cladograms," in *The Origin of the Higher Groups of Tetrapods,* edited by Hans-Peter Schultze and Linda Trueb. Cornell University Press, 1991. Used by permission of the publisher.)

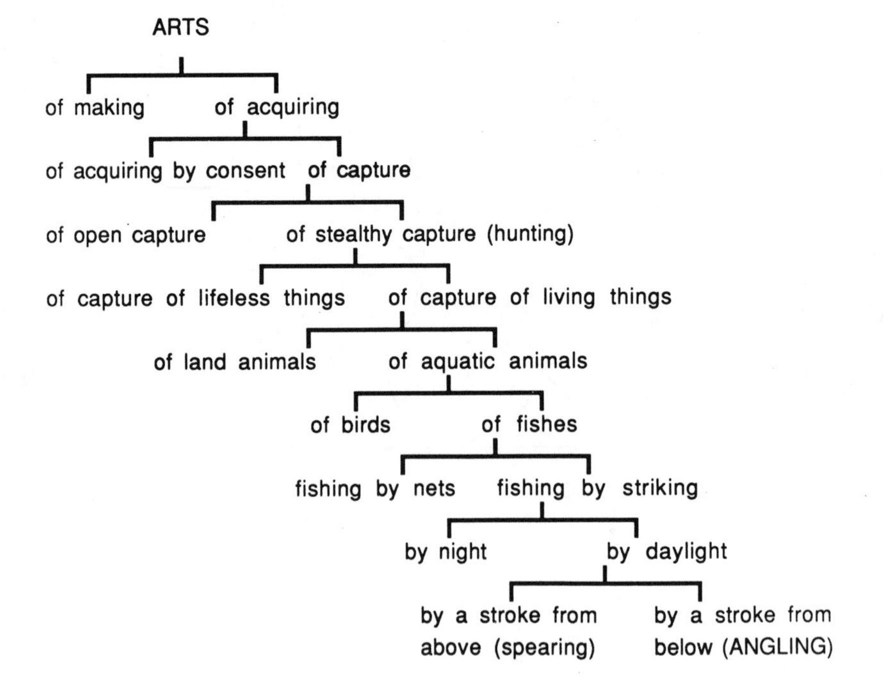

the natural ordering of organisms. This distinction probably represented the early views of Lamarck.

In the early nineteenth century, however, logical division, based on the tree of Porphyry, became associated with the concepts of "higher" and "lower" organisms in animal classification, thus emphasising the conceptual connection of the tree and the *scala naturae*. We shall see in Chapter 3 that the tree, in various guises, had a similar role well into the second half of the twentieth century in reinforcing the idea of progress in evolution, again taking over, as it were, from the more naive *scala*. The early nineteenth century association arose from the embryological speculations of the German school of *Naturphilosophie,* paralleled by the work of the French transcendental

Fig. 2.4. The "Tree of Porphyry", with placement of the "taxon" (*Individuum*) Socrates in a dichotomous classification. (From A.L. Panchen. "The Early Tetrapods: Classification and the Shapes of Cladograms," in *The Origin of the Higher Groups of Tetrapods,* edited by Hans-Peter Schultze and Linda Treub. Cornell University Press, 1991. Used by permission of the publisher.)

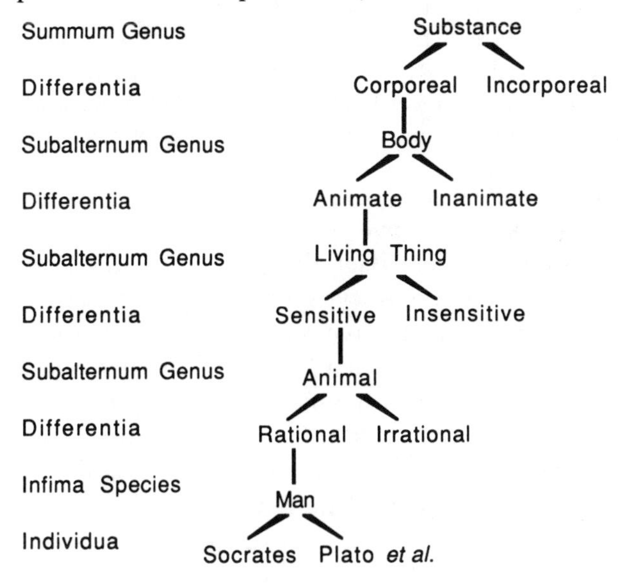

morphologists led by Etienne Geoffroy Saint-Hilaire. We have already seen that Oken, one of the most prominent of *Naturphilosophen,* erected the parallel between ontogeny and the *scala naturae* and that this was extended as a threefold parallel, including the fossil record, by Agassiz and Chambers.

Karl Ernst von Baer was brought up in the tradition of *Naturphilosophie* (Gould 1977b, pp. 52–63) but set out to refute the *scala naturae* interpretation of ontogeny of Oken and others. In order to attack the concept of the *scala naturae* in ontogeny, von Baer, in the first volume of his great *Entwickelungsgeschichte der Thiere* (1828), established the principle of *divergence.* His views on the pattern of classification were closely similar to those of Cuvier, with the Animal Kingdom divided into four *embranchements,* and, like Cuvier, he was keenly aware of the adaptive significance of taxonomic characters. His views, however, seem to have been developed independently. Furthermore, as one brought up in the tradition of *Naturphilosophie,* he saw taxonomic characters as a nested set of "*Baupläne*" (fundamental body plans) within each *embranchement.* His views on the divergent nature of the ontogeny of related animals are set out as four laws of development (from Gould 1977b, p. 56):

(1) The general features of a large group of animals appear earlier in the embryo than the special features.
(2) Less general characters are developed from the most general, and so forth, until finally the most specialised appear.
(3) Each embryo of a given species [*Thierform*], instead of passing through the stages of other animals, departs more and more from them.
(4) Fundamentally therefore, the embryo of a higher animal is never like [the adult of] a lower animal, but only like its embryo.

Thus according to von Baer the pattern of development of (particularly vertebrate) embryos should be represented by a dendrogram showing successive subdivision, with shared early ontogenetic characters succeeded by divergent later ones, and so on. A classification based on ontogenetic features would then be represented by a dendrogram of a similar form (Fig. 2.5). The use of ontogeny in classification was discussed specifically in a famous essay by Henri Milne-Edwards (1844), which included a much reproduced diagram of the classification of vertebrates presented in the form of a Venn diagram of inter-nested ovals (Fig. 2.6). When one considers that Milne-Edwards's classification was based on von Baer's principles, it shows two curious features. First, shorn of its terminal taxa it is undoubtedly in tree-of-Porphyry- or Hennigian-comb-like form. Second, Milne-Edwards draws a series of fine lines connecting taxa both within and between the larger taxa in which they are grouped. The first feature suggests that Milne-Edwards had departed from the purity of von Baer's original principles to incorporate the concepts of "higher" and "lower" vertebrates in his scheme. Von Baer, on the other hand, insisted, like Cuvier, that any given organ system was adaptive, and that while one might be able to arrange (say) the locomotor characters of a series of animals in a morphocline (higher and lower), the order of animals would not coincide with a series arranged according to feeding adaptations.

The second feature of Milne-Edwards's diagrammatic classification is of great importance to our theme. When he connects two or more taxa within a single larger taxon, various ideas may be implicit. The connections may represent a *scala naturae,* as with "Amphioxus", "Cyclostomes", and "Chondropterygians" within fish; or "Rongeurs", "Insectivores", "Quadrumanes", and "Bi-

Fig. 2.5. The pattern of development and relationships of vertebrates based on von Baer's laws. (From Chambers 1844, after Carpenter 1839.)

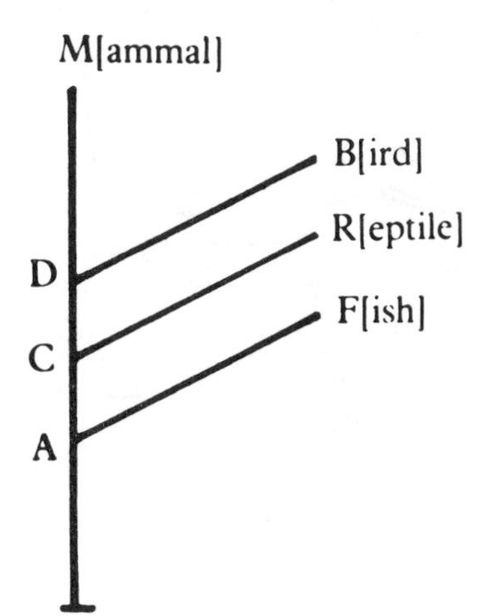

manes" within "mammals with a discoidal placenta". These might of course also be interpreted as phyletic sequences, which would violate the inter-nested pattern of the classification no more than does a *scala naturae*. However, when he connects taxa across the boundaries of more inclusive taxa that contain them, he is making explicit the most important problem of classification: that of homoplasy. Thus the amphibian caecilians are connected to reptilian snakes, because of their legless wormlike form; urodeles are connected to lizards ("*saurians*"), and anurans to chelonians. Chelonians (turtles and tortoises) are connected to both birds and monotremes to express the shared character of a beak, and, most striking of all, *Chrondropterygians* (sharks and their allies) are connected to *Cetacés* (whales) amongst the mammals because of their

Fig. 2.6. Milne-Edwards's diagram of vertebrate relationships as a Venn diagram, but with lines indicating analogy. (From Milne-Edwards 1844.)

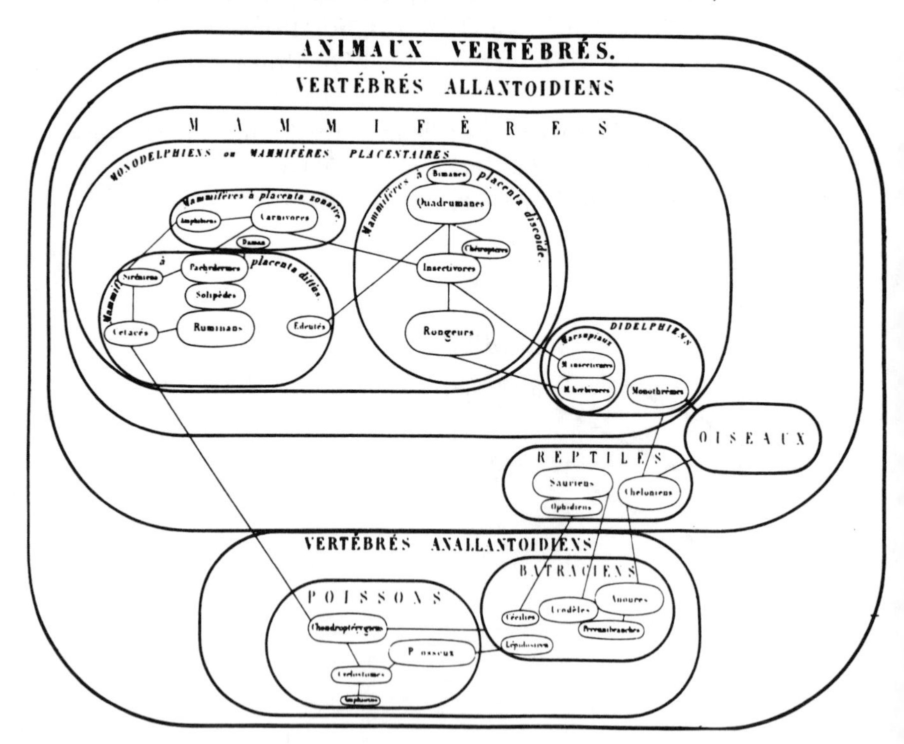

similarity of body form. In most cases it is possible to distinguish these resemblances as analogies, in contrast to homologies which define groups and characterise *Baupläne* (a distinction usually credited to Richard Owen), but in some cases that distinction is difficult or impossible to make. The inherent difficulty of distinguishing homology and analogy appears in an odd episode in the history of classification.

III. Quinarian circles and family trees

This episode is usually associated with the names of MacLeay and, *inter alia,* his disciple and publicist William Swainson. William Sharp MacLeay was a British entomologist who eventually emigrated to Australia and who, while resident in Sydney, was a considerable influence on the young Thomas Henry Huxley, naturalist aboard H.M.S. *Rattlesnake* (Winsor 1976). MacLeay's principal work on classification, *Horae Entomologicae* (1819, 1821), was presented in two volumes, the first a discussion of the classification of scarab beetles, the second an elaboration of his ideas on homology and analogy. He regarded both these latter phenomena as important intrinsic features of a natural classification. Affinity (recognised by homology) was expressed by arranging a number of groups in a linear series, or (in recent terminology) a morphocline, but according to MacLeay for any such series there is always a parallel series of an equal number of taxa which can be arranged parallel to it (a *scala* with rungs!). The rungs are represented by *analogous* features; an example of MacLeay's is the ten orders of insects then recognised. His third principle, after the nature of affinity and analogy, was the proposition that the affinities of any series of related taxa would be discovered to form a circle so that a "morphocline" A-B-C-D-E should be arranged so that E is also adjacent to A, and similarly for its parallel series A′-B′-C′-D′-E′. Furthermore, all groups of taxa forming a taxon of the next higher rank had five members, and hence the term "quinarian" applied to MacLeay's scheme of classification and to him and its other advocates.

Thus MacLeay's system of classification was a valiant attempt to reconcile the principle of plenitude (the closed rings) with Cuvier's insistence on a non-linear pattern of classification and on the adaptive nature of characters, together with an incorporation of both homology and homoplasy, and an attempt to find numerical regularity in the division of groups. He divided the Animal Kingdom into five major groups – Radiata, Acrita (splitting Cuvier's Radiata into echinoderms and polyps, etc., respectively), Mollusca, Vertebrata, and Annulosa – and arranged these diagrammatically in a circle. Each group was then to be sub-divided into five subgroups on the same principles. Each sub-group was supposed to correspond by analogy to each of those in the other major groups in an analogous position, the whole being regarded as part of the regular pattern of Creation. The final feature of MacLeay's system, to take account of plenitude, which he regarded as compromised by his five separate animal groups, was to have adjacent circles representing those groups "inosculate", touching one another in his diagram. At each of these points of contact, a small extrinsic taxon is inserted to bridge the taxonomic gap. Thus Cirreped(i)a (barnacles) was the "osculant" group between Radiata and Annulosa, and Cephalopoda (octopus, squid, etc.) between molluscs and vertebrates, but these groups were members of neither.

William Swainson was an ornithologist and biogeographer who attempted to extend and publicise MacLeay's system in popular works on natural history (Swainson 1834, 1835), but was in part also responsible for bringing it into disrepute. He was more concerned with the numerical aspects of the system than MacLeay had been, suggesting first that ten was the natural number (to incorporate the osculent groups) and later three, appearing to be five. This latter opinion was because each group was said to consist of one *typical* group (e.g., "quadrupeds" – mammals – within vertebrates), one *sub-typical* group (birds), and three *aberrant* ones (reptiles, amphibia, fish). However, the aberrant groups formed a trio at the next lower rank. Typical groups are generalised, embodying characters also possessed by other groups; aberrant groups are specialised; whereas sub-typical groups are

> always those which are most powerfully armed, either for inflicting injury on their own class, for exciting terror, producing injury, or creating annoyance to man. . . . They are in short symbolically

> the types of evil . . . scorpions, *Acari,* spiders, and all those repulsive insects, whose very aspect is forbidding, and whose bite or sting is often capable of inflicting serious bodily injury.
>
> (*Swainson 1835, pp. 245–6*)

Similarly the "wild, revengeful" [and sub-typical] bison which shows "an innate detestation of man" is to be contrasted with the "useful, docile and tameable" [and typical] ox.

The work of MacLeay and his fellow quinarians enjoyed a tremendous vogue, particularly in England, from the early 1820s until a little after 1840 when their ideas went out of fashion (Ospovat 1981, pp. 101–4). Alfred Russel Wallace (1855) attributed their decline in popularity to the method of classification proposed by Strickland (1841). Strickland's method of classification was *synthetic* (agglomerative: see Sneath and Sokal 1973) in that he took any given species (A) and asked what other two (or more) species (B and C) were most closely related to it, thus producing a diagram of affinity, B – A – C. Other species were then added to give a map of affinities. This was by no means always a linear series, and he cites a simple arrangement of five species thus:

```
B – A – C – D
        |
        E
```

By such a method Strickland built up complex diagrams of the arrangement of bird genera within families. The type of diagram produced by Strickland has two features: (1) it is a *minimal spanning tree* in which the taxa are connected directly to one another (Page 1987) rather than a *minimal Steiner tree* in which additional points are added so that no taxon is directly connected to another; (2) it is unrooted, and thus cannot be converted directly into a classification or, in other words, be represented by a dendrogram (phenogram or cladogram) (see below). A dendrogram may be used to represent a hierarchical classification, as does the type of figure, essentially a Venn diagram, used by Milne-Edwards and reproduced above. The essential feature of both dendrograms and Venn diagrams is that the terminal taxa (species if the classification proceeds from first principles) are grouped together to give a smaller number of taxa at the next rank and so on. We have already noted one form of such a dendrogram in the tree of Porphyry ("Hennigian comb") which is dichotomous and asymmetrical. Among dichotomous dendrograms the other extreme is a completely symmetrical one.

Given further information an unrooted tree can be *rooted*. In the case of a spanning tree this implies direction in the relationships expressed so that in Strickland's example the "affinity" (relationship) B → A may not be the same as A → B. The asymmetry of relationship means that A and B, although both may be species, would represent taxa of different rank if the spanning tree is rooted and thus converted into a dendrogram. The asymmetry of relationship may also be compared with that represented by true relationships such as father–son or mother–daughter, suggesting that the pattern of a dendrogram may be compared to a family tree. Thus a spanning tree, when rooted, can have real species at the nodes as well as comprising the "terminal taxa". As well as placing species at different ranks, this implies, if the dendrogram represents phylogeny, then one species is ancestral to another. The methods of analysis used in both phenetic and cladistic classification (see next section and Chapter 7) result in taxa at the ends of the branches only ("operational taxonomic units" or "terminal taxa" respectively), so that phenograms and cladograms are rooted Steiner trees.

We now see, however, that a dendrogram can represent phylogeny as well as the pattern of classification; this is the hypothesis that the apparent relationships of natural classification, between one species and another, are real relationships. To propose this is to propose a theory that evolution has occurred (Chapter 3). This is in fact what Charles Darwin did in his *Origin of Species . . .* (1859), so it is of some interest to discover to what extent an irregular dendrogram was regarded as a fair representation of natural classification in the years before the *Origin*.

Bowler (1976) presents the case that in the 1840s and 1850s the concept of divergence of organisms, particularly as recorded in the fossil record, became well established. For a general acceptance of the concept three factors seem to be involved: first, von Baer's embryological principles as described above; second, a considerable improvement in the fossil record (Agassiz's great work on fossil fish appeared between 1833 and 1844); and third, an attempt to reconcile the functional anatomy of Cuvier with the idealistic morphology of the *Naturphilosophen*. We have already noted the influence of von Baer and Milne-Edwards and also Agassiz's adoption of the threefold parallel of embryology, progression in palaeontology, and classification. But, whatever von Baer's original intention, both of the latter authors imparted a strong directional component to the parallel, so that, as *pattern,* a tree-of-Porphyry-like arrangement rather than an undirected dendrogram would represent their views on (at least vertebrate) classification.

Bowler and also Rudwick (1976, chap. 5) cite the prize essay of Heinrich-Georg Bronn (1858) as summarising a consensus of palaeontologists on the pattern of life, as seen just before the publication of *On the Origin of Species.* Bronn accepted a progressive advance in organisation of organisms, seen as fossils, with time (but not, or not until much later, a theory of transmutation). However, he also accepted that extrinsic factors produced adaptation, causing divergence of groups of organisms in anatomy as well as pattern. Bronn represented his views in a strikingly treelike diagram (Fig. 2.7) with a strong central axis ("trunk"). The vertical dimension represents both geological time and level of organisation, so that the side branches are taken to continue through time but not necessarily to have reached a higher grade. The central axis suggests linear progress towards man, but Bowler claims that this was not necessarily Bronn's intention.

Another important contributor to naturalists' perceptions of the pattern of classification about the middle of the nineteenth century was Richard Owen. He is to be credited with the reconciliation of the idealistic morphology of the *Naturphilosophen* and Geoffroy Saint-Hilaire with the functional anatomy of Cuvier to give "functional morphology". In establishing the distinction between homology and analogy referred to above, Owen (1843) was making the principal move in establishing the reconciliation. For every taxonomic group there was a Platonic *archetype* (the *Bauplan*): members of the taxon represented variants of the archetype brought about by adaptive necessity. Those furthest removed from the archetype were highest within the taxon. Owen's use of the idea of homology is further discussed in Chapter 4, but in the present context it can be seen that his views are not inconsistent with a hierarchical classification. More strongly, however, the innate principle, the archetype, is a conservative force and is not directional as the *scala naturae* is, so that all the differences among organisms are adaptive.

Thus while Owen's scheme incorporates the concepts of "higher" and "lower" as in a *scala* or a tree of Porphyry, the factor that determines "height" – that is, adaptation – is also the factor that produces diversity. It therefore seems reasonable to conclude that to Owen a natural classification of organisms could have been represented by a dendrogram that was not a tree of Porphyry and certainly not a *scala*. Furthermore, it would have an irregularity of pattern because divergence would be dictated by too many contingent features of the environment, not by any innate organising principle.

Owen's principal presentation of his ideas on functional morphology was in his *On the Archetype and Homologies of the Vertebrate Skeleton* (1848). Charles Darwin embarked on his greatest tax-

Fig. 2.7. Bronn's "tree" of animal relationships, with geological time and level of organization conflated along the axis. (After Bronn 1858, from Bowler 1976.)

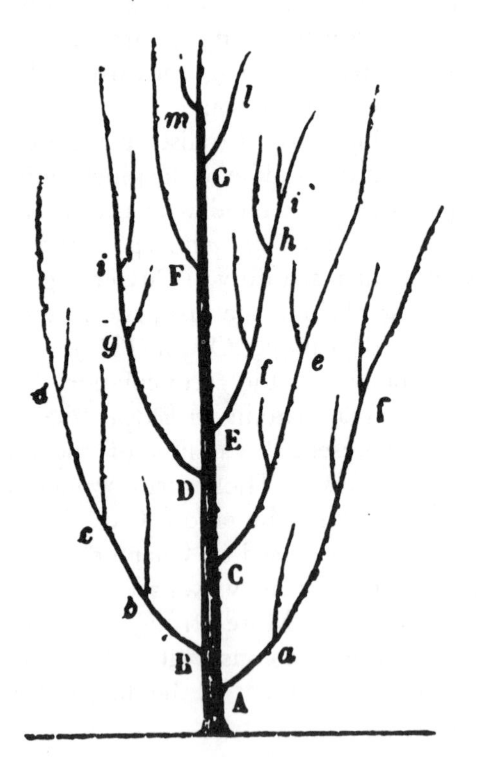

onomic work, the classification of living and fossil barnacles, in 1846 (F. Darwin 1887, vol. 1, p. 80). Two volumes were published on the fossils in 1851 (see 1851b) and 1854 (1854b), respectively, and the two parts of the monograph on the living forms in the same years. It has been suggested by Smith (1965) that Darwin's barnacle work was undertaken at least in part to refute MacLeay's quinarian system which he would have seen as anti-evolutionary. One can cite a passage from the second volume of the living cirripede volume (Darwin 1854a, p. 16) in support of this view:

> I may venture to express strongly my opinion, that the group [cirripedes] is formed on a distinct type; as different from the other three or four main Crustacean groups . . . as these differ from each other; the difference, moreover, being of the kind considered by the highest authorities on this subject, as the most important. It should be observed that there is no special blending at either end of the Cirripedial series, towards any one of the other main groups of Crustacea: it is hardly possible to take some one Cirripede, and say that it leads more plainly than some other Cirripedes, onto ordinary Crustaceans. Moreover, a great range of structure, as we shall soon briefly show, is included within the group . . .

Cirripeda (*sic*) was an "osculent" group between crustaceans (of the Annulosa) and echinoids (of the Radiata) in MacLeay's system; thus Darwin's assertion of their independence could just have the significance suggested by Dr. Sydney Smith, particularly when added to the evidence that Darwin's interest in the group was roused by an abnormal (and apparently "inosculating") form discovered on the *Beagle* voyage. The barnacle work was begun two years after Darwin had completed his *Essay* "On the Variation of Organic Beings Under Domestication; and on the Principles of Selection" (in Darwin and Wallace 1958), and Ghiselin (1969) suggests, more reasonably, that what was to have been an account of a single abnormal cirripede turned into a series of monographs on the whole group in the light of Darwin's *almost* complete (Ospovat 1981) evolutionary theory and his excitement at the significance of the

> truly wonderful fact – the wonder of which we are apt to overlook from familiarity – that all animals and all plants throughout all time and space should be related to each other in group subordinate to group. . . . The several subordinate groups in any

> class cannot be ranked in single file but seem rather to be clustered round points, and these round other points, and so on in almost endless cycles. On the view that each species has been independently created, I can see no explanation of this great fact of classification of all organic beings; but, to the best of my judgement, it is explained through inheritance and the complex action of natural selection, entailing extinction and divergence of character . . .

This quotation from near the end of Chapter IV of the *Origin* (Darwin 1859) is followed by the long final paragraph in which the metaphor of a tree is made explicit, to represent both classification and phylogeny (see Chapter 3, Section II). It is also made clear that this is no tree of Porphyry, nor yet a dendrogram determined by mathematical rules, but that the irregularity of the tree is a true and important property of the contingent nature of phylogeny.

IV. Grades, phenograms, and cladograms

By the time of the publication of the *Origin,* the form of classification as a hierarchical clustering was established, and soon after, it probably would have been accepted by most taxonomists that a dendrogram, drawn in the style of a human genealogy but with successively higher ranks instead of more and more distant generations, could represent a natural classification. We defer what is meant by the phrase "natural classification" until Chapters 3 and 6, but in most cases, following Darwin, the apparent genealogical arrangement would have been regarded as real, so that the dendrogram would represent phylogeny.

The fashion for genealogical dendrograms, or phylogenetic trees, representing real taxa, started with Haeckel (1866), who tried to match in pictorial form the vividness of Darwin's tree metaphor. We shall consider trees as phylogenies more fully in Chapter 3, but meanwhile there are important points to be made about the post-*Origin* rash of trees that soon appeared, in terms of pattern and their claim to represent natural classification.

Most made the claim, implicitly or explicitly, to represent both classification and phylogeny, and most, again either explicitly or implicitly, had a vertical axis that represented more than a sequence of ranks. Agassiz had invented a (according to him) non-phylogenetic type of diagram, now known as a spindle diagram, in which the vertical axis represented geological time, and the "spindles" groups of organisms, with the width of the spindles providing an impression of numbers or diversity. But the horizontal order and distance of the spindles, particularly in modern phylogenetic versions, often represent "morphology" or anatomical difference among groups, so that two parameters are conflated on the horizontal axis. It will be recalled that in Bronn's tree, time and morphology were conflated along the *vertical* axis.

In an excellent review of the history of classification and phylogeny reconstruction in teleost fish, Patterson (1977) contrasts, *inter alia,* two nineteenth-century trees. The first is that of Gill (1872) (Fig. 2.8). This is set out like a human genealogical tree (without the marriages!) but its important features are that (1) it is a Steiner tree with the taxa being classified only at the tips; (2) even as *genealogy* the vertical axis cannot represent geological time; as *classification* it represents the pattern of inter-nested taxa; as *phylogeny,* only the sequence of speciation events culminating in the terminal taxa; (3) the horizontal axis is not a measure of morphology or anything else; (4) most of the branchings at each "node" are dichotomous. In all these respects it is comparable to the cladograms produced by modern cladistic analysis. In contrast, I shall take the second of the three of Haeckel's fish phylogenies illustrated by Patterson (Haeckel 1889): "Haeckel simplified his trees and complicated his taxonomy, but introduced no new factual justification for his schemes". The diagram (Fig. 2.9) is superficially similar to Gill's, with a majority of dichotomous furcations, but in two lines particularly, those leading to Teleostei/Physostomi and to Amphibia, a whole series of taxa appear successively along an unbranched line. No time axis can be placed parallel to these lines because the taxa are a mixture of extinct (Leptolepides, Thrissopides, Ctenodipterini) and extant (Amiades, Monopneumones, Dipneumones) groups. Further the higher-ranking Dipneusta appears distal to "Crossopterygii(?)". (Note also Perennibranchia–Cryptobranchia–Salamandrina–Anura within the Amphibia: all are extant groups.)

These vertical series of taxa can be seen only as *scalae naturae* in which the individual taxa were presumably seen as representing different grades of organisation with the whole series as (in modern

Fig. 2.8. "A quasi-genealogical tree" of vertebrates after Gill 1872. (From Patterson 1977.)

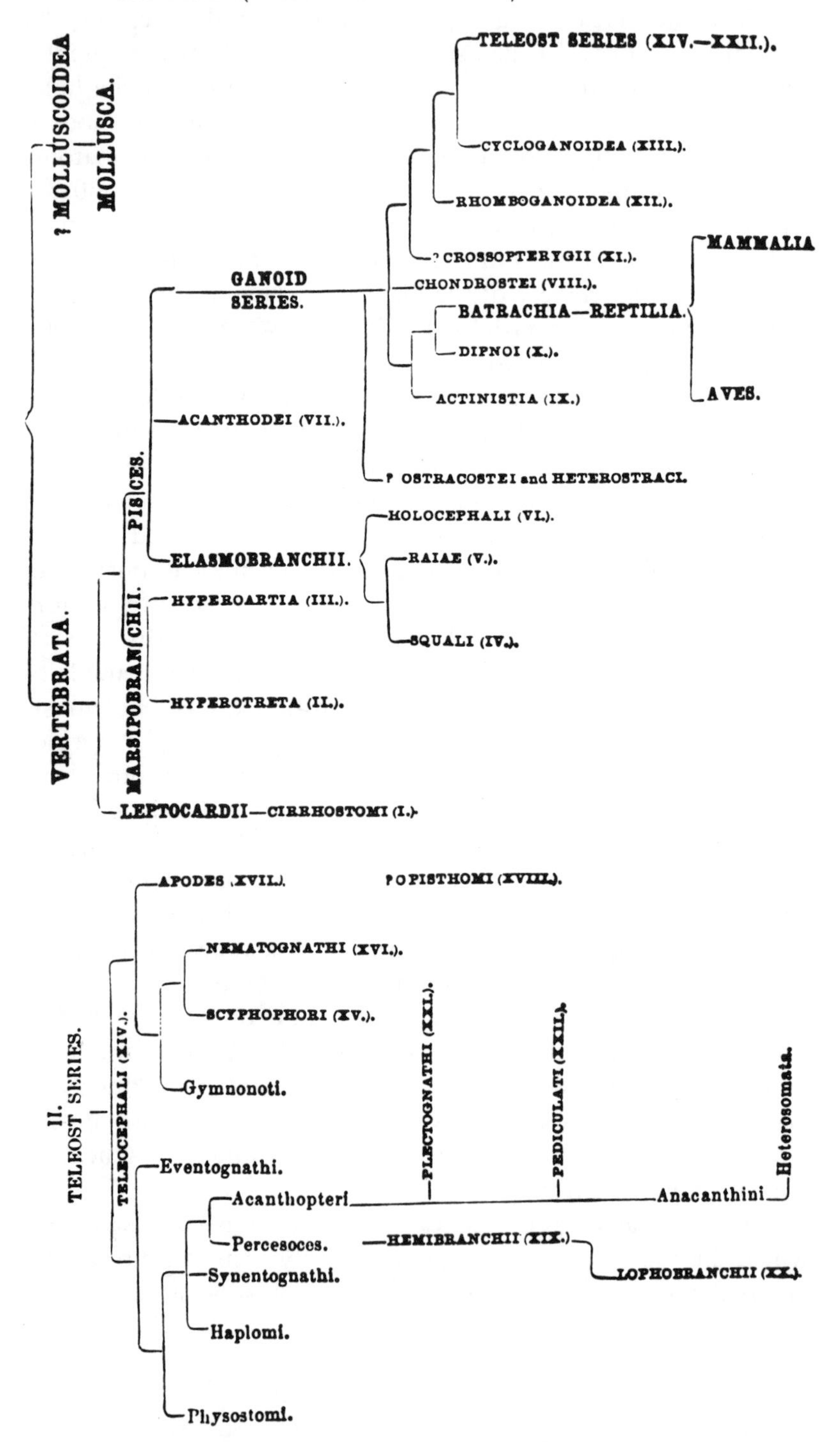

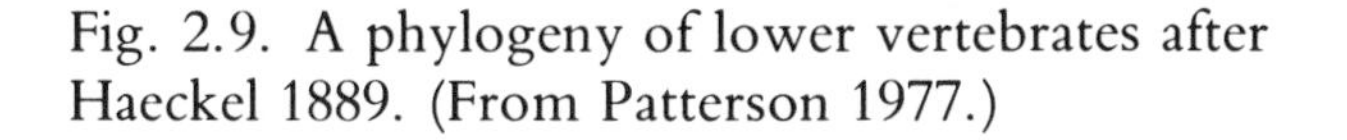

Fig. 2.9. A phylogeny of lower vertebrates after Haeckel 1889. (From Patterson 1977.)

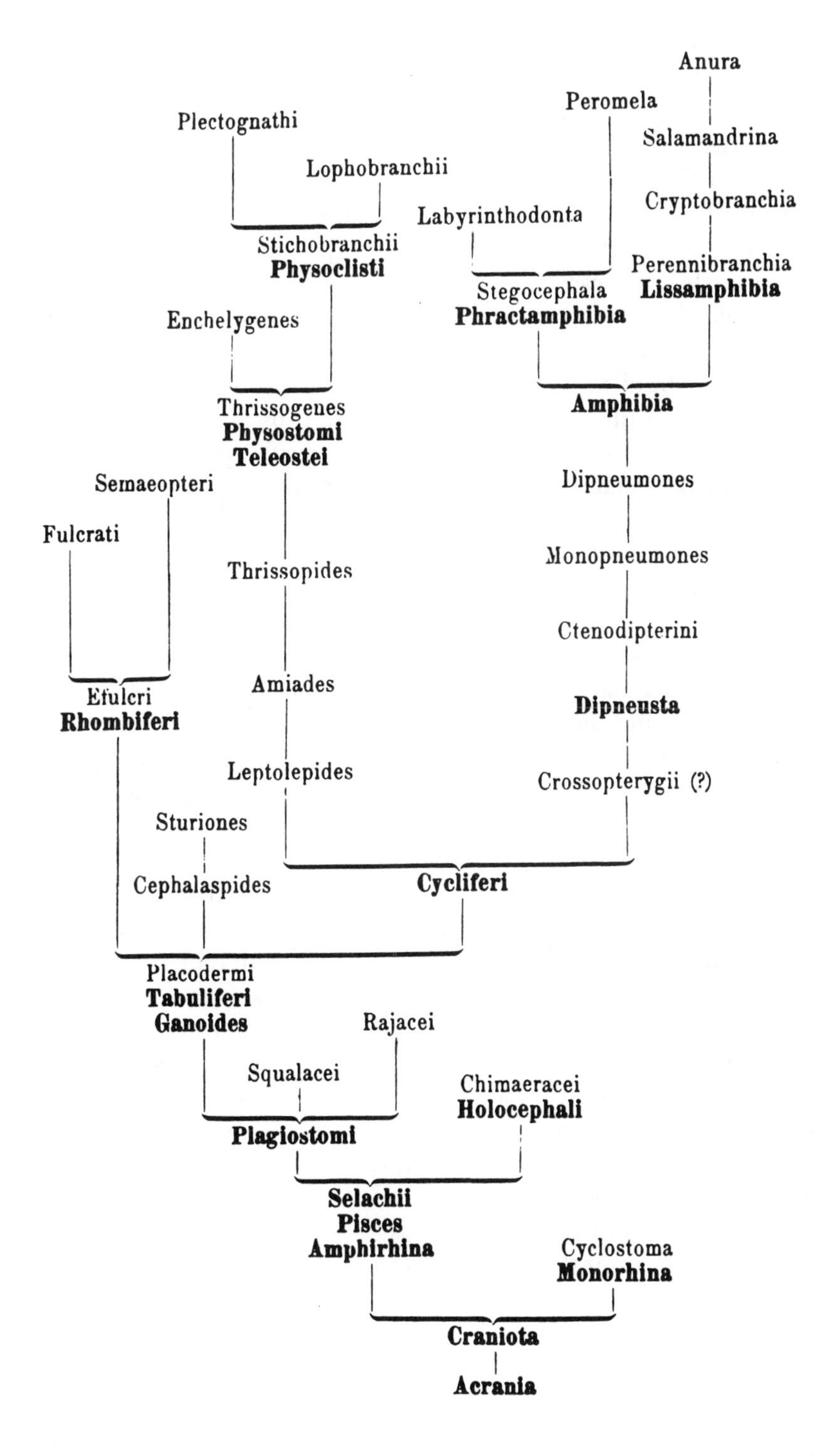

terminology) a *morphocline*. Interpreted as phylogeny, each series would have to be interpreted as an ancestor–descendant series. But one supra-specific taxon cannot be ancestral to another: if (e.g.) the taxon Leptolepides included the ancestors of Amiades, then the latter should be a sub-group of the former. Nevertheless, such grade groups persisted into this century. Patterson cites an example within fish in Romer (1945, fig. 83). There is also another way in which supposed grade differences have been allowed to affect the pattern of taxonomy. The two most famous examples, championed by Mayr (1969, 1974, respectively), represent established tradition. In each case they involve the separation as a taxon of high rank of an "advanced" group from the more inclusive group in which it should be contained, leaving the more "primitive" members behind as a separate (to the cladists, *paraphyletic;* i.e., incomplete) group. In the first example, the advanced taxon *Homo sapiens* (family Hominidae) is removed from the great apes (Pongidae). In the second, birds (class Aves) are removed from the class Reptilia.

In both examples, if a diagram were drawn from the pattern of classification, the groups would appear as branches from the same node, but a diagram of phylogeny would be quite different. In this case the same phylogenetic diagram serves for both (Fig. 2.10). Mayr's claim is that taxonomic clustering should represent "genes in common", and that the latter is represented by proximity along the horizontal axis. In the first case, B would represent the orang-utan; C, chimpanzees plus gorilla; and D, man. In the second case, C would represent the crocodilians; B, other diapsid reptiles; and D, birds. Seen as a phylogeny, A represents the common ancestor of B, C, D in each case: seen merely as a representation of classification, it represents the primitive condition. In each case the claim is that D should be placed in a taxon different from B + C. Patterson (1981b) has pointed out that in the first case at least the premises are false: all biochemical and genetic tests show that man is more closely related to the African apes than either is to the orang-utan (Andrews and Cronin 1982). My purpose here is merely to note that in both cases the arrangement is to justify placing D in a major taxon different from B + C *because it is taken to have reached a higher grade*.

I conclude this historical review by describing the types of pattern of classification that result from two rival modern schools of taxonomy: *phenetics* and *cladistics*. Their methods are described in Chapter 7. Merely considered as pattern, the dendrograms produced by these schools – phenograms and cladograms, respectively – have a number of features in common: (1) if fully "resolved", that is, if the analysis has been successfully brought to completion, they are dichotomous throughout – that is, all nodes form the base of two branches; (2) they are minimal Steiner trees with the objects, such as species, being classified at the tips of the branches and uniformly occupying the lowest rank; (3) they are "rooted", and the point at which they are rooted represents the highest rank and thus the all-inclusive taxon (we shall see in Chapters 7–9 that both methods can produce unrooted trees, but these are usually rooted subsequently).

There are fundamental differences in both the methods and the significance of the results of phenetics and cladistics. Phenetics is grouping by "overall" (aggregate: Charig 1982) similarity; the nodes represent all the characters shared by their dependent branches and, ultimately, original taxa [*operational taxonomic units* (OTUs) to the pheneticist]. Cladistics is grouping by uniquely

Fig. 2.10. Presumed phylogeny used to advocate a "grade" classification by Mayr and others. Horizontal axis: "morphology". In the first case, B = orang-utan, C = gorilla plus chimpanzees, D = *Homo sapiens* (Mayr 1969). In the second case, B = other diapsid reptiles, C = crocodilians, D = birds (Mayr 1974); A = common ancestor.

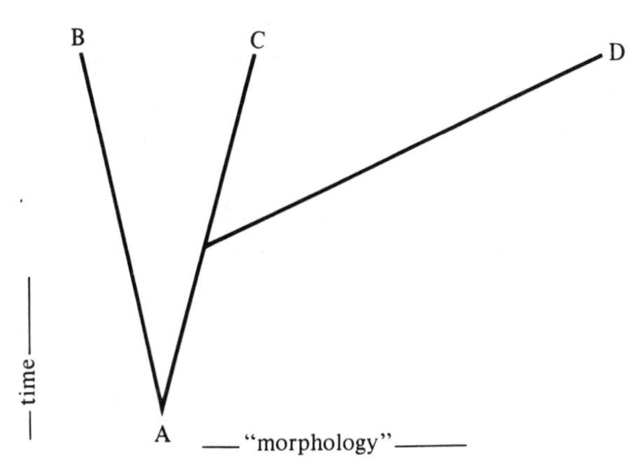

shared derived characters ("*synapomorphies*": Chapter 4); the nodes represent those characters unique to the taxa (including the "*terminal taxa*": OTUs) descendant from them. But the differences between phenograms and cladograms *as pattern* are concerned mainly with the significance of the (conventionally) vertical axis – which I shall term the "axis of rank". In the final phenogram, with all the OTUs arranged at the same level, their relative positions and the horizontal distances between them do not have any significance outside that of the branching pattern that connects them (unlike Mayr's phyletic diagrams above!). Thus in a phenogram the horizontal axis is not a linear measure of "morphology" or "phenetic difference" or anything else. But the vertical axis is. A pair of OTUs (analogous to the "sister-groups" of cladistics; see below) will have a small vertical distance between themselves and the node or branch-point that connects them exclusively if they are closely similar, but a greater distance if they are less so (Fig. 2.11). Thus an upright phenogram has a vertical axis which is a measure of relative difference between OTUs and combinations of OTUs. This is useful as a graphic representation of the degree of confidence (as a reciprocal of branch length) that can be reposed in any pairing, but it means that no branch-point is necessarily at the same horizontal level as any other. If they were, as they are in conventional cladograms of cladistic taxonomists, then they could be used as a guide to ranking, those groups arising at the same level being given the same rank. In order to cope with the absence of clear ranks in phenograms, Sneath and Sokal (1962; 1973, pp. 294–6) suggest a concept of *phenons,* which are analogous to taxa, but are produced by drawing lines across the phenogram at predetermined levels representing percentage similarity. Lines at different levels would then represent different ranks and thus delimit a hierarchy of phenon categories. The system, however, has not been generally adopted, and there are a number of objections to it (Chapter 7).

Cladistic methods are discussed more fully in Chapter 7. For our present purposes it is the form of the resulting cladograms (*s.s.*) that is important. As with phenograms a fully resolved Hennigian cladogram is dichotomously branched at every node. Tricho- or polytomies are regarded as incompletely resolved dichotomies whose lack of resolution is due to insufficient or contradictory evidence. In claiming to represent phylogeny, the early cladists claimed only to represent the pattern of cladogenesis or speciation but not anagenesis or the phyletic evolution of clades. For this reason neither the horizontal nor the vertical axis of a Hennigian cladogram represents any measurable quantity. The horizontal axis does not represent any measure of "morphology", nor does the vertical. Similarly, the vertical axis does not represent any measure of similarity (as it does in a phenogram), nor yet evolutionary progress, nor yet time; except that in a tree of Porphyry pattern, if it is interpreted as phylogeny, there is an *ordinal* time axis. For that reason the results of cladistic analysis can be represented equally well by Venn diagrams of nested binary sets similar to Milne-Edwards's diagram (Fig. 2.6) (Hennig 1966, fig. 18; Charig 1982, fig. 1). Because of this there is considerable artistic freedom in the presentation of cladograms (Fig. 2.12). There is, however, a category of exceptions to the statement that cladograms have no

Fig. 2.11. A phenogram with "phenons" (see text). (After Sneath and Sokal, *Numeral Taxonomy*. W. H. Freeman & Co. 1973. Used by permission of the publisher.)

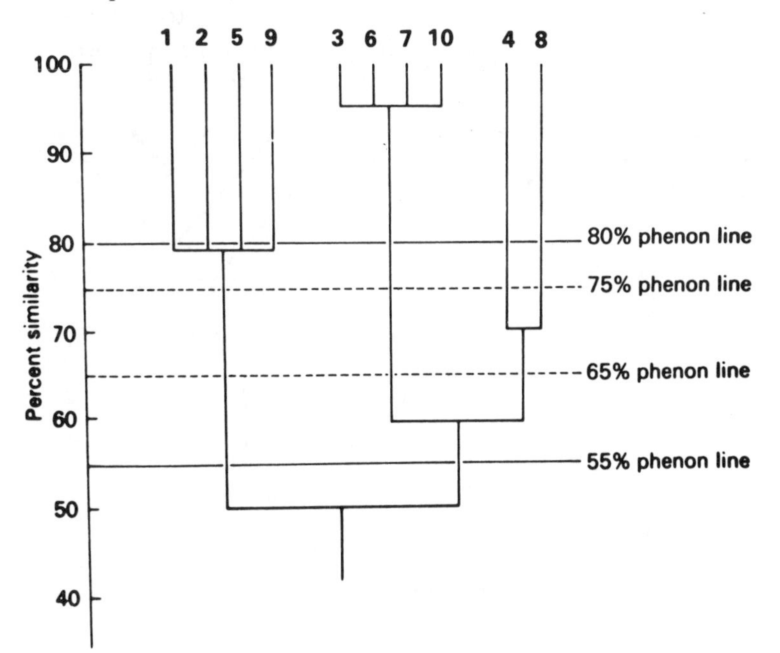

measure of distance incorporated in them. The methods of numerical cladistics use algorithms in which the branch lengths on their dendrograms represent the number of uniquely shared characters possessed by the taxon (terminal or otherwise) at the end of the branch. The algorithm is an attempt to minimise the total branch length (the principle of parsimony). When the final cladograms are published, however, it is usual to show only the pattern of branching (see Chapter 9).

V. Conclusions

In this chapter I have attempted to begin to develop the theme that a theory that evolution had occurred depends on a perceived arrangement of organisms that is in some way regarded as "natural". I have attempted to show, all too briefly, the dom-

Fig. 2.12. Cladograms: (a) as in phylogenetics with circles at interior nodes, usually taken to represent common ancestors, that is, a "tree"; (b) without "ancestors" at interior nodes, usually taken to indicate a "synapomorphy scheme"; (c) "toasting-fork" pattern, now usual, derived from computer cladistics.

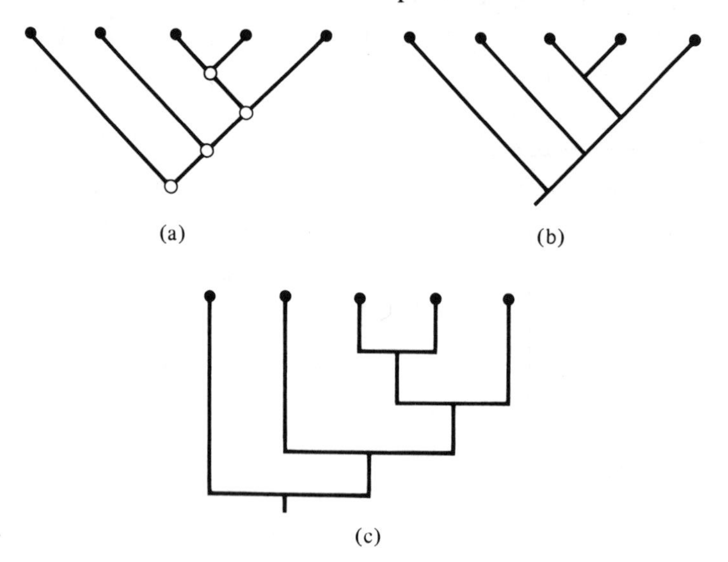

inance of the concept of the *scala naturae* from ancient times forward to the early nineteenth century. But there was a conflict between the *scala*, seen as the natural order of organisms, and the results of classification by logical division which lead to an inclusive hierarchy. This conflict is explored by Rieppel (1988). Until the aim of classification was not primarily to follow the correct procedure, logical division, but to achieve the natural pattern, the picture of phylogeny as having the same pattern could not emerge. Lamarck's original theory of evolution was simply to explain progression along (or rather up) the *scala naturae:* it explained progression in the history of life but not the results of classification.

I spent some time on the quinarians, usually regarded as a taxonomic joke in rather poor taste, because at the time I believe their aim was a legitimate and important one. Without a theory of evolution there need be no presumption of the contingent nature of the natural order of organisms arising from an awareness of their history. It was scientifically proper, therefore, to look for some regular organising principle, analogous to the Periodic Table of chemical elements, to explain the natural pattern of classification. The quinarians were a byway of taxonomic history, but their failure emphasises contingency and thus the expectation of irregularity in the pattern of natural classification.

The method of logical division, which we shall look at in more detail in Chapter 6, yields a divergent inclusive hierarchy of "groups within groups". The idea that it could also yield a natural pattern depended on rejection of the *scala naturae* in comparative anatomy, as by Cuvier and Agassiz and the later Lamarck, and in embryology by von Baer. The divergent nature of the hierarchy was also powerfully supported by the distinction between homology and analogy, understood and misinterpreted by the quinarians, but highlighted as a phenomenon in the pattern of classification by Owen. Thus Darwin (and Wallace; see Chapter 3) was able to take for granted the irregular, and thus contingent, nature of the pattern of classification to be explained by the theory of evolution.

The role of palaeontology, developed in the next chapter, particularly in the interpretation of the tree of Porphyry as phylogeny, seems to me to have been equivocal in all this. The contingent nature of the diversity of organisms was emphasised by the fossil record, but the paralleling of "progression" in the fossil record, ontogeny, and the *scala naturae* by the anti-evolutionist Agassiz and the mystic evolutionist Chambers kept the *scala* concept going until later in the nineteenth century, when palaeontologists started to pay more attention to the discovery of ancestor-descendant sequences in the fossil record than to deciphering the pattern of life. We saw this in the pattern of classification produced by Haeckel. Thus the *scala naturae* was not defeated after the acceptance of evolution. I also drew attention to the distinction between spanning and Steiner trees. A *rooted* spanning tree, as represented by the linear sequences of taxa in Haeckel's figure (whether interpreted as phylogeny or not), implies the concept of *grade* and hence harks back to the *scala*. But grade is also implicit in the Steiner trees used by Mayr to justify his separation of man from the great apes and birds from reptiles as taxa of equal rank to that of the remnant group. Phenograms and cladograms are also Steiner trees, but their horizontal axes are not measures of taxonomic distance or of grade. We shall see in the next chapter, however, that the *scala* can appear in cladistics in the form of the Hennigian comb, linear descendant (to coin a phrase) of the tree of Porphyry.

3

Patterns of phylogeny

It is a characteristic feature of elementary biology texts (e.g., English "A-level" texts such as Roberts 1987, Simkins and Williams 1989) that they almost always include their one or more chapters on evolution near the *end* of the book. It is traditional to include a section on "evidence for evolution", with the principal evidence cited as from Comparative Anatomy, Embryology, Physiology, Biochemistry (including immunology), and sometimes Behaviour. But all these are evidence for the phenomenon of *natural classification*. The fossil record may be discussed, but more as an abbreviated history of the biosphere than as a basis for a logical argument, and there is usually a brief treatment of the evidence from geographical distribution; but the missing ingredient is any clear statement of the scientific reason for proposing a theory of evolution.

One of the most important points to be emphasised in this book is that phylogeny is the theory proposed to explain natural classification. The pattern of evolution, or phylogeny, and particularly common ancestry, is the *explanans;* the phenomenon of natural classification is the *explanandum* (Brady 1985). In other words, the theory that evolution has occurred is the theory that the apparent relationships of classification are real relationships – that is, the pattern of classification is in fact a family tree. So this chapter will explore patterns of phylogeny in the light of the patterns of classification seen in Chapter 2.

I. The *scala naturae* as phylogeny

When we consider theories of evolutionary mechanism in Chapters 11 and 12, we shall see that in almost every case they are compounded of two classes of factors: intrinsic or internal factors, and extrinsic or environmental ones. Lamarck's theory was the first fully worked-out theory of evolution and included as a primary mechanism what Lovejoy (1936) called the "temporalizing of the Chain of Being". Thus, according to Lamarck, some internal force produced evolutionary progress up the *scala naturae*. His theory ("*marche de la nature*") was introduced to a lecture audience at the Museum d'Histoire Naturelle in 1800. In 1786 he had suggested that separate *scalae* were necessary for the Plant and Animal Kingdoms, rejecting the popular eighteenth-century idea (Ritterbush 1964) that "zoophytes" were intermediate between the two. He did, however, retain two current ideas, that "subtle fluids" (electricity) were the driving force and that spontaneous generation produced the raw material on which his innate, or orthogenetic, mechanism acted:

> [O]nce the difficult step [of accepting spontaneous generation] is made, no important obstacle stands in the way of our being able to recognise the origin and order of the different productions of nature.
>
> (*Lamarck 1802, transl. Burkhardt 1977*)

But Lamarck seems to have had a fluctuating but dwindling faith in the adequacy of a simple *scala* to represent even the unconstrained pattern of animal evolution. In 1802, the faith appeared absolute:

> Ascend from the simplest to the most complex; leave from the simplest animalicule and go up along the scale to the animal richest in organisation and facilities . . . then you will have hold of the true thread that ties together all of nature's productions, you will have a just idea of her *marche* and you will be convinced that the simplest of her living productions have successively given rise to all the others.
>
> (*Burkhardt's transl.*)

By the time of his *Philosophie Zoologique* (Lamarck 1809), his position was contradictory. He was still sufficiently convinced of the integrity of the animal *scala* to make the distinction between *classification* and *distribution générale* noted in Chapter 2 (Section I), but his diagram of the natural arrangement of animals belies this (Fig. 3.1):

> The table on p. 179 may facilitate the understanding of what I have said. It is there shown that in my opinion the animal scale begins by at least two separate branches, and that as it proceeds it appears to terminate in several twigs in certain places.
>
> (*Lamarck 1809: Elliot's 1914 transl. p. 178*)

The two branches are retained in Lamarck's definitive phylogeny, in the *Histoire Naturelle des Animaux sans Vertèbres* (1815–22). The first, "*série des animaux inarticulés*", included infusorians, polyps, and radarians, plus ascidiens and molluscs; the second, *"série des animaux articulés", vers* (primitive "worms"), then a dichotomy

Fig. 3.1. Phylogeny of animals from Lamarck (1809). (In Elliot's 1914 translation the sparse dotted lines are replaced by solid lines.)

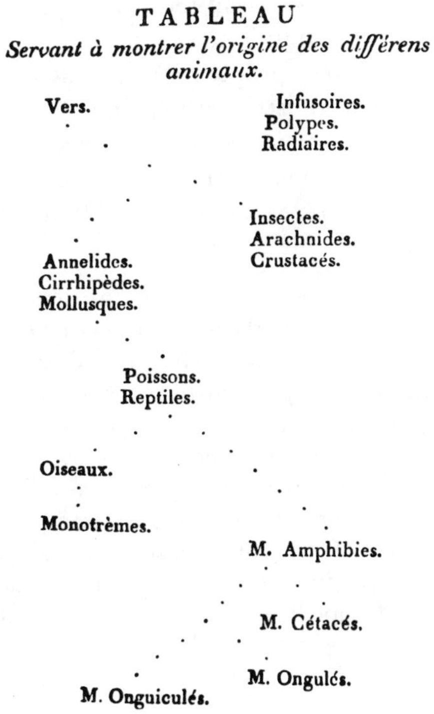

TABLEAU

Servant à montrer l'origine des différens animaux.

Vers.

Infusoires.
Polypes.
Radiaires.

Insectes.
Arachnides.
Crustacés.

Annelides.
Cirrhipèdes.
Mollusques.

Poissons.
Reptiles.

Oiseaux.

Monotrèmes.

M. Amphibies.

M. Cétacés.

M. Ongulés.

M. Onguiculés.

into annelids on the one hand and insects and other arthropods on the other. But the vertebrates were no longer attached to either ramus and reflected another new feature. All animals were divided into three grades: *animaux apathiques,* said to lack a nervous system; *animaux sensibles,* with nerves; and *animaux intelligents,* including only vertebrates. The grades, first introduced in 1812, represent stages on the evolutionary *scala,* while the branching represents the environmentally induced deviations from it.

Robert Chambers, in the *Vestiges of the Natural History of Creation,* also adopted the dual arrangement of *scala* and branches as well as the paralleling of (human) ontogeny, phylogeny, and the fossil record which we saw in Chapter 2. In the first edition (1844), he expressed dissatisfaction with a single *scala,* and later moved to a belief in the multiple spontaneous generation of animal life, so that by the ninth edition (1851) he was able to announce that the problem was solved: a plurality of series must have occurred with at least three invertebrate *scalae* and the probable polyphyletic origin of fish from all three. Later, parallel evolutionary *scalae* also appear, notably in the phylogenetic ideas of Thomas Henry Huxley. Before Darwin's publication of the *Origin,* Huxley was an advocate of the doctrine of "persistence" in the fossil record. In this he was siding with the geologist Charles Lyell in opposing "progressionism" as advocated by Buckland and other catastrophists (see Chapter 5). But even long after the publication of the *Origin,* Huxley believed that major animal groups had differentiated before the known fossil record. The following outburst is quoted by Desmond (1982) in a study of the scientific and social context of Huxley's beliefs:

> I confess it is as possible for me to believe in the direct creation of each separate form as to adopt the supposition that mammals, birds, and reptiles had no existence before the Triassic epoch. Conceive that Australia was peopled by kangeroos and emus springing up ready-made from her soil and you will have performed a feat of imagination not greater than that requisite for the supposition that the marsupials and great birds of the Trias had no Palaeozoic ancestors belonging to the same classes as themselves. The course of the world's history before the Trias must have been strangely different from that which it has taken since, if some of us do not live to see the fossil remains of a Silurian mammal.
>
> (*Huxley 1869*)

Huxley persisted in this belief, which is expressed in his 1880 "arrangement" of mammals in which each order is represented as having passed, in parallel with all the others, through a series of grades until it reached its definitive grade. Even in this century, palaeontologists, such as Henry Fairfield Osborn, with mammals (see Chapter 11) and D. M. S. Watson, with the "labyrinthodont" amphibians of the Paleozoic and Triassic (Watson 1956), tried to show that a number of clades went through parallel and synchronous evolutionary changes driven by some internal mechanism. A more recent case, at least as far as the pattern is concerned, is Jarvik's (1980) arrangement of the major groups of vertebrates, and, more specifically his proposal of the polyphyletic origin of tetrapods.

II. Family trees

Charles Darwin returned from the voyage of H.M.S. *Beagle* on 2 October 1836. In late June or early July 1837, he opened his first Transmutation Notebook (Notebook B: Darwin 1987); an early entry (Darwin 1987, p. 176) reads as follows: "organised beings represent a tree irregularly branched. – Hence Genera"; to which he subsequently added "as many terminal buds dying as new ones generated".

On subsequent pages in the same notebook Darwin doodled irregular trees (Fig. 3.2), and a hypothetical phylogenetic tree was the only illustration in *On the Origin of Species* (Fig. 3.3). In his explanation of the last, Darwin (1859, pp. 116–17) explains that

Fig. 3.2. An irregular phylogenetic tree from Darwin's first transmutation notebook. (From Darwin 1987.)

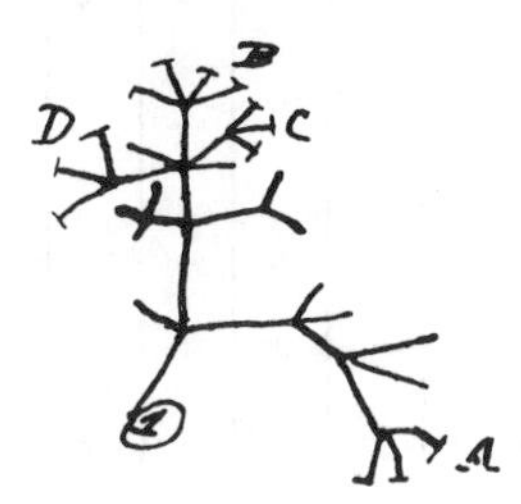

Fig. 3.3. A hypothetical time-based phylogeny of the ancestral species A to L "of a genus large in its own country"; the only illustration in *On the Origin of Species* (Darwin 1859).

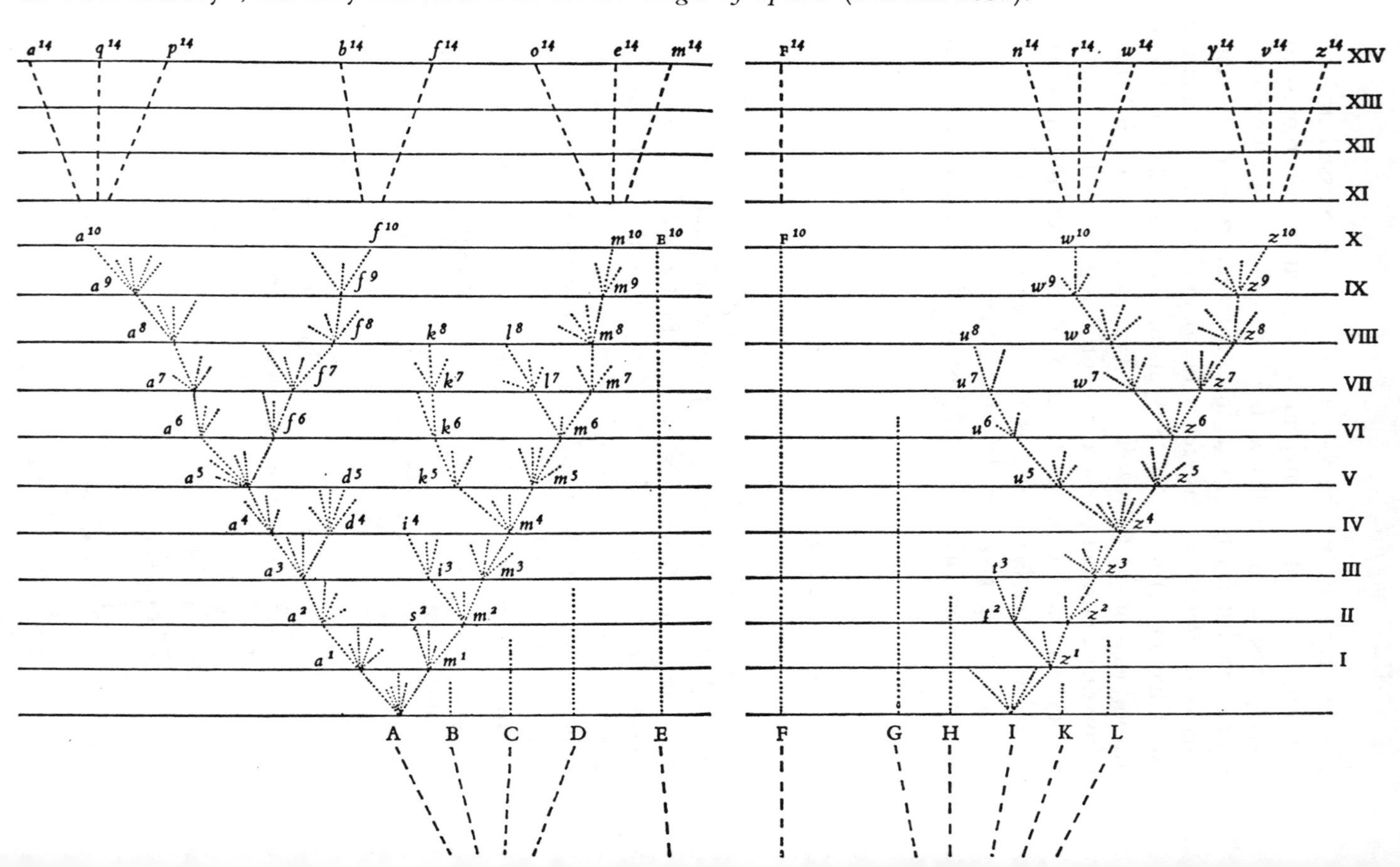

the vertical axis represents geological time – "The intervals between the horizontal lines in the diagram, may represent each a thousand generations" – and that the horizontal axis represents morphology – "these species are supposed to resemble each other in unequal degrees, as is so generally the case in nature and is represented in this diagram by the letters standing at unequal distances".

We saw in the previous chapter that Darwin, in Chapter IV of the *Origin,* specifically explained the pattern of classification in terms of phylogeny and that he saw that pattern as an irregular, inclusive, divergent hierarchy. The last long paragraph of Chapter IV, following that quoted in Section IV, begins as follows:

> The affinities of all the beings of the same class have sometimes been represented by a great tree. I believe this simile largely speaks the truth. The green and budding twigs may represent existing species; and those produced during each former year may represent the long succession of extinct species . . . and this connexion of the former and present buds by ramifying branches may well represent the classification of all extinct and living species in groups subordinate to groups.

One might well ask who before had used the simile of "a great tree". One answer is certainly "Alfred Russel Wallace". In 1855, three years before he sent an outline of his theory to Darwin, precipitating their joint communication to the Linnean Society on 1 July 1858, Wallace published his paper, written in Sarawak, "On the Law Which Has Regulated the Introduction of New Species". I shall consider that law, and the series of propositions on which it is based in Chapter 5 (Section III), but in the paper Wallace speaks of the "branching of the lines of affinity, as intricate as the twigs of a gnarled oak or the vascular system of the human body", and goes on:

> Again if we consider that we have only fragments of this vast system, the stem and main branches being represented by extinct species of which we have no knowledge, while a vast mass of limbs and boughs and minute twigs and scattered leaves is what we have to place in order, and determine the true position each originally occupied with regard to the others, the whole difficulty of the true Natural System of classification becomes apparent to us.

Apart from the tree diagram of Bronn, not published until 1858 (see Chapter 2, Section IV), I know of no other naturalist before Darwin and Wallace who used the simile of the irregular tree. Leaving aside the question of whether Darwin copied any features of his elaborate simile from Wallace, it seems more than mere coincidence that Darwin and Wallace had cited the simile of the tree to picture the pattern of classification (the *explanandum*) as a vast irregular dendrogram and had both proposed a theory of phylogeny (the *explanans*) to explain it. To them the correct method of classification yielded the natural and real pattern of classification. The latter was an empirical and contingent phenomenon crying out for explanation. The explanation was that the apparent relationships of classification were real.

After the publication of *On the Origin of Species,* tree production became a growth industry among naturalists (to coin a multiple play upon words). Most famous of treelike representations of both classification and phylogeny were the illustrations produced for several of his books by Ernst Haeckel. Haeckel's famous "woody trees" appear first in the second volume of his *Generelle Morphologie* (1866) as eight consecutive plates. The first is a general tree showing in outline a "*monophyletische Stammbaum der Organismen*" divided into three major branches for the three kingdoms (Plantae, Protista, Animalia). There then follows a tree for all plants, then separate ones for the principal animal phyla (*Coelenteraten oder Acephalen, Echinodermen, Articulaten, Mollusken, Wirbelthiere*), and finally "*Stammbaum der Saugethiere mit Inbegriff des Menschen*". All are labelled "*entworfen und gezeichnet von Ernst Haeckel, Jena 1866*", but only the echinoderm and vertebrate trees are claimed to be "*palaeontologisch begründet*". These two are also unique in having a vertical axis of the standard geological periods (*Laurentische, Cambrische, Silurische . . . Caenolithische, Recent*) with an interesting variant: each period is preceded by a corresponding anteperiod. In the echinoderm tree, these anteperiods are scaled as of relatively modest length, and, with the exception of the *Antetrias,* relatively little of phylogenetic importance appears to have happened in them; but in the vertebrate tree (Fig. 3.4), each anteperiod appears to have been longer than its corresponding period, and most important cladogenetic events appear to have occurred near or within them.

As *pattern* this reminds us of the *scala naturae* sequences in the dendrogram of classification of Haeckel's which we saw in the last chapter. However, in fairness to Haeckel, it must be noted that in *Anthropogenie* (1874; the fifth edition was translated as *The Evolution of Man;* the quotation is from the 1910 issue), he makes the point that

> you must never confuse direct descendants with collateral branches, nor extinct forms with living. . . . When for instance, we say that man descends from the ape, this from the half-ape (lemur), and the lemur from the marsupial, many people imagine we are speaking of the living species of these orders of mammals which they find stuffed in our museums. . . . All these living forms have diverged more or less from the ancestral form; none of them could engender the same posterity that the stem-form really produced thousands of years ago.

Haeckel is often thought of as the promoter of the "missing-link" fallacy; specifically that the common ancestor of apes and man would have been a morphological average of the respective living forms, an intermediate rung on the *scala naturae.* He did indeed introduce the generic name (a *nomen nudum*) of *Pithecanthropus* in the classification in his *Generelle Morphologie* (1866, vol. 2, p. CLX); and later in his *Natürliche Schöpfungsgeschichte* (1868), he described *Pithecanthropus alalus,* his speechless ape man. The name *Pithecanthropus* was subsequently appropriated by Dubois (1894), who eventually found the first specimens of the hominid now known as *Homo erectus* (Reader 1981).

The theoretical basis for all of Haeckel's trees was his best-known contribution to evolutionary theory, his "fundamental law of organic evolution", the *Biogenetic Law.* As Gould (1977b, p. 76) points out, the law of recapitulation was "discovered" many times in the decade following 1859, and it had a long history before that (Chapter 2). It is, however, always associated with the name of Haeckel, who expresses it in various ways including the aphoristic "ontogeny recapitulates phylogeny":

> In view of these facts, we may now give the following more precise expression to our chief law of biogeny: – The evolution

Fig. 3.4. Palaeontologically based phylogeny of the vertebrates; inset (in the original) "grade-tree" of vertebrates (see Fig. 3.5). (From Haeckel 1866.)

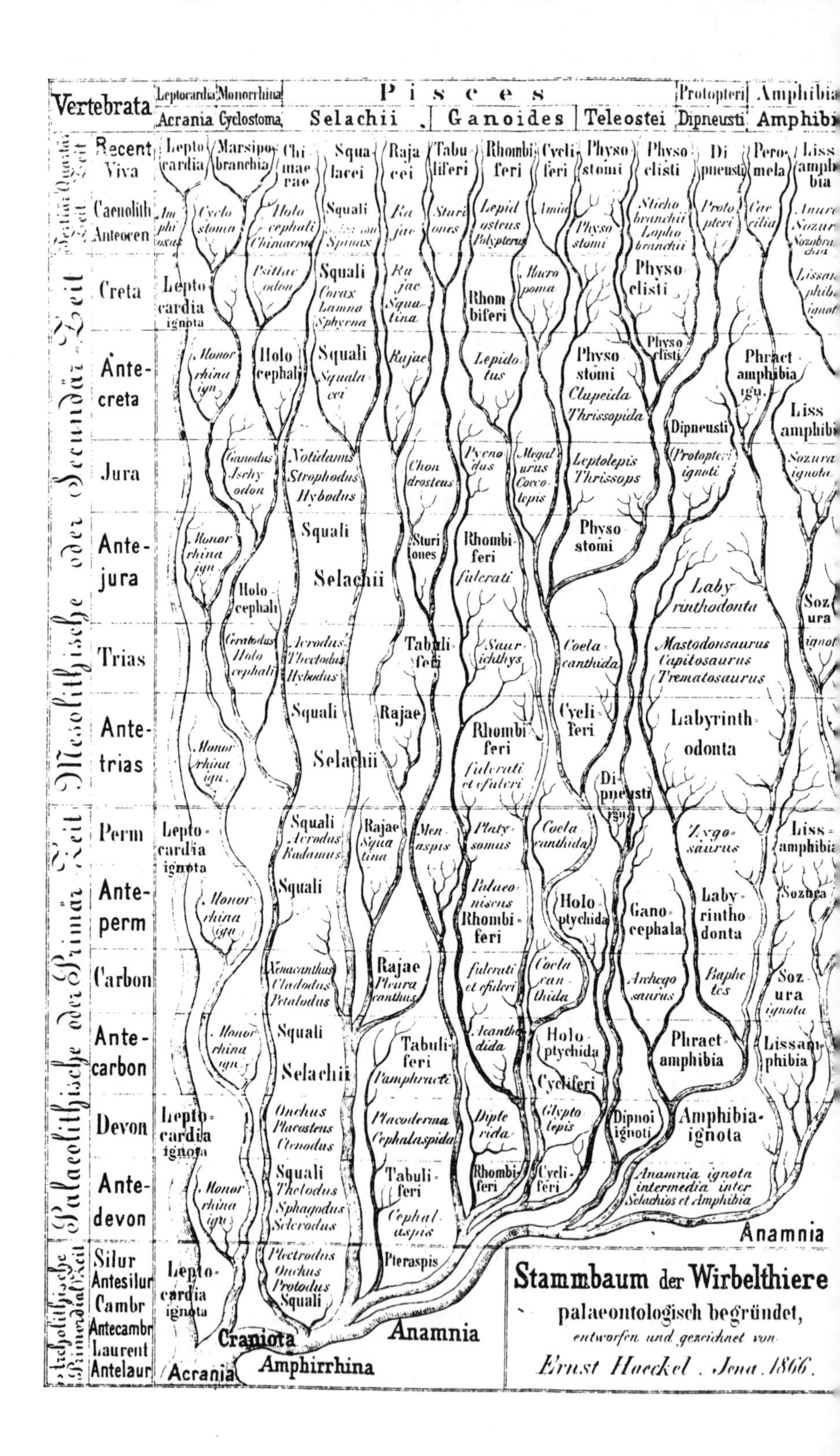

Vertebrata
Leptocardia
Monorrhina
Pisces
Protopteri
Amphibia
Acrania
Cyclostoma
Selachii
Ganoides
Teleostei
Dipneusti
Amphibia
Recent Viva
Caenolith Anteocen
Creta
Ante-creta
Jura
Ante-jura
Trias
Ante-trias
Perm
Ante-perm
Carbon
Ante-carbon
Devon
Ante-devon
Silur
Antesilur
Cambr
Antecambr
Laurent
Antelaur
Mesolithische oder Secundär-Zeit
Palaeolithische oder Primär-Zeit
Leptocardia
Marsipobranchia
Chimaerae
Squalacei
Rajacei
Tabuliferi
Rhombiferi
Cycliferi
Physostomi
Physoclisti
Dipneusti
Peromela
Lissamphibia
Holocephali
Chimaera
Squali
Spinax
Rajae
Sturiones
Lepidosteus
Polypterus
Amia
Physostomi
Stichobranchii
Lophobranchii
Protopteri
Caecilia
Anura
Sozura
Leptocardia ignota
Psittacodon
Squali
Corax
Lamna
Sphyrna
Rajae
Squatina
Rhombiferi
Macropoma
Physoclisti
Monorrhina ign.
Holocephali
Squali
Squalacei
Rajae
Lepidotus
Physostomi
Clupeida
Thrissopida
Phractamphibia ign.
Dipneusti
Lissamphibia
Ganodus
Ischyodon
Notidanus
Strophodus
Hybodus
Chondrosteus
Pycnodus
Megalurus
Coccolepis
Leptolepis
Thrissops
Protopteri ignoti
Sozura ignota
Monorrhina ign.
Squali
Selachii
Sturiones
Rhombiferi fulcrati
Physostomi
Holocephali
Labyrinthodonta
Ceratodus
Holocephali
Acrodus
Thectodus
Hybodus
Tabuliferi
Saurichthys
Coelacanthida
Mastodonsaurus
Capitosaurus
Trematosaurus
Squali
Selachii
Rajae
Rhombiferi fulcrati et efulcri
Cycliferi
Labyrinthodonta
Monorrhina ign.
Dipneusti ign.
Leptocardia ignota
Squali
Acrodus
Radamus
Rajae
Squatina
Menaspis
Platysomus
Coelacanthida
Zygosaurus
Lissamphibia
Monorrhina ign.
Squali
Palaeoniscus
Rhombiferi
Holoptychida
Ganocephala
Labyrinthodonta
Sozura
Xenacanthus
Cladodus
Petalodus
Rajae
Pleuracanthus
fulcrati et efulcri
Coelacanthida
Archegosaurus
Baphetes
Sozura ignota
Monorrhina ign.
Squali
Selachii
Tabuliferi
Pamphracti
Acanthodida
Holoptychida
Cycliferi
Phractamphibia
Lissamphibia
Leptocardia ignota
Onchus
Placosteus
Ctenodus
Placoderma
Cephalaspida
Dipterida
Glyptolepis
Dipnoi ignoti
Amphibia ignota
Monorrhina ign.
Squali
Thelodus
Sphagodus
Sclerodus
Tabuliferi
Cephalaspis
Rhombiferi
Cycliferi
Anamnia ignota intermedia inter Selachios et Amphibia
Anamnia
Leptocardia ignota
Plectrodus
Onchus
Protodus
Squali
Pteraspis
Craniota
Anamnia
Acrania
Amphirrhina
Stammbaum der Wirbelthiere
palaeontologisch begründet,
entworfen und gezeichnet von
Ernst Haeckel. Jena 1866.

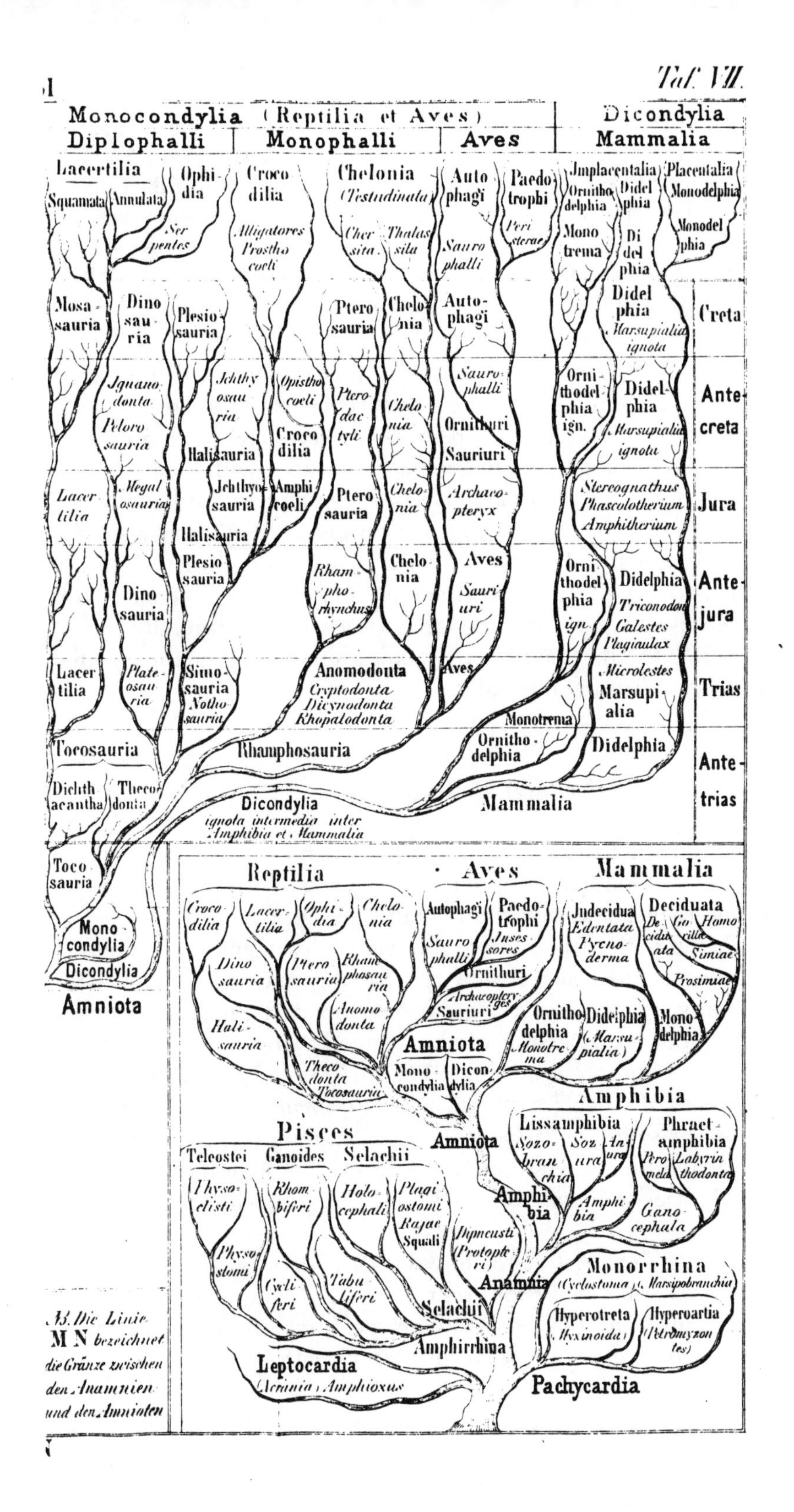
Taf. VII.
Monocondylia (Reptilia et Aves)
Dicondylia
Diplophalli
Monophalli
Aves
Mammalia
Lacertilia
Squamata
Annulata
Ophidia
Serpentes
Crocodilia
Alligatores
Prosthocoeli
Chelonia
(Testudinata)
Chersita
Thalassita
Autophagi
Paedotrophi
Peristerae
Sauro phalli
Jmplacentalia
Ornithodelphia
Didelphia
Placentalia
Monodelphia
Monotrema
Didelphia
Monodelphia
Mosasauria
Dinosauria
Plesiosauria
Pterosauria
Chelonia
Autophagi
Didelphia
Marsupialia ignota
Creta
Jguanodonta
Pelorosauria
Jchthyosauria
Opisthocoeli
Crocodilia
Pterodactyli
Chelonia
Saurophalli
Ornithuri
Sauriuri
Ornithodelphia ign.
Didelphia
Marsupialia ignota
Antecreta
Halisauria
Lacertilia
Megalosauria
Jchthyosauria
Amphicoeli
Pterosauria
Chelonia
Archaeopteryx
Stereognathus
Phascolotherium
Amphitherium
Jura
Halisauria
Plesiosauria
Dinosauria
Rhamphorhynchus
Chelonia
Aves
Saururi
Ornithodelphia ign.
Didelphia
Triconodon
Galestes
Plagiaulax
Antejura
Lacertilia
Plateosauria
Simosauria
Nothosauria
Anomodonta
Cryptodonta
Dicynodonta
Rhopalodonta
Aves
Microlestes
Marsupialia
Monotrema
Trias
Tocosauria
Rhamphosauria
Ornithodelphia
Didelphia
Antetrias
Dichthacantha
Thecodontia
Dicondylia
ignota intermedia inter Amphibia et Mammalia
Mammalia
Tocosauria
Monocondylia
Dicondylia
Amniota
Reptilia
Aves
Mammalia
Crocodilia
Lacertilia
Ophidia
Chelonia
Autophagi
Paedotrophi
Insessores
Saurophalli
Jndecidua
Edentata
Pycnoderma
Deciduata
Decidua
Gorilla
Homo
Simiae
Prosimiae
Dinosauria
Pterosauria
Rhamphosauria
Ornithuri
Anomodonta
Archaeopteryges
Sauriuri
Ornithodelphia
Monotrema
Didelphia
(Marsupialia)
Monodelphia
Halisauria
Amniota
Thecodonta
Tocosauria
Monocondylia
Dicondylia
Amphibia
Pisces
Amniota
Lissamphibia
Sozobranchia
Sozura
Anura
Phractamphibia
Peromela
Labyrinthodonta
Teleostei
Ganoides
Selachii
Physoclisti
Rhombiferi
Holocephali
Plagiostomi
Rajae
Squali
Dipneusti
(Protopteri)
Amphibia
Amphibia
Ganocephala
Physostomi
Cycliferi
Tabuliferi
Anamnia
Monorrhina
(Cyclostoma (Marsipobranchia)
Hyperotreta
(Myxinoida)
Hyperoartia
(Petromyzontes)
Selachii
Amphirrhina
Leptocardia
(Acrania) Amphioxus
Pachycardia
N.B. Die Linie MN bezeichnet die Gränze zwischen den Anamnien und den Amnioten

of the foetus (or *ontogenesis*) is a condensed and abbreviated recapitulation of the evolution of the stem (*phylogenesis*); and this recapitulation is the more complete in proportion as the original development (or *palingenesis*) is preserved by a constant heredity; on the other hand, it becomes less complete in proportion as a varying adaptation to new conditions increases the disturbing factors in the development (or *cenogenesis*).

(*Haeckel 1910, p. 8*)

As far as the pattern of Haeckel's trees is concerned, however, the causality of the biogenetic law is reversed. The early trees, as in the first edition of *Generelle Morphologie* (1866), represent a hierarchical arrangement in general terms and, with the exception of those two that are *palaeontologisch begründet,* could be redrawn as cladograms. By the first edition of *Natürliche Schöpfungsgeschichte* (1868), Haeckel had shifted his focus of attention more specifically to the ancestry of man, a theme continued in *Arthropogenie* (1874) and represented by a majestic tree whose massive trunk dominates (Fig. 3.5). In this case the early stages represent ancestors created by Haeckel from embryology.

III. *Stufenreihen* and cladograms

Haeckel's two terms *palingenesis* and *c(a)enogenesis,* introduced in the quotation above, are the analogues in ontogeny of the internalist and environmentalist components of phylogeny which we noticed in talking about the phylogenies of Lamarck, Chambers, and Haeckel himself. Arising out of his theory, we must now reintroduce a concept developed in Chapter 2. This is the phylogenetic version of the sequence of taxa in the tree of Porphyry, known after Abel (1911, 1924, 1929) as a *Stufenreihe* ("step-series"). Abel distinguished *Stufenreihen* from *Ahnenreihen* ("ancestor-series"). Palaeontologists had long realised that it was improbable that any simultaneously morphological and chronological series of fossils they had (i.e., a morphocline that was also

Fig. 3.5. A tree to show human ancestry. Note that the vertical axis (trunk) represents grade (but conflated with time) and that ancestors on the trunk are reconstructed from embryology. (From Haeckel 1874.)

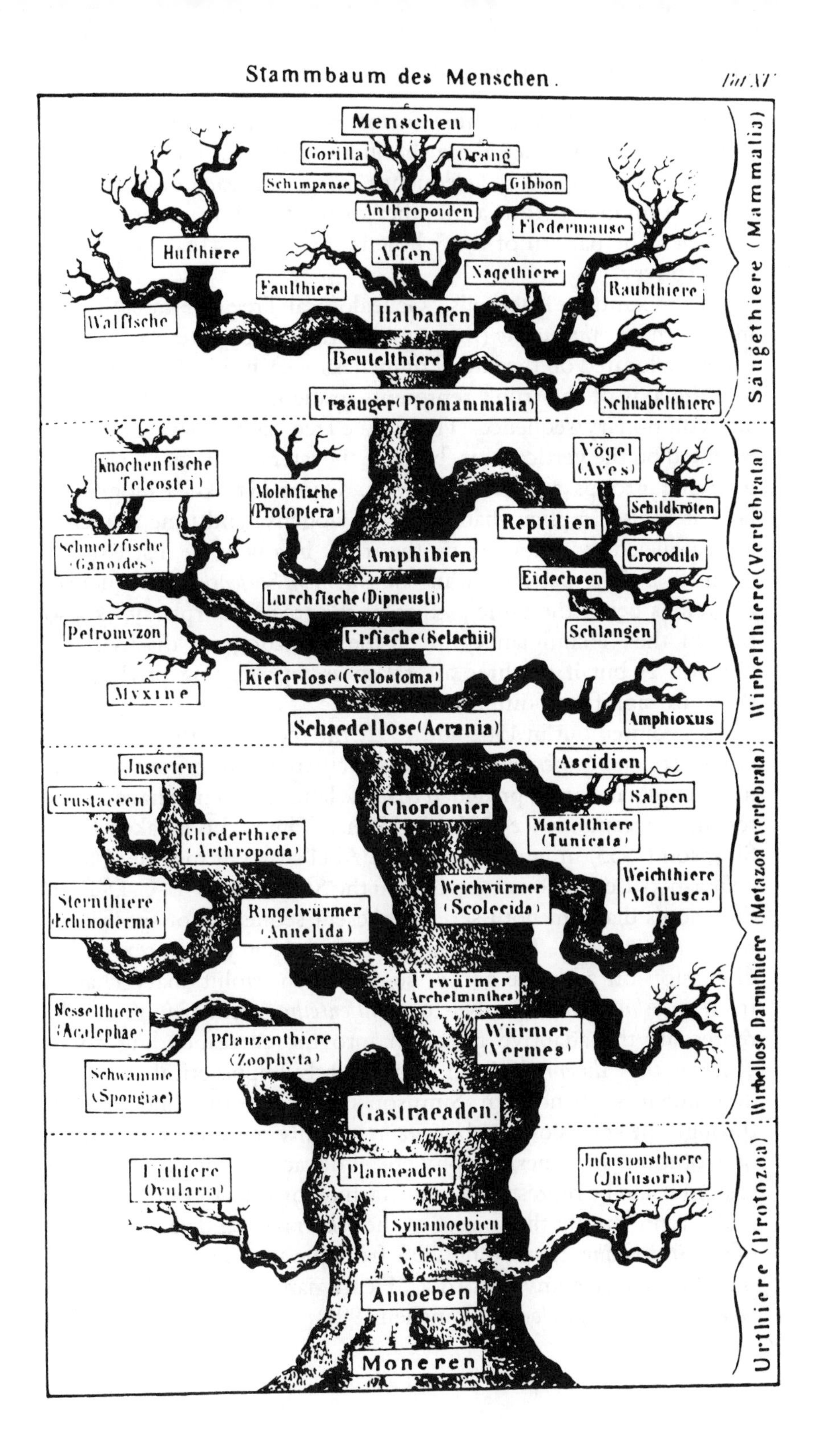
Stammbaum des Menschen.
Menschen
Gorilla
Orang
Schimpanse
Gibbon
Anthropoiden
Fledermause
Hufthiere
Affen
Nagethiere
Faulthiere
Raubthiere
Walfische
Halbaffen
Beutelthiere
Ursäuger (Promammalia)
Schnabelthiere
Säugethiere (Mammalia)
Knochenfische (Teleostei)
Vögel (Aves)
Molchfische (Protoptera)
Schildkröten
Reptilien
Schmelzfische (Ganoides)
Amphibien
Crocodile
Eidechsen
Lurchfische (Dipneusti)
Petromyzon
Urfische (Selachii)
Schlangen
Kieferlose (Cyclostoma)
Myxine
Amphioxus
Schaedellose (Acrania)
Wirbelthiere (Vertebrata)
Ascidien
Insecten
Crustaceen
Salpen
Chordonier
Mantelthiere (Tunicata)
Gliederthiere (Arthropoda)
Weichthiere (Mollusca)
Sternthiere (Echinoderma)
Weichwürmer (Scolecida)
Ringelwürmer (Annelida)
Urwürmer (Archelminthes)
Nesselthiere (Acalephae)
Würmer (Vermes)
Pflanzenthiere (Zoophyta)
Schwamme (Spongiae)
Gastraeaden.
Wirbellose Darmthiere (Metazoa evertebrata)
Planaeaden
Infusionsthiere (Infusoria)
Synamoebien
Amoeben
Moneren
Urthiere (Protozoa)

a chronocline) was a genuine ancestor-descendant sequence even at the species level, let alone the individual one. Thus the series would be interpreted as a sequence of species, each divergent by a speciation event from a member of the true but unknown *Ahnenreihe*. The pattern of the whole is the same as that of the tree of Porphyry.

As an example, Haeckel's "Genealogical Tree of Humanity" is in pattern an *Ahnenreihe* (the trunk) paralleled by two *Stufenreihen* (the taxa located on the lateral branches on each side). In Haeckel's tree the axis of the trunk represents grade (*scala naturae*) and also an evolutionary sequence. The whole tree cannot have time represented on the vertical axis, because the lateral branches terminate in living groups at different levels. Given Haeckel's biogenetic law, it is also implicit that grade, phylogeny, and time sequence are paralleled by ontogeny up the trunk in a way we discussed in Chapter 2. The pattern of an *Ahnenreihen/Stufenreihen* sequence, as well as being the same as that of the tree of Porphyry, is also that of the "Hennigian comb" type of cladogram discussed in Chapter 2, but if all three are taken as representing phylogeny, there are significant differences.

As I pointed out in 1982 and 1991, the reconstruction of an *Ahnenreihe* from a *Stufenreihe* involves exactly the assumptions implicit in a Hennigian comb, provided that the latter is taken to have a one-to-one correspondence with phylogeny. Figure 3.6 is taken from Simpson (1953) in which he revives Abel's two terms: the *Ahnenreihe* a-b-c-d-e is reconstructed from the *Stufenreihe* A-B-C-D-E. In Simpson's diagram both series represent sequences of species, or rather populations. The diagram is presented in the context of his discussion on adaptive zones, so that each evolutionary change from an *Ahnenreihe*-member to a *Stufenreihe*-member (e.g., a--A) represents the adaptive radiation into the corresponding zone, whereas the *Ahnenreihe* itself is an entirely hypothetical ancestor-descendant sequence. In Simpson's diagram the line passing through the stages of the *Ahnenreihe* actually represents the course of evolution (anagenesis or phyletic evolution in this case), but the system could be represented in the fashion of a cladogram, with the nodes representing the species of the *Ahnenreihe* and each member of the *Stufenreihe* terminating a line from the axis. The whole would then be exactly equivalent to a "Hennigian comb", *if the latter is taken to represent phylogeny* (Panchen 1991).

However, the Hennigian comb, like any other cladogram, is also (or exclusively) a representation of character-distribution giving a hierarchical clustering: as a representation of phylogeny it is a tree (in the cladist sense). As we pointed out in the previous chapter (Section II), a Hennigian comb cladogram is identical not only in pattern but also in logical structure to the tree of Porphyry, except that in the latter, as I pointed out, the terminal taxa arising from each node (the members of the *Stufenreihe*!) are not specified. The Hennigian comb is particularly characteristic of the work of palaeontological cladists.

Its rationale is explained by Patterson and Rosen (1977). It was one of Hennig's *dicta* that of a pair of sister-groups of extant orga-

Fig. 3.6. Abel's concept of the *Stufenreihe*, as used by Simpson (1953). The *Ahnenreihe* a-b-c-d-e represents phyletic evolution. The *Stufenreihe* members A to E represent evolution by cladogenesis into a succession of adaptive zones (clear cells). (From A. L. Panchen, "The Early Tetrapods: Classification and the Shapes of Cladograms", in *The Origin of the Higher Groups of Tetrapods,* edited by Hans-Peter Schultze and Linda Treub. Cornell University Press, 1991. Used by permission of the publisher. After Simpson.)

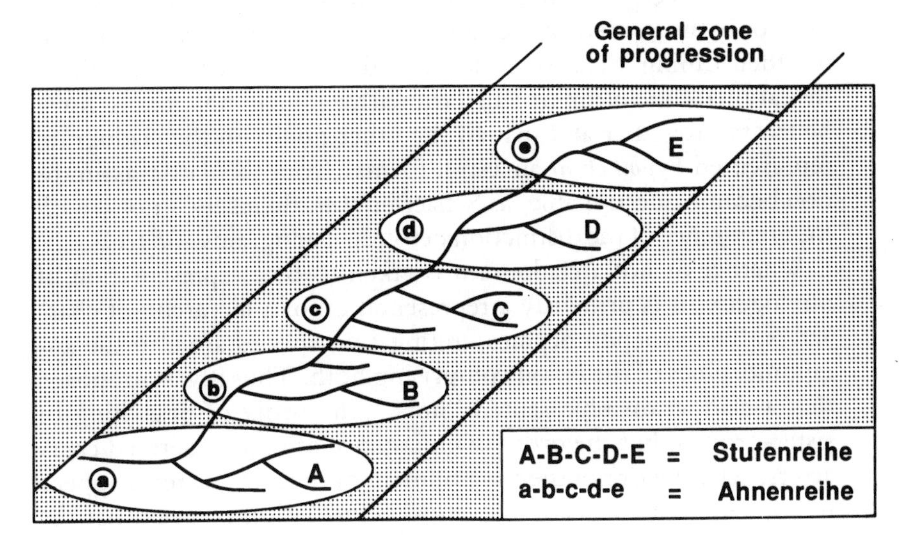

nisms one would be found to have more unique (autapomorph) characters than the other more primitive one (the "deviation rule"). Patterson's and Rosen's work was on the Teleostei (advanced bony fish) whose living sister-group they decided is the Halecomorphi, represented by the primitive extant freshwater fish *Amia*. Having agreed on this pairing of two extant groups, their concern was then to place a series of fossil fish that either fell within the teleosts as diagnosed at that time (i.e., had all the teleost autapomorphies) or possessed some but not all teleost autapomorphies. These latter fish were arranged in a *scala naturae* of cumulative teleost characters (Fig. 3.7), but if they are to be good taxa they also have autapomorphies of their own. Thus the two extant taxa (Halecomorphi and Teleostei) are the living sister-groups of each other. The series of fossil forms are "stem-group" teleosts (Hennig 1966), and the Teleostei are the "crown-group" (Jefferies 1979). Because it is hierarchical, each terminal taxon in a Hennigian comb has a different rank, but to avoid the necessity of assigning each to a different category, the fossil taxa are labelled "plesions" so that they do not upset the categories of living taxa in the final classification.

The incorporation of fossil species into cladistic analysis was a perplexing problem in the early stages of the development of Hennigian cladism. At first, fossils were sometimes placed at a lower position on the cladogram than extant forms, on the assumption that the vertical axis represented "absolute time" (e.g., Brundin 1968); however, it soon became clear that cladistic analysis could yield only the *sequence* of speciation events leading to extant species, not their timing. Schaeffer, Hecht, and Eldredge (1972) then drew the rational conclusion that all terminal taxa, including fossils, must be treated together and that all the other nodes on the cladogram represented *hypothetical* ancestral forms. Even this latter assumption became questionable, as Schaeffer *et al.* themselves note, with the emergence of the distinction between cladograms and trees.

Whether or not a cladogram is taken to be a representation of phylogeny, it is primarily a representation of a hierarchy of characters. The synapomorphies uniting two species are also the autapomorphies of the taxon comprising both. Thus it was proposed that a cladogram represented "not . . . the order of branching of sister-groups, but the order of emergence of unique derived characters, whether or not the development of these characters happens to coincide with speciation events" (Hull 1980); or even that the

Fig. 3.7. Cladogram of extant and some fossil teleost taxa. Extant taxa: double lines; the numbers in circles and their double arrowhead lines give the range of possible placements of fossil taxa *incertae sedis*. (From Patterson and Rosen 1977.)

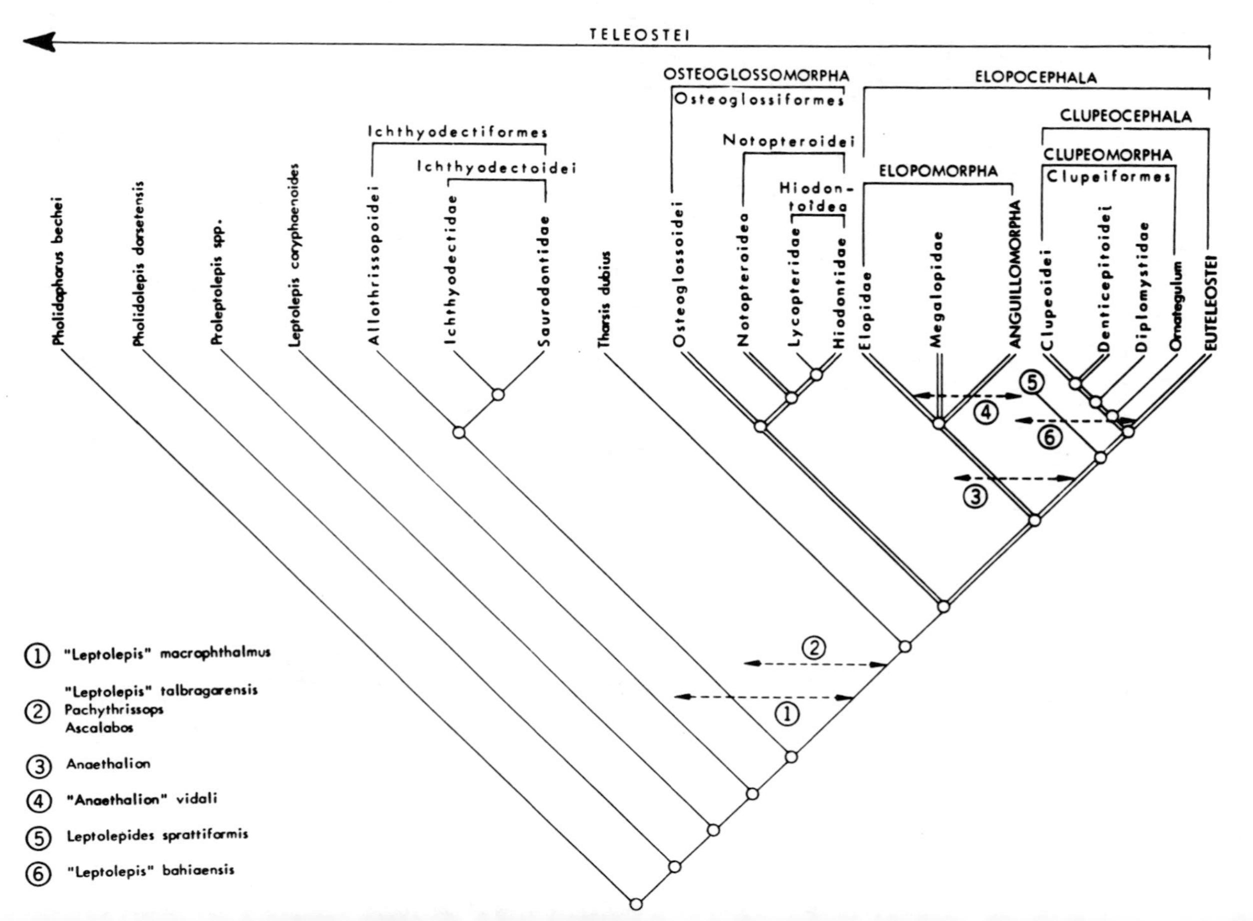

cladogram was "an atemporal concept... a synapomorphy scheme" (Nelson 1979). If a cladogram is interpreted as phylogeny, it then becomes a tree in the restricted sense in which cladists use that term. As defined by Eldredge (1979a), a cladogram is "*a branching diagram depicting the pattern of shared similarities thought to be evolutionary novelties ("synapomorphies") among a series of taxa*", and a tree (*"phylogenetic tree" or "phylogram"*) is "*a diagram* (not necessarily branching!) *depicting the actual pattern of ancestry and descent among a series of taxa*" (italics as in the original).

Most modern cladists would remove any reference to phylogeny (e.g., the phrase "thought to be evolutionary novelties") from the definition of a cladogram to minimise the assumptions inherent in cladistics as a method of classification. There is, however, less unanimity about the relationship of cladograms and trees. One school of thought believes that there is only one valid tree corresponding to any given cladogram – that which is isomorphic with it – whereas another (the so-called transformed cladists) holds that there is a series of trees corresponding to every cladogram. This controversy is further discussed in Chapter 8.

IV. Reticulate phylogeny

It is well known that amongst plants new species can arise by hybridisation (e.g., Wagner 1968, 1969). Furthermore, it is possible to test the nature of such "intertaxa" by breeding a similar form from the putative parent species (Wagner 1983) or by other methods (Wanntorp 1983). Rosen (1978, 1979) has also suggested that populations of the fish genera *Heterandria* and *Xiphophorus* are intra-generic hybrids and should be interpreted as such in both taxonomic and area cladograms in an exercise in vicariance biogeography (Chapter 5).

The recognition of hybrid species has a profound effect on the methodology of phylogeny reconstruction. That effect is considered in Chapter 10; here we are concerned only with pattern. In a phylogenetic tree, "intertaxa" must obviously be represented by lines joining the hybrid to its putative parents so that the pertinent part of the tree has a convergent as well as a divergent pattern. In other words it is *reticulate* (Humphries 1983; Nelson 1983; Wagner 1983; Wanntorp 1983). If speciation by hybridisation between sister species were as common as speciation by cladogenesis producing sister species, the "trees" (in the cladist sense) would appear as meshes or networks like fishing nets or chicken wire, rather than representing a divergent hierarchical pattern. Because the existence of the phenomenon of species production by hybridisation is known, whereas its extent is not, how much reticulation the "true" tree should contain cannot be known. There is, moreover, a further complication: hybrid species may result from the pairing of species more distantly related than sister-groups (see the works cited in this section) so that the pattern of the tree is more complex.

If cladograms are to be distinguished from trees, it is axiomatic that the former be divergent throughout (e.g., Patterson 1980a). However, Wagner (1983), Bremer and Wanntorp (1979), and Wanntorp (1983) all suggest that cladograms should be reticulate if cladistic analysis suggests that a taxon is hybrid, whereas Nelson and Platnick (1980, 1981) state that no cladogram should show any reticulations. According to the latter authors, a hybrid that favours neither parent species over the other (i.e., that shares an equal number of apparent synapomorphies with each parent species) will appear with both in an unresolved trichotomy. Otherwise it will appear as sister-group of one or the other. If the hybridisation is between non–sister species, a polytomy will result, which can be resolved as an "Adams consensus tree" (Adams 1972; Nelson 1983).

V. Conclusions

At the beginning of this chapter, I was at pains to emphasise that patterns of classification of organisms, produced by taxonomists, were logically prior to patterns of evolution, so that patterns of phylogeny were hypotheses to explain the views of their authors on the nature of patterns of classification. This does not mean, however, that there is any justification for the *a priori* assumption that the pattern will be a divergent hierarchy. As we saw in the previous chapter, the founders of modern taxonomy, notably Linnaeus, continued the Aristotelian tradition of classification according to the principles of logical division (Chapter 6) without assuming that the natural order of organisms was represented by a hierarchical clustering; hence Lamarck's distinction between *classification* and *distribution générale* (Chapter 2, Section I). This traditional separation is upheld today by the pheneticists, who produce phenograms that are divergent and hierarchical but invoke a different series of techniques for the reconstruction of phylogeny (Chapter 9). In cladistics, Hennig's original intention was that the patterns of classification and phylogeny should be isomorphic, whereas the "transformed cladists" have divested themselves of the assumption that they are guided by the nature of phylogeny. The transformed cladists are unique, however, in that they reject evolution as an *a priori* assumption, but accept as axiomatic the assumption that the natural order of organisms is a divergent hierarchy.

Lamarck, as we have seen, became an evolutionist in the taxonomic tradition of Linnaeus, but modified his views on the pattern of phylogeny when he was no longer able to accept a single *scala* each for plants and animals. It is worth noting, however, that the literal arrangement of Lamarck's original pattern based on a simple *scala* would have been a series of parallel phyletic lines identical to one another as far as they were complete (Fig. 3.8). The reason is that spontaneous generation was supposed to have occurred throughout the history of life so that any taxon was polyphyletic to the highest possible degree (Winsor 1976). This complication did not, how-

Fig. 3.8. The steady-state theory of evolution as originally envisaged by Lamarck. Parallel oblique lines represent phylogeny in parallel of separate, spontaneously generated organisms. Once the earliest of these reaches the human condition, the fauna of the world (in types, if not in relative numbers) remains constant.

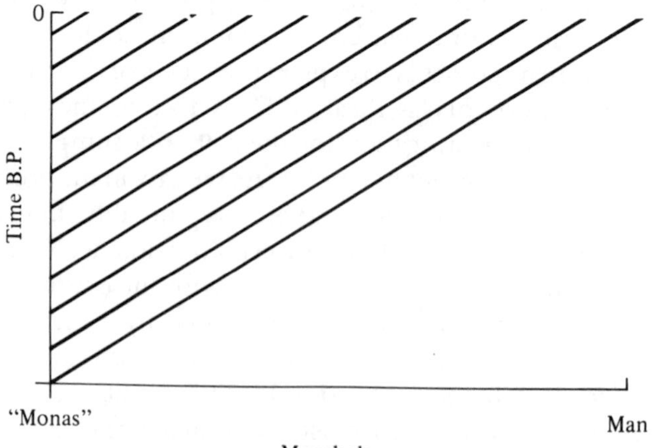

ever, appear in Lamarck's diagrams. Whatever Lamarck's views on phylogeny, his classification is more one of grades than clades, but not consistently either. This situation prevailed in taxonomy until the widespread acceptance of cladistics. Most authors reconstructed phylogeny to yield some semblance of a cladistic pattern but deliberately conflated grade and clade differences in constructing a classification – a procedure justified by Mayr by his "genes-in-common" criterion, which is essentially a phenetic one.

We must turn now to patterns of phylogeny that are both divergent and hierarchical and thus derived from the precedent set by Darwin and Wallace. If these reconstructed patterns have real taxa only at the ends of the branches – that is, as terminal taxa – they are Steiner trees. Thus cladistics yields only Steiner trees, and the tree of Porphyry/*Stufenreihe*/Hennigian comb pattern is a Steiner tree. As a *Stufenreihe* the terminal taxa represent both a chronocline and a (less than perfect) morphocline. The successive nodes represent an ancestor-descendant sequence, the *Ahnenreihe,* which is wholly hypothetical. It is also, of course, a phyletic *scala naturae.* The same pattern produced as a cladogram (Hennigian comb) is in addition a hierarchical structure in which each terminal taxon has a different rank from those adjacent to it (Panchen 1991). As used by palaeontologists, the Hennigian comb has, or should have, an extant taxon at each end of the series; the intermediate ones may all be fossil. The diagrams of phylogeny produced by noncladist palaeontologist were, and still are, a mixture in various proportions of Steiner and spanning trees. In some cases of "spindle diagrams" (Chapter 2, Section IV), it is often difficult to say which type of tree any part of the diagram represents.

In this chapter and in Chapter 2, it can be seen how difficult it is to disentangle patterns of classifications and patterns of phylogeny. The historical and logical priority of the former is there in theory and is insisted on by the transformed cladists. Unfortunately, their whole methodology depends on the assumption that the natural order of organisms is a divergent hierarchy. If it were not so, they, like the pheneticists, would continue to produce dichotomous dendrograms because their method would conceal the true pattern. It is ironical that the only feature of phylogenetic pattern that is corroborated beyond reasonable doubt is the reticulation produced by hybrid species in plants. That feature of pattern should be reflected back from phylogeny to taxonomy, but cladistics cannot do other than reject it.

4

Homology and the evidence for evolution

In Chapters 2 and 3, I set out to suggest that the theory that evolution has occurred is the *explanans* of the phenomenon of natural classification (the *explanandum*). We also saw that the hierarchical pattern of classification, apart from being the normal way in which classifications of objects are presented, is the result, at least historically, of an accepted method. That method, described in Chapter 6, is logical division. But since the time of Darwin and Wallace, the resulting hierarchy has been regarded as the natural pattern. Most methods of classifying organisms are therefore designed to produce a divergent hierarchy not just for convenience but because it is presumed to correspond to a real phenomenon in nature.

But classifications depend on characters possessed by the objects to be classified. Thus if the natural arrangement of organism is a divergent hierarchy, there must be an underlying divergent hierarchy of characters, and there should be something more to a character correctly uniting two species than a coincident resemblance. Behind the idea that one is dealing with the "same" character in two species, or higher taxa, is the concept of homology. Thus as natural classification is logically prior to phylogeny, homology is perhaps to be regarded as logically prior to natural classification.

It is not the case, however, that the concept of homology arose before that of natural classification in the history of systematics. The development of the homology concept as the basis of systematic biology took place in the early years of the nineteenth century.

It is no coincidence, except in a literal sense, that systematic theories of phylogeny were first proposed at the same time.

In this chapter and the next, we shall look at the traditional "evidence for evolution", including the evidence for "affinities". However, bearing in mind the relationship between classification and evolution, we must ask whether there is any "evidence for evolution" that is not in fact taxonomic. We shall see that all taxonomic evidence, apart from being evidence for classification rather than for evolution, is based on the concept of homology. That is not to say, however, that *all* cases of homology are the raw material of taxonomy.

I. Homology

The classic account of the history of the concept of homology is E. S. Russell's *Form and Function* (1916). There is also a recent and excellent historical review in Rieppel (1988). Russell claims to find the idea of homology and even, implicitly, the distinction between homology and analogy in the *Historia Animalium* and *De Partibus Animalium* of Aristotle. He also draws attention to Belon's (1555) figures of the skeleton of a bird and a mammal showing homologous bones, but much more dubiously infers a distinction between homology and analogy in Buffon.

However, the formalisation of the concept of homology and the enunciation of the criteria by which it is identified arose within the tradition of transcendental morphology, whose most important figure was Etienne Geoffroy Saint-Hilaire. Before that Goethe (1807) had successfully applied the idea of homology in his interpretation of the leaves, sepals, petals, and stamens of flowering plants as modifications of a common type of appendage. He was also the first advocate of the vertebral theory of the vertebrate skull (Goethe 1817–24).

Geoffroy developed his ideas in a series of *mémoires* from 1796 onwards, culminating in his *Philosophie Anatomique* (1818). In his theory of homology, confusingly termed *théorie des analogues,* he sets out his principle of the unity of composition where, for example, all vertebrates are constructed from the same series of units or "organic materials". *"Analogues"* (i.e., homologues) are then identified using the *principe des connections* – that is, the principle of topographic similarity and relationships to surrounding structures. Geoffroy further employed a method of designating an animal in which a particular structure was most fully developed as a standard of comparison for homologues of that structure. Thus, in his studies of the vertebrate sternum, he took the plastron of chelonians as his standard, identified the nine elements that composed it, and then attempted to identify those elements in its proposed homologues. In doing this, he employed his third principle (after those of composition and connections), that of the *loi de balancement* – the idea that over-development of one component, or whole structure, was compensated for by the under-development of others.

Unfortunately, Geoffroy's attempt at a unifying theory of pure morphology over-reached itself and attracted the all-too-ready ridicule of his colleague and former friend Cuvier. Geoffroy had already adopted a division of the Animal Kingdom into four major taxa, corresponding to those of Cuvier, thus:

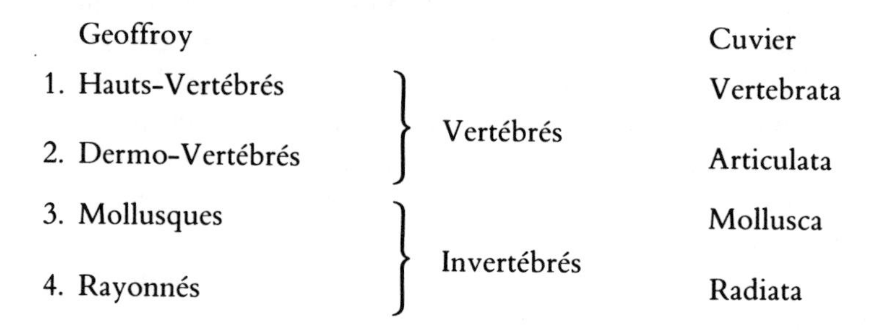

Geoffroy		Cuvier
1. Hauts-Vertébrés	Vertébrés	Vertebrata
2. Dermo-Vertébrés		Articulata
3. Mollusques	Invertébrés	Mollusca
4. Rayonnés		Radiata

The important differences between the two, however, were that whereas Cuvier regarded his four *embranchements* as distinct creations, whose form was related only to function, Geoffroy regarded them as derived from a common plan, with function as the explanation of the variations on that plan. Later in life, Geoffroy, when speculating on the possibility of transmutation, adopted a radical environmentalism, with adaptive evolutionary change dependent on direct action of the environment on embryos (Chapter 11).

Geoffroy's grouping of articulates (annelids and arthropods) and vertebrates together as *Vertébrés* is a symptom of his views. The exoskeleton of a lobster segment, representing one, was the homologue of a fish vertebra, representing the other. The lobster segment rotated through a right angle would give the vertebral centrum with its neural and haemal arches and spines; the viscera would then have to be rotated through a further right angle to bring the nerve cord to a dorsal position and the gut to a ventral one.

Not satisfied with this concept, he then wanted to bring his *Invertébrés* into the same system and welcomed the comparison, by Meyranx and Laurencet (1830), of a cephalopod with a vertebrate bent back at the umbilicus. This idea was ridiculed by Cuvier (1830), and thus began the famous, and increasingly acrimonious, Cuvier–Geoffroy debate (Appel 1987).

An attempt to reconcile the morphology of Geoffroy and the German *Naturphilosophen* with Cuvier's "functional anatomy" was made by Richard Owen. He combined the teleological approach of Cuvier, Geoffroy's principle of connections, and the concept of serial homology, taken from Oken and Goethe. In Owen (1843), we find the first clear definitions distinguishing analogy and homology. (It should be noticed, however, that in Owen's definitions comparable structures could be both analogous and homologous, similar in function *and* derivation.) Thus:

> *Analogue:* A part or organ in one animal which has the same function as another part or organ in a different animal.
> *Homologue:* The same organ in different animals under every variety of form and function.

Although Owen is usually credited with distinguishing the two for the first time, at least in English, we have already seen (Chapter 2, Section III) that MacLeay (1819, 1821) had introduced the concepts of affinity, based on homology, and parallelism, based on analogy in the classification of his ten orders of insects.

In his major work on comparative anatomy, *On the Archetype and Homologies of the Vertebrate Skeleton* (1848), Owen developed a theory of the derivation of the whole skeleton, skull and appendicular skeleton included, from modified vertebrae. The archetype was an idealised vertebra of which all the "vertebrae" of the skeleton were modifications. Thus Owen (1849) was able to distinguish three types of homology: (1) *special homology,* as in the comparison of "the same organ" within animals of different species; (2) *general homology,* as in the comparison of the structure in an animal species with that of the archetype; (3) *serial homology,* that among vertebrae, whole segments, and so forth, within one individual. The corresponding parts, he referred to as *homotypes.*

In the same publication, as Russell notes, Owen declared the primacy of comparative anatomy and the principle of connections over mode of embryological development in establishing homologies:

> There exists doubtless a close general resemblance in the mode of development of homologous parts; but this is subject to modification, like the forms, proportions, functions, and very substance of such parts, without their essential homological relationships being thereby obliterated. These relationships are mainly, if not wholly, determined by the relative position and connection of the parts, and may exist independently of form proportions, substance, function and similarity of development.
>
> (*Owen 1849, p. 6*)

This approach can, however, lead to problems: popular examples noted by Rieppel (1988) concern the descended testes of higher mammals and the pelvic girdle of advanced teleost fish. The latter may be situated in the same transverse plane as, or even in front of, the pectoral girdle. In both cases it may be claimed that homology of the derived condition of the structure with the primitive one demands a knowledge of its ontogeny to be unambiguous.

Gegenbaur (1870), writing post-*Origin,* had an evolutionary concept of homology (see below), but insisted on the use of ontogeny for the determination of homology, while emphasising that the interpretation of ontogeny depended ultimately on comparative anatomy. Gegenbaur also distinguished between homologous structures, regarding them as morphologically equivalent, and analogous structures, which were only "physiologically equivalent" (i.e., convergent in evolutionary terms), as had Owen, but Gegenbaur unfortunately used the terms "general" and "special" homology in a different way. Gegenbaur's *special* homology referred to the same phenomenon as Owen's, but the definition and interpretation were different: the English edition of Gegenbaur (1874), states that it is the name given "to the relations which obtain between two organs [each from a different organism] which have had a common origin, and which have also a common embryonic history". His definition was thus an evolutionary one, but nevertheless he was insistent that common descent was to be inferred from homology, and thus comparative anatomy, and not vice versa. Gegenbaur's *general* homology referred not to comparison with an archetype but rather was an extension of Owen's serial homology concept.

At roughly the same time, Ray Lankaster (1870) wrote "On the Use of the Term Homology in Modern Zoology, and the Distinction between Homogenetic and Homoplastic Agreements". (Note that the correct term is "homoplastic" – the adjective corresponding to the noun "homoplasy" – rather than "homoplasic" or even "homoplasious"!) Lankaster advocated rejection of the term "homology" altogether as tainted by Platonic idealism. Corresponding to modern usage, all functional resemblances were analogies. Those resemblances between "structures which are genetically related, in so far as they have a single representative in a common ancestor" are *homogenous,* while all others were *homoplastic:* "Homoplasy includes all cases of close resemblance of form which are not traceable to homogeny, all details of agreement not homogenous, in structures which are broadly homogenous, as well as in structures having no genetic affinity". Thus to Lankaster, serial homology was a case of homoplasy, as well as the result of parallel and convergent evolution.

We must now consider briefly the impact of evolutionary ideas on the concept of homology. Russell notes that Lamarck "arrived at a roughly accurate distinction between homologous and analogous structures. More importance, he thought, was to be attributed in classifying animals to characters which appeared due to the 'plan of Nature' than to such as were produced by an external modifying cause". Thus homology to Lamarck (although he did not use the term) may be taken to be a result of his first law of 1815, that of the innate *marche* of Nature, whereas analogy resulted from convergent adaptations due to the "inheritance of acquired characters" (see Chapter 11). The distinction between homology, resulting from phylogenetic relationship, and analogy, resulting from convergent adaptation, was also made by Darwin (1859) and Haeckel (1866).

As with classification in general, Darwin was clear that homology was part of the *explanandum* of which his theory of transmutation by natural selection was the *explanans:*

> What can be more curious than that the hand of a man, formed for grasping, that of a mole for digging, the leg of the horse, the paddle of the porpoise, and the wing of the bat, should all be constructed on the same pattern, and should include similar bones, in the same relative positions? Geoffroy St Hilaire has insisted strongly on the high importance of relative connexion in homologous organs: the parts may change to almost any extent in form and size, and yet they always remain connected together in the same order. . . . Hence the same names can be given to the homologous bones in widely different animals. . . . . . . If we suppose that the ancient progenitor, the archetype as it may be called, of all mammals, had its limbs constructed on the existing general pattern, for whatever purpose they served, we can at once perceive the plain signification of the homologous construction of the limbs throughout the whole class.

As we have seen, Gegenbaur and Lankaster both essayed an evolutionary ("genetic": Russell) definition of homology. Gegenbaur, in a passage quoted by Russell (1916, p. 265), appears to favour an evolutionary definition but then makes it clear that he regards establishment of homology as a prerequisite for hypotheses of phylogeny:

> Homology . . . corresponds to the hypothetical genetic relationships. In the more or the less clear homology, we have the expression of the more or less intimate degree of relationship. Blood-relationship becomes dubious exactly in proportion as the proof of homologies is uncertain.

Thus appears to have begun the practice of defining homology in terms of phylogeny, but using non-evolutionary criteria for establishing it. Russell (1916, p. 355ff.) gives an account of the way in which Oscar Hertwig expressed his dissatisfaction at this state of affairs, but it persists, perhaps as the majority view, to the present day. But in seeking the relationship between homology and evolution, homology as *explanandum* or evidence, we must revert to pre-evolutionary explanations of homology.

Earlier in this section, we saw that Owen's general homology was based on comparison of a single structure with its *archetype*. In the case of the vertebrate skeleton, each skeletal segment was compared to an idealised, or *archetypal* vertebra. This is precisely what Plato would have understood by an *Idea* (Chapter 6). Thus Owen's approach to homology is characterised by the stance of philosophical idealism: individual homologies are seen as transformations (not in the evolutionary sense) of the archetype. Terms related to archetype are "morphotype" and "*Bauplan*". Archetypes can be structures or whole organisms, such as the "generalised vertebrate" of elementary texts.

Patterson (1982) defines an archetype as "an idealisation with which features of organisms may be homologised by abstract transformations which entail no hypotheses of hierarchic grouping". He describes it as "conceived as Platonic essence". There is, however, an important distinction to be drawn between *Idea* and essence, and the embodiment of essence is represented by Patterson's definition of morphotype as "a list of the homologies (synapomorphies) of a group". This is exactly as the term "essence" would have been understood by Aristotle (Chapter 6). *Bauplan* does not seem to be distinguishable from the archetype as applied to a whole organism, but tends to be used with reference to the ground plan of phyla or other taxa of high rank (e.g., Valentine 1986).

II. The classification of homology

In this section I shall develop a new classification of types of homology. I do not believe that the categories I use form a hierarchy, but for convenience I shall give that classification a hierarchical pattern. Following the review by Patterson (1982), the primary division is into *Transformational* and *Taxic* homologies, a dichotomy introduced for evolutionary studies by Eldredge (1979a). Transformational homologies concern comparisons between two structures that differ in some significant way but are thought to be built on a common plan. If it is hypothesised that a structure (or structures) in one individual is a transformational homologue of a structure in another, then that is *singular* homology. Such a hypothesis need not be evolutionary: homology between the ear ossicles (incus and malleus) of mammals and the bones of the jaw articulation (quadrate and articular, respectively) of reptiles and other jawed vertebrates (the cliché example!) was proposed by Reichert (1837) without any phylogenetic implications.

I shall build on Owen's distinction between *general* and *special* homology to divide singular homologies. I use the term "general homology" when a structure in an individual is compared to that in a hypothetical reference, which could be either a hypothetical ancestor, or perhaps an archetype. When a comparison is made between corresponding structures in two real individuals, then the hypothesis in one of special homology. This will be the case, whether or not one of the individuals is thought to show the ancestral condition relative to the other.

Singular homology is opposed to *Iterative* homology, which is transformational homology in which a structure is interpreted as a member of a series. Iterative homology is discussed by Roth (1984, 1988). She makes the useful point (Roth 1984) that, analogous to the hierarchical arrangement of taxic homologues (i.e., homology as synapormorphy; see below), there is a hierarchy of similarity in serial homologues (iterative homologues in a single individual). One example given is:

VERTEBRAE
Cervical Vertebrae
Cervical Vertebrae Exclusive of Axis and Atlas
– vertebra c5 –

Paralogy in molecular biology (Chapter 9) is also cited as a case in which it is possible to homologise characters within individuals.

Roth (1984) makes one point about the serial homology of vertebrae which seems to me to lead to an important conclusion:

> One can use this example [homology of the fore and hind limbs of a single individual] to discuss an alternative definition for serial homology. Riedl (1978) [cited here as 1979] defined serial homologues as structures derived from structures which were identical in the common ancestor. Although as Bateson (1913) pointed out, the terminal members of a string of serial homologues are usually distinctive – the atlas was probably never identical to other vertebrae, though arguably one might consider it a serial homologue – nonetheless – this difficulty in Riedl's definition might be remedied by recognising degrees of serial homology.

I suggest that Roth's examples may be used to distinguish between Owen's general homology and special homology, for both singular and iterative homology. In the case of general homology, the principle evoked is pure idealism. Owen did not believe that his hypothetical vertebrae forming the skull ever existed as real vertebrae; the cranial "vertebrae" were taken to be homologues of anterior vertebrae in the archetype. Hence the significance of Roth's (and Bateson's) point about the atlas vertebra (whether true or not!). Patterson (1982) states that all transformational homology is validated by reference to archetypes, but this need not be the case. Two other possibilities occur to me.

The first possibility is where two structures are compared without any necessary phylogenetic assumption, as noted above for Reichert's original homologisation of ear ossicles and jaw bones. No archetype need be implicit. The criterion used, apart from Geoffroy's principle of connections, is, as suggested by Roth, similar developmental pathways in ontogeny. The second possibility is where a phylogenetic scenario is referred to overtly. This case may again be subdivided. First, it may be proposed that the condition in one organism, as might be the case with Reichert's homology for the jaw bones, is the plesiomorph or primitive state, and that in the other (as with ear ossicles) it is the apomorph or derived state. A hypothesis of phylogenetic transformation from one to the other, supported by comparative anatomy, embryology, and the fossil record, involves no invocation of an archetype. Second, the proposal might be that the conditions in both organisms, while different from each other, are said both to be derived from that of a common ancestor. The condition of a third organism might be cited as representing the ancestral one, without the implication that the organism represents the ancestral species; or the ancestral condition might be a hypothetical one. But it is my contention that in neither case is an archetype invoked. To sustain this argument it is necessary to show that the proposition of a common ancestor differs from that of an archetype. An example might be the famous conclusion of Goodrich (1916, 1930) that the heart and aortic arches of reptiles and birds (Sauropsida; see Huxley 1870) on one hand and those of mammals (Theropsida) on the other could not be derived one from the other but were divergent modifications of a condition at the same grade as that of living Amphibia. It is improbable that the nearest common ancestor of sauropsids and theropsids had an arrangement of heart and aortic arches identical to that of any living amphibian (Foxon 1955), but its condition (if one accepts the former existence of the ancestor) was not archetypal.

Hypothetical ancestors, whatever their claims to reality, are logically different from archetypes. The point is made by Roth's comment about the atlas vertebra: if the atlas vertebra was always different in all vertebrates from those that followed it, then the hypothetical ancestor of all vertebrates had a distinctive atlas. The archetypal vertebrate, in Owen's scheme, had a series of identical vertebrae. Ancestor and archetype are different concepts.

The second major category is that of *Taxic homology,* which concerns the diagnostic characters used to unite taxonomic groups. Because a natural hierarchy of groups is taken, by cladists at least, to imply a natural hierarchy of characters, the categories of taxic homology are also hierarchical. *Autapomorphies* characterise a taxon; *Synapomorphies* unite two (or more) taxa in a taxon of higher rank; *Symplesiomorphies* are non-diagnostic; they are apomorphies at a rank higher than that under consideration.

I now set out the classification:

I TRANSFORMATIONAL HOMOLOGY

A. SINGULAR HOMOLOGY: that between structure(s) in an individual to an "anatomical singular" in another (reference) individual.

1. **General Homology:** comparison to a hypothetical reference.
   a. Comparison to an ancestral condition:
      (1) Single structure to single structure in hypothetical ancestor;
      e.g., Heart and aortic arches of mammal to ancestral tetrapod condition,
      Orthology in molecular taxonomy (Chapter 9) in part.
      (2) Replicated structure to single structure in hypothetical ancestor; *e.g., Paralogy in molecular taxonomy (Chapter 9) in part.*
   b. Comparison to an archetypal condition:
      (1) Single structure to single structure in archetype; *? example.*
      (2) Replicated structure to single structure in archetype; *? example.*
2. **Special Homology:** comparison to another individual.
   a. Single structure to single structure in another organism; *e.g., Heart and aortic arches in a mammal to those of a urodele, Orthology in part.*

b. Replicated structure to single structure in another organism;
e.g., Paralogy in part.

B. ITERATIVE HOMOLOGY (HOMONOMY *sensu* Riedl 1979 and Patterson, 1982): interpretation of a structure as one of a series.

1. **General Homology:** comparison to a hypothetical reference.

a. Comparison to an ancestral condition:

(1) Single structure as derived from member (or members) of a series in hypothetical ancestor; *e.g., derivation of girdle and limb from branchial arch skeleton (Gegenbaur).*

b. Comparison to an archetypal condition:

(1) Single (or replicated) structure as derived from a member (or members) of a series in the archetype; *e.g., the vertebral theory of the skull (Goethe, Owen), ? floral parts as leaf homologues (Goethe).*

2. **Special Homology:** comparison within an individual (or between individuals):

a. Single structure to single structure within an individual; *e.g., humerus and femur, paralogy in part.*

b. Single structure as a member of a series within an individual – serial homology;
e.g., Serial homology of a vertebra to all others.

c. Homology of all replicated parts in an individual; *e.g., "Haversian systems, enucleate erythrocytes, or xylem vessels" (Patterson 1982).*

II TAXIC HOMOLOGY

A. AUTAPOMORPHY: homology of corresponding structures in members of a taxon, thus constituting a "character" or part of the morphotype of that taxon;
e.g., the (?penta-)dactyl limb characterising Tetrapoda.

B. SYNAPOMORPHY: homology of corresponding structures uniquely shared by members of two taxa recognised as "sister-groups", thus constituting a "character" or part of the morphotype of the higher-rank taxon of which the sister-groups are the only members;
e.g., the (?penta-)dactyl limb, uniting Amphibia and Amniota.

C. SYMPLESIOMORPHY: homology of corresponding structures of two or more taxa not uniquely shared, thus constituting a "character" of a taxon of some higher rank; *e.g., the (?penta-)dactyl limb shared by Aves and Mammalia, but not unique to them.*

A few comments are necessary about the classification. The first point is that I have been unable to think of convincing cases of singular general homology where reference is made to an archetype. If this is an empty class, that may be of significance. The second concerns taxic homology. Patterson (1982) wanted to synonymise "homology" and "synapomorphy", but in fact, while rejecting transformational homology as "empty transformations which lead to no new hypotheses of grouping", he retained the term "homology" to characterise them. Later Patterson (1987) relaxed his censorious attitude somewhat. Ax (1987), however, did not even accept the equation of taxic homology and synapomorphy. Negative features (e.g., secondary absence of wings in some pterygote insect orders) are autapomorphies of those orders (or synapomorphies of their constituent suborders) but not homologies, according to Ax; Hennig (1966) made a similar point. Patterson (1982) discussed the point and disagreed with both: suppression of wings is apomorph *and* homologous. I accept Patterson's view, amended to say that *taxic* homology is synonymous with (*both* aut- and syn-) apomorphy. Symplesiomorphy is simply apomorphy of a rank higher than that under consideration: taxic homologies form a hierarchy (or are *assumed* to axiomatically!) as do taxa. This is reinforced by Bock's (e.g., 1969, 1974, 1977) insistence, echoed by Ghiselin (1966), that statements of homology are meaningless unless accompanied by a "conditional phrase" stating the rank of the homology. Bock (1974) gives the example "the wings of birds and the wings of bats are homologous as the forelimbs of tetrapods". They are thus homologous as derivatives of the tetrapod limb, but only analogous as wings. I have tried to make the point about the ranking of taxic homologies by using the same example in each category.

Thus the case can be made that taxic homologies (or apomorphies) form a hierarchy that is logically prior to the hierarchical pattern of natural classification and are thus a component of the *explanandum* of evolution. They are therefore not evidence for evolution. There are objections to be made to the proposition of logical priority. Rieppel (1988) discusses the principal one:

> Kowalevsky and von Baer selected different axes of symmetry in their comparison of ascidian larvae with vertebrates. The choice of the axis of symmetry resulted from the acceptance of a particular hypothesis of grouping, the ascidians being related either to vertebrates (Kowalevsky) or to molluscs (von Baer). The example demonstrates not only the conventional or arbitrary choice of the frame of reference within which homology must be analysed, but it illustrates furthermore that the acceptance of the frame of reference may depend on a conjecture of grouping based on other characters. In other words, homology is no inductive clue to phylogenetic relationships; instead it must be established deductively in the light of a hypothesis of grouping . . . . (Rieppel 1984)

Patterson (1982) considered the demarcation of (taxic) homology from convergence (i.e., homoplasy). He suggested three tests:

1. *Similarity,* effectively Geoffroy's principles of anatomical (compositional) and topographical similarity. Similarity is normally taken to be the test of any sort of homology, although Patterson cites Bjerring's (1967) homologising of the polar cartilage of chondrichthyan and actinopterygian fishes and birds with the subcephalic muscle of coelacanths which fails anatomical similarity.
2. *Congruence,* a development of the cladistic principle of parsimony; this assumes the taxonomic hierarchy *a priori* (Chapters 7 and 8).
3. *Conjunction,* in which, if the two proposed homologues, each from a different organism, occur together in a third organism, the test is failed. As an example, Panchen and Smithson (1990) proposed that the ischium, a bone of the tetrapod pelvic girdle, is the homologue of the bony scute that occurs behind the pelvic fin in a group of fossil fish. We note that finding an organism with both ischium and scute would refute the homology.

According to Patterson all three tests are valid for taxic homology, but transformational homologies either fail, or are not tested by, at least one of them. Thus Bjerring's homology fails anatomical similarity; congruence does not test transformational homology (or, for that matter, symplesiomorphy), and all iterative homologies plus paralogy fail conjunction. Autapomorphies and synapomorphies, on the other hand, pass all three tests.

Part of the purpose of this chapter was to consider homology as "evidence for evolution"; we are now equipped to do so. We shall be considering taxic homology in relation to taxonomy and classification in Chapters 6–8. Meanwhile, however, I shall reiterate the important point that if the natural hierarchy is taken *a priori,* then the recognition of taxic homologies is part of the methodology of classification and thus is part of the process that yields the *explanandum* of which phylogeny is the *explanans*. Even if one were to take the diametrically opposed view that the phenomenon of phylogeny must be taken *a priori* to validate classification (Charig 1982; Ridley 1986), taxic homology is still not evidence for evolution. Hennig's (1966) position, as the founder of cladistics, was to use the divergent pattern of phylogeny and the mode of speciation to validate his technique of classification. Homology (synapomorphy), therefore, could not be evidence for phylogeny or speciation. The remaining question is then whether transformational homology yields such evidence. I believe that the answer is no, unless other data are invoked.

All homology involving reference to a hypothetical ancestor presupposes evolution and therefore cannot be evidence for it. Reference to an archetype is reference to something in the immutable world of Platonic idealism. If homology is proposed between the skull of a living organism and the first few vertebrae of the archetype, the comparison is between something that is transient and whose characters are thus contingent, and something (if its existence is admitted at all) that is outside time and transcendent. There can be no hypothesis of transmutation in time from a timeless abstraction.

We now turn to special homology, in the sense in which, following Owen, I have used the term. The example given above, in the case of singular homology, was that of the heart and aortic arches of a mammal as compared to those of a urodele amphibian (newt or salamander). The condition in the latter is regarded as the nearest extant approximation to the primitive tetrapod condition. We have also noted Goodrich's conclusion that the condition in sauropsids (extant reptiles and birds) and that in mammals cannot be derived one from the other, but that both must be derived from the amphibian grade of organisation (Fig. 4.1). This is disputed by Holmes (1975) and Gardiner (1982), but it does not affect the principle of the argument.

Fig. 4.1. "Diagrams of *heart* and *aortic arches* in an Amphibian, A; a Mammal, B; a Reptile . . . C; and a Crocodile, D". In ventral view; stylised and diagrammatic. (From Goodrich 1930.)

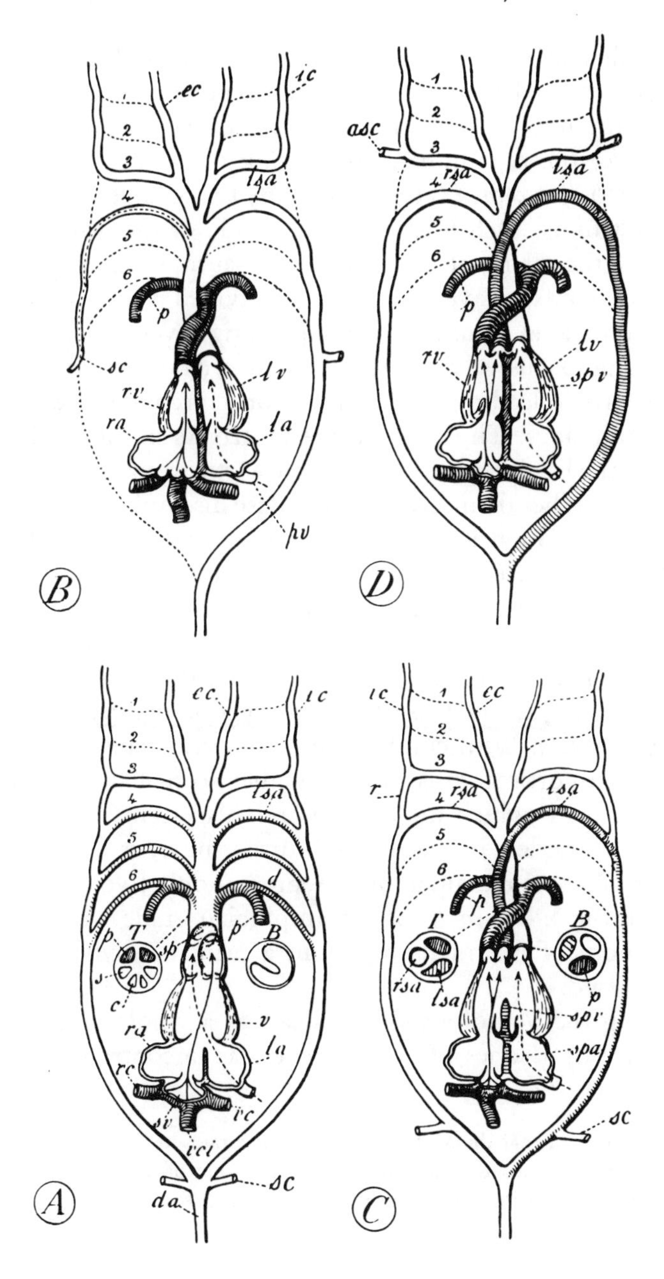

The orthodox conclusion is that of the three paired aortic arches posterior to the carotid arch in a urodele: no. 3 in Fig. 4.1 (nos. 1 and 2 are taken to be present in some fish); the single (left) aorta of mammals is in part the homologue of the left systemic (no. 4).* The right systemic of urodeles is then taken to be the homologue of the subclavian artery of mammals. Paired arches no. 5 are not represented in mammals, and arches no. 6 are retained only as pulmonary arteries, the *ducti Botalli* having disappeared. Furthermore, the common trunk leaving the single ventricle in the urodele (*bulbus cordis*) is represented by the separate pulmonary and aortic trunks, each leaving its own side of the now divided ventricle, in the mammal. In the reptilian sauropsids, on the other hand, three trunks leave the partially divided ventricle, a pulmonary trunk (as in mammals) leaving the right ventricle, a middle trunk leading exclusively to the left aortic arch, and a trunk from the left ventricle leading to the right aortic arch and the carotid arch on both sides. Thus the latter two trunks in sauropsids are taken to be the homologues of the single aortic trunk of mammals.

The aortic arches themselves are likewise dissimilar in mammals and sauropsids. Goodrich (1930) gives a figure (his fig. 587C: our Fig. 4.1C) for "a Reptile (Chelonia, Lacertilia, Ophidia, Rhynchocephalia)" representing the primitive sauropsid condition. He shows the condition as essentially the same in "a Crocodile" (D), apart from an almost complete separation of the ventricles in the latter, together with the loss of any connection between the third and fourth arches. With completion of the ventricular septum in birds, the left aortic arch is lost altogether so that the remaining aortic arch, from which the carotid arch also stems, is the right in birds, but the left in mammals. Furthermore, it is taken that the two trunks that remain in birds represent only two of the three into which the *bulbus cordis* of the amphibian grade was divided, whereas those of mammals represent a primary division of the *bulbus* into two.

Accepting the homology of the heart as a whole, and at least the presence of a series of aortic arches, as uncontroversial – in

*It might be argued that the aortic arches in vertebrates are serial homologues of one another and thus that our example is not one of singular homology. However, we are concerned here with their individual fate, not their collective origin.

fact, as a taxic homology of all vertebrates – the interest in Goodrich's thesis is in divergent transformations of a perceived basic plan. Although there is no doubt that he saw the transformations as phylogenetic, no reference need be made to phylogeny in order to validate the homology. The criteria are the compositional and topographic ones of Geoffroy, together with that of ontogeny.

We consider the use of ontogeny as evidence for evolution in the next section, and the fossil record in Chapter 5, but there is a further point to be made here about our example of the heart. In looking at the two proposed transformations, there is an important difference (as presented) between the amphibian → mammal one and the amphibian → bird one. The first is comprised of only two states, which are "polarised" (Chapters 7–8) by the statement that the amphibian condition is "primitive" (plesiomorph within tetrapods) and the mammalian one "derived" (apomorph). In the sauropsid case, however, the sequence is amphibian (state) → "reptile" → crocodile → bird. The hearts in this series are arranged in a morphocline (Maslin 1952), polarised as before, but with more possibility of error (Eldredge and Cracraft 1980). There is, however, the possibility of corroboration of the sequence of a multistate morphocline by ontogeny. There is also the possibility, but not in the case of the heart and aortic arches, of corroboration by discovery of a chronocline from the fossil record. Our example of the ear ossicles of mammals is a case in point. We shall discuss the validity of the use of morphoclines and chronoclines together with ontogeny as evidence of evolution later in Chapter 5 and their combined use to validate specific arrangements of organisms in Chapters 7 and 8. Meanwhile, however, we must turn to a special case of comparative anatomy that has long been regarded as potent evidence for the truth of evolution.

III. Vestigial organs

That special case is the phenomenon of "vestigial organs": I place the phrase in quotation marks, for not to do so would be to prejudge the issue. Both Wallace (1855) and Darwin (1859) claimed "rudimentary organs", as both termed them, as evidence for evolution:

> Another important series of facts, quite in accordance with, and even necessary deductions from the law now developed [see Chapter 5, Section III], are those of *rudimentary organs*. That these really do exist, and in most cases have no special function in the animal oeconomy, is admitted by the first authorities in comparative anatomy. The minute limbs hidden beneath the skin in many snake-like lizards, the anal hooks of the boa constrictor, the complete series of jointed finger-bones in the paddle of the Manatus and whale, are a few of the most familiar instances. . . . If each species has been created independently, and without any necessary relations with pre-existing species, what do these rudiments, these apparent imperfections mean?
>
> (*Wallace 1855*)

> On my view of descent with modification, the origin of rudimentary organs is simple. . . . I believe that disuse has been the main agency; that it has led in successive generations to the gradual reduction of various organs, until they have become rudimentary, – as in the case of the eyes of animals inhabiting dark caverns, and the wings of birds inhabiting oceanic islands, which have seldom been forced to take flight, and have ultimately lost the power of flying.
>
> (*Darwin 1859, Chapter XIII*)

The argument using vestigial organs as evidence for evolution is a familiar one and need not be discussed in detail. It consists of demonstrating the homology between the proposed vestigial organ in one species with the (usually larger) organ of known function in another. The demonstration is then followed by the assertion that the organ in the first species is either functionless or has a trivial function different from that of its homologue in the second species.

Scadding (1981) objects to this argument and asserts that " 'vestigial organs' provide no evidence for evolutionary theory". He points out the impossibility of asserting that any particular organ or structure is functionless, having characterised vestigial organs by use of that assertion. Thus, according to Scadding, no organ can be identified as vestigial and the argument fails, or rather, as I would put it, reverts to the status of homology-as-evidence-for-evolution. Scadding's point is well taken, but I do not believe that an organ or structure must be patently functionless to be regarded as vestigial. The reduced wings below sealed elytra in some ground beetles (one of Darwin's examples), or those of flightless birds, may retain some function in ontogeny and may also have acquired a new one in the adult, the "exaptation" of Gould and Vrba (1982). Provided, however, that the implicit transformational homology can be polarised with reasonable certainty, it seems to me that a phylogenetic interpretation of the transformation is eminently the most reasonable one; the proposed homology (*fide* Patterson!) has no taxonomic content and thus is not part of the *explanandum* for which phylogeny is the *explanans,* and therefore, vestigial organs are acceptable evidence for evolution.

Another category of evidence, related to that from vestigial organs, is I think more potent. It is that from genetic potential no longer expressed. Gould draws attention to it in his well-known popular essay (Gould 1980b) and in his book entitled *Hen's Teeth and Horse's Toes* (1983). Victorian naturalists, including Darwin, were fascinated by the phenomenon of "atavism", in which primitive or supposed ancestral characters appeared inexplicably in individual organisms. Gould draws attention to Marsh's (1892) compilation of cases of polydactyly in horses. These were of two sorts. The first was polydactyly by duplication, common in mammals and including hexadactyly in man. In this case, an extra digit represented abnormal development of the single (third) digit of the horse. Duplication was identifiable because the splint bones representing the vestiges of the second and fourth digit co-existed with the abnormal extra toe. In polydactyly by atavism, however, one of the splints itself was abnormally developed to produce a variably developed extra toe, a presumed reversion to an ancestral condition and plausible evidence for evolution.

The hen's teeth of Gould's essay refers to the work of Kollar and Fisher (1980). In this they showed that if mesenchyme from the molar tooth region of an embryo mouse was cultured alone, spongy bone would be produced but no tooth vestige. If, however, the mesenchyme was cultured with epithelial tissue from the first and second branchial arches of five-day chick embryos, the latter in some cases formed enamel and induced dentine in the mouse mesenchyme. The induction was mutual: mouse mesenchyme induced enamel in the chick epithelium; chick epithelium induced dentine in mouse mesenchyme. In four of their fifty-five grafts, recognisable teeth were produced, but these lacked any molar cusp pattern; conceivably, they were to be regarded as "hen's teeth".

Retention of inductive potential in the chick from tooth-bearing ancestors seems the only plausible explanation – again potent, non-taxonomic, evidence for evolution.

Similar to evidence from vestiges and from atavism and also, like them, evidence from transformational homology is the phenomenon inelegantly labelled "exaptation" by Gould and Vrba, noted above. Again the title of one of Gould's (1980a) books of essays has made one example famous. Members of the mammalian carnivore family, the Ursidae (bears) lack the first digit in the hand; the giant panda, now known to be ursid, has the "Panda's thumb" as a pseudo-digit formed from an enlarged and modified radial sesamoid bone of the wrist. Again evolution is the only plausible explanation.

IV. Ontogeny and molecular biology

We have seen in Chapters 2 and 3, and also in the previous section, how ontogeny, particularly that of vertebrates, has contributed to speculations about the pattern of classification and the pattern of evolution, and to discussions about the nature of homology. Since the publication of von Baer's work, there has been a tension between his view of the evidence from embryos about the ordering of adult animals as a divergent hierarchy, and that of recapitulation theory suggesting a *scala naturae*. We have developed Haeckel's ideas on recapitulation in some detail (Chapter 3, Section II), but said less about the original "pre-phylogenetic" recapitulation theory generally known as the Meckel–Serres Law (Russell 1916).

Serres was a disciple and colleague of Geoffroy, whose anatomical studies were concerned chiefly with human embryos and particularly with abnormalities. He also elaborated the parallel, as had a number of the German *Naturphilosophen* including Oken and Meckel, between a *scala naturae* arrangement of adult organisms and the embryology (usually) of man. Serres's particular contribution was his studies of abnormal foetuses, in which, for instance, the head was missing. This suggested to him that in some respect development had been arrested, in this case at a molluscan stage, and led to his theory of developmental arrests (Gould 1977b): "If the formative force of man or the higher vetebrates is arrested in its impulse, it reproduces the organic arrangements of lower animals. . . . These cases of pathological anatomy are only a prolonged embryogeny" (Serres 1830: quoted by Gould). Thus Serres took his study of "Teratology" as corroboration of recapitulation.

We will consider recapitulation as a basis for polarising characters in Chapter 8. I have already noted the importance of ontogeny in establishing homologies; this is reviewed by Roth (1988). What I want to do in the present context is to suggest as a thought experiment the question, If the Meckel–Serres Law were true (or more correctly highly corroborated), would the phenomenon of recapitulation which it embodies, of itself, constitute evidence for evolution? This clearly could not be the case for Haeckel's biogenetic law which assumes phylogeny as one of its premises, but the Meckel–Serres Law merely assumes that organisms, or at least animals, can be arranged along a *scala naturae*.

My answer to the question is no. But, if the triple parallel of *scala*, ontogeny, and the fossil record were corroborated as Chambers took it to be in the early editions of the *Vestiges* (see Chapter 3, Section I), then probable explanations of the phenomenon would narrow to progressionism and evolution. If there were then evidence to favour evolution rather than progressionism, a powerful case could be mounted. We shall consider that in the next chapter.

Meanwhile we must turn to von Baer's more accurate view of comparative embryology to ask, in the same spirit as our thought experiment, whether the observations that yielded von Baer's laws constitute evidence for evolution. Once again I believe the answer to be no. Von Baer's four laws of development are set out in Chapter 2 (Section II). Richard Owen was impressed by the importance of von Baer's laws and drew parallels between those laws and his knowledge of the fossil record (Ospovat 1976). Owen (1840–45) particularly drew parallels between the ontogeny of mammalian teeth and the dentition of primitive mammals from the fossil record. A primitive artiodactyl, *Anoplotherium,* had the full complement of mammalian teeth, forty-four in number, and was said by Owen to have a generalised, or archetypal, tooth pattern with respect to extant artiodactyl ungulates. As quoted by Ospovat:

> [A]lmost every kind and degree of variety, save that of increased number of teeth, has been superinduced in later and existing forms of hoofed mammals upon the primitive Anoplotherian formula, which may therefore be regarded as the type of perfect standard of the dentition of the great natural group of *Ungulata*.

We pointed out in Chapter 2 (Section III) that Owen appears to have had a hierarchical view of the arrangement of organisms that was a combination of the idealism of morphology and the functional anatomy of Cuvier, so that both divergence from the archetype of a group and advance upon it were related to adaptation. However, his paralleling of a divergent hierarchy rather than a *scala naturae* in morphology with the fossil record yields corroborated evidence of a pattern of classification. If we wish to retain the distinction between *explanandum* and evidence that I have maintained so far, Owen's work cannot be cited as evidence for evolution, even with the ontogeny–palaeontology parallel which he draws.

An analogous argument applies to comparative molecular biology (Chapter 9). Once again the evidence from the various techniques of biochemical and molecular taxonomy yield patterns of classification which may then be interpreted as phylogeny; they are part of the *explanandum*. Once again, also, these patterns are frequently combined with data from the fossil record, but this time not to demonstrate a common pattern. Much of the interest and a great deal of the controversy surrounding molecular taxonomy arise from the fact that it is normal for the taxonomic patterns to be calibrated by means of agreed palaeontological markers, thus allowing their presentation as phylogenies ("trees") on a base of geological time.

5

Geological and geographical evidence

It is assumed by most laypeople who concern themselves with evolution, and by many professional biologists, that the fossil record constitutes the primary and most important evidence that evolution has occurred. This also seems to be the assumption of many "creationists", who attempt to refute all aspects of evolutionary theory in favour of their own religious beliefs. Thus creationist tracts are usually aimed at discrediting the evidence from the fossil record and, of necessity, the methodology of geological dating (Morris 1974; Gish 1979; see Kitcher 1982 and Newell 1982 for commentaries on "Creation Science").

I. The fossil record

A position opposite to that of assuming that the fossil record is the primary evidence for evolution is taken by Kitts (1974). His review is concerned with the relationship of the fossil record to theories of evolutionary mechanism, but in the preamble he is concerned with the nature and interpretation of fossils:

> Palaeontologists often claim that fossils tell us something. But fossils, by themselves tell us nothing; not even that they are fossils. . . . When a palaeontologist decides whether or not something found in the rock is the *remains* of an organism he decides, in effect, whether or not it is necessary to invoke a biological event in the explanation of that thing. Sometimes a judgement is made without hesitation because the object in question resem-

> bles some part of a living organism. . . . But . . . [t]here is no structural feature which serves to distinguish fossils from entities of every other kind. . . . An object ceases to be regarded as a fossil when a physical explanation of its origin is accepted. The "Jellyfish" of the Nankoweap group are rejected when Cloud's (1960, p. 43) claims that they are gas bubble markings is accepted.

Steno (Niels Stensen) is usually credited with establishing the status of fossils, as the remains of once-living organisms and not some parallel creation placed initially within strata. He did this in two publications, a "digression" from a study of the dissection of the head of a shark (Steno 1667) and *The Prodromus to a Dissertation concerning Solids Naturally Contained within Solids* (Steno 1669). His inferences about the nature of fossils depended on two principles, termed by Gould (1983) *the principle of sufficient similarity* and *the principle of moulding.* Steno invoked the first principle by recognising that the common "tongue stones" (*glossopetrae*) of northern Mediterranean strata were similar in external appearance and histology to the teeth of living sharks. The second principle comprised his recognition that glossopetrae found within sediments had impressed their form on the surrounding rock matrix without themselves taking up the form of pre-existing cavities within the rock. The fossils must therefore have pre-dated consolidation of the sediments in which they were found.

If one is to construct an argument for the fossil record as evidence for evolution, Steno's conclusion, that fossils are the remains of once-living organisms, must be the first step. The second, denied by the creationists, is that data from fossils do constitute a historical record that can be deciphered. There are two other possibilities, both patently false: either that nothing historical can be inferred from the record or that all organisms now represented by fossils were entombed in quick succession or simultaneously.

During the eighteenth century, the idea of arrangement of sediments as data for a historical (stratigraphic) record slowly emerged, with the principle of superposition implicit (Rudwick 1976). The principle of correlation, that contemporaneous sediments could be recognised by the similarity of their fossil biota, is usually associated with the names of William Smith in Britain and Cuvier and Brongniart in France. Smith began recording his ideas on characterising strata by fossils in 1796. One of his first stratigraphic maps, of the area around Bath, was completed in 1799, and the earliest surviving geological map of the whole of England and Wales was produced by Smith in 1801. This was followed by his illustrated *Strata Identified by Organised Fossils,* of which four parts were published between 1816 and 1819.

Cuvier and Brongniart published their map of the geology of the Paris Basin, together with a memoir explaining their technique of correlating strata by fossils, in 1811. There have been rather pointless debates about their priority over Smith or vice versa, but the main point for our argument is that the principles of superposition and correlation were understood and applied by geologists from early in the nineteenth century.

With the acceptance of fossils as once-living organisms and of the fossil record as a historical document, the next stage of the argument is to demonstrate that there were organisms alive in the geological past that are no longer extant today and, contrariwise, that most living species and at least some higher taxa extend back into the past only to a limited degree. The first is easy to demonstrate. It is in the highest degree improbable that living representatives of such spectacular fossils as any of the known species of dinosaur are alive today in our well-explored world, although the discovery of the coelacanth *Latimeria* in 1938 shows that the deep sea can still produce surprises. Similarly, it is highly improbable that Huxley's "Silurian mammal" (Chapter 3, Section I) will ever be found. Two important points arise from these two types of negative evidence: the first is the reality of extinction; the second the phenomenon of "progression" in the fossil record.

It is to Cuvier that we owe the acceptance of the reality of extinction:

> It was Cuvier who first recognised clearly that this question . . . would never be resolved decisively except by using the large terrestrial quadrupeds as a 'crucial experiment'. . . . If a study of these fossil bones [*Megatherium* and mastodons], using the powerful new methods of comparative anatomy, could prove that they had belonged to species distinct from any known alive, the reality of extinction would be proved almost beyond dispute.
> (*Rudwick 1976, p. 107*)

Cuvier's early conclusion was that one great extinction event and a single "world anterior to ours" explained these spectacular extinct species. Later, however, as a result of stratigraphic work in various parts of Europe, notably that by his colleague Faujas de St. Fond on the Chalk of Maastricht, Cuvier revised his opinion. In a lecture in Paris in 1801, he declared that "the most remarkable and most astounding result" of his and his colleagues' work was that "the older the beds in which these bones are found, the more they differ from those of animals we know today" (Rudwick 1976, p. 107). Then followed the work by Cuvier and Brongniart on the stratigraphy of the Paris Basin in which they arrived at the principle of correlation by fossils. They distinguished seven major formations above the Cretaceous chalk and thus established, in the post-Cretaceous Cenozoic, the same principle that William Smith had demonstrated in the "Secondary" (Mesozoic Era).

Cuvier and Brongniart, however, reached another conclusion, which is important to our argument. They concluded that the strata of the Paris Basin represented alternate freshwater and marine conditions with fossil molluscs appropriate to the particular environment. Cuvier concluded from this that the history of the area had been marked by a series of marine incursions that extinguished the terrestrial and freshwater fauna, so that the next, and different, terrestrial fauna must have colonised the area from elsewhere. Thus he built up a picture of Earth history in any particular locality as successive periods of stability punctuated by revolutionary events that destroyed one flora and fauna, to be subsequently replaced by another. He did not envisage the revolutions as world-wide catastrophes, but other people did. Notably in England, "catastrophism" was combined with theology to resurrect the idea of the Deluge as the last of Cuvier's revolutions. This was done in Robert Jameson's (1813) very free translation of Cuvier's work and was proposed as a theory by William Buckland (1823).

The relevance of this discussion to our argument is that if systematic change, and more arguably progress, is seen in the fossil record, there are two possible explanations – either independent creation and extinction, eventually of every species of organism throughout geological time, or "descent with modification" of one species from another. The next stage of the argument is to enhance the probability of the second over the first.

If it were established that Earth history was marked by a series of catastrophes that extinguished the whole existing biota, then (assuming "progression" – or change of the biota with time – in the fossil record) creation of a new fauna and flora after each catastrophe would seem the inevitable explanation. This was the position taken by Buckland, despite the fact that in his *Bridgewater Treatise* (Buckland 1836; see Chapter 11, Section II) he recanted on the Deluge as the last catastrophe. If, on the other hand, it could be shown that species appear in the record at times other than those immediately following world-wide catastrophes and become extinct at times other than at the onset of such catastrophes – in fact, that the origin and extinction of species occur throughout the fossil record – then (given "progression") evolution might seem the more probable explanation. This would still be the case even if it were shown, as is now generally accepted (Stanley 1987), that there have been a number of large-scale extinctions throughout the fossil record. That case would be strengthened if it were shown that even major extinction events did not eliminate all species and also (see below) that chronoclines continued through extinction events. The first can be shown by the history of various taxa through an extinction event. It is also emphasised by "Lazarus taxa", which disappear from the record around the time of the event, but reappear subsequently, presumably representing a reduction in numbers with subsequent recovery (Jablonski 1986b).

An early and vivid demonstration that origination and extinction occurred throughout the fossil record was a by-product of Lyell's (1830–3) method of dating Tertiary strata. He used fossil molluscs as his stratigraphic reference, but did not use the technique of designating characteristic fossils or suites of fossils to identify each zone in the way that Smith and Cuvier and Brongniart had pioneered. Lyell's method was statistical. Very few of the numerous species of fossil mollusc found at the beginning of the Tertiary period are extant. At the present time, by definition, all species are. Then even if one makes no assumptions about species characterising strata, and none about more recent forms being "advanced", the geological age of any stratum should be inversely related to its percentage of extant species. Lyell divided the Tertiary into four epochs: the Eocene, immediately after the Cretaceous, with about 3 per cent of extant species; the Miocene with about 20 per cent; the older Pliocene with from a third to over a half;

and the newer Pliocene with about 90 per cent (Gould 1987). It is thus implicit in the success of his method that at no time in the Tertiary is it possible to draw a line through the stratigraphic record when the whole fauna was extinguished to be replaced by a new and different one.

There is, however, a double irony about using Lyell's successful methodology to demonstrate this truth. I have used his data to suggest that, given a choice between catastrophism and evolution as explanation of progression in the fossil record, evolution is the incomparably more probable hypothesis. But Lyell, in the first edition of the *Principles of Geology* (1830–3), rejected both progression and evolution. He claimed that the physical environment changed, but not directionally. There was, he believed until late in life, no irreversible change in the physical and biological world. He, like Huxley, predicted the discovery of Paleozoic mammals, invoking the incompleteness of the fossil record to explain that none had been found, but in the first edition he made an even more startling prediction, this time about the future history of the Earth. Past environmental conditions could recur:

> Then might those genera of animals return, of which the memorials are preserved in the ancient rocks of our continents. The huge iguanodon might reappear in the woods, and the ichthyosaur in the sea, whilst the pterodactyle might flit again through umbrageous groves of tree ferns.
>
> (*Lyell 1830, p. 123*)

Thus Lyell rejected evolution until late in life, when the failure of his predictions about the fossil record and the influence of Darwin's theory caused him reluctantly to change his mind.

But Lyell was in a minority in the first half of the nineteenth century in rejecting progression in the fossil record. It was accepted by the *Naturphilosophen,* by Cuvier, by Buckland, and by Agassiz, although Agassiz's progression was internalist and later he was to reject it as perhaps implying evolution (Gould 1977a). The evidence for progression was almost always the vertebrate fossil record. It was, and is, more difficult to make progressionist claims about the invertebrates after the first appearance of most Metazoan phyla in the Cambrian period at the beginning of the Phanerozoic. Progression in the vertebrate record is most easily accepted if one accepts a simple *scala naturae* picture of the diversity of vertebrates through geological time.

II. Fossils and transmutation

We have so far established that fossils are the remains of once-living organisms, that the fossil record is truly historical, and that the biota has changed with time. Extinction is a reality, and the origin of species at different times during the geological past, whether this is interpreted as progression or not, is beyond reasonable doubt. Catastrophism in the Tertiary is refuted, and similar refutations could be demonstrated for the other geological eras of the Phanerozoic. We must now search for convincing evidence of transmutation in the fossil record, which can then reasonably be interpreted only as evolution.

There are two aspects of this search, which we have discussed already in Chapters 2 and 3, and in the pages of the last chapter. The first aspect is cladistic; the second is concerned with transformational homology. We have already seen that it is unreasonable to expect to find ancestor-descendant sequences in the fossil record but that we might well expect to find an arrangement of characters in a group of organisms containing both fossil and extant forms which gave a cladogram at least part of which was in Hennigian-comb or tree-of-Porphyry-like form.

As a demonstration of the first aspect, we shall take the cliché example of the evolution of horses (family Equidae). Speculation about the origin of extant horses began with Gaudry and Kovalevsky, but the simplified picture of horse evolution as a progressive line from Eocene to Recent we owe to Marsh's (1879) work on North American equids. Marsh's series was first announced in 1874. Two years later Huxley predicted the discovery of an earlier Eocene form. It was discovered and named *Eohippus* (now correctly *Hyracotherium*) within two months of Huxley's prediction.

Subsequent work by Cope, Matthew, Osborn, Stirton, and others showed that there was no single undeviating line from *Eohippus* to *Equus,* the only extant genus. Simpson (1951, 1953) was at pains to emphasise this:

> The Equidae had no trends that: (1) continued throughout the history of the family in any line, (2) affected all lines at any one time, (3) occurred in all lines at some time in their history, or (4) were even approximately constant in direction and rate in any time for periods longer than on the order of 15 to 20 million years at most. . . . The whole picture is more complex, but also more instructive, than the orthogenetic progression that is still being taught students.

In fact in the early Miocene, grazing horses first appear in the record and are separable into a large number of genera, but there also appears to have been a radiation of "primitive" browsing horses into a number of genera, with *Anchitherium* as a typical representative, shortly before this. This pattern was charted by Simpson (1953), and a similar "phylogeny" is given by MacFadden (1985). In both, the pre-Miocene history is represented by a time-based *scala naturae* of genera: *Hyracotherium – Orohippus – Epihippus – Mesohippus – Miohippus,* although Simpson emphasises that this Haeckelian trunk to the tree includes many specific lineages.

In order to judge the status of the fossil record of horses as evidence for evolution, we shall take a small number of representative genera to produce a simplified classification representative of the whole clade. Figure 5.1a was first presented as an exercise in an Open University course book (Gass *et al.* 1972) with drawings taken from Simpson (1951). Students were asked to "attempt to establish a greatly simplified 'evolutionary tree' for the horse family". If, however, all the represented genera are well characterised – that is, have unique (autapomorphous) characters – they cannot be ranged in an ancestor-descendant sequence but can be ranged as a cladogram (Fig. 5.1b). The pattern of the cladogram is that of a Hennigian comb with each successive node adding a new synapomorphy until P and Q share all the advanced horse features. To use the cladogram as evidence for evolution, we can then interpret it as phylogeny and look for congruence between the order of the terminal taxa (the fossil genera) and their stratigraphic dates.

From the dates of the terminal taxa, each node on the cladogram can be assigned a date (see Fig. 5.1b) that represents a minimum date for the origin of the synapomorph character(s) at that node. The date is the oldest for any genus originating from the node. I then suggest that if the ages of the terminal taxa (from E to Q in

Fig. 5.1. (a) "Drawings of skulls, forelimbs and upper molar teeth of several genera of fossil horses" with suggested phylogeny, from Open University course book (Gass *et al*. 1972. Published by permission of the authors and the publisher, The Open University, United Kingdom.) (b) Cladogram derived from the data in (a); dates of specimens allow the nodes to be dated so that cladogram is also a tree.

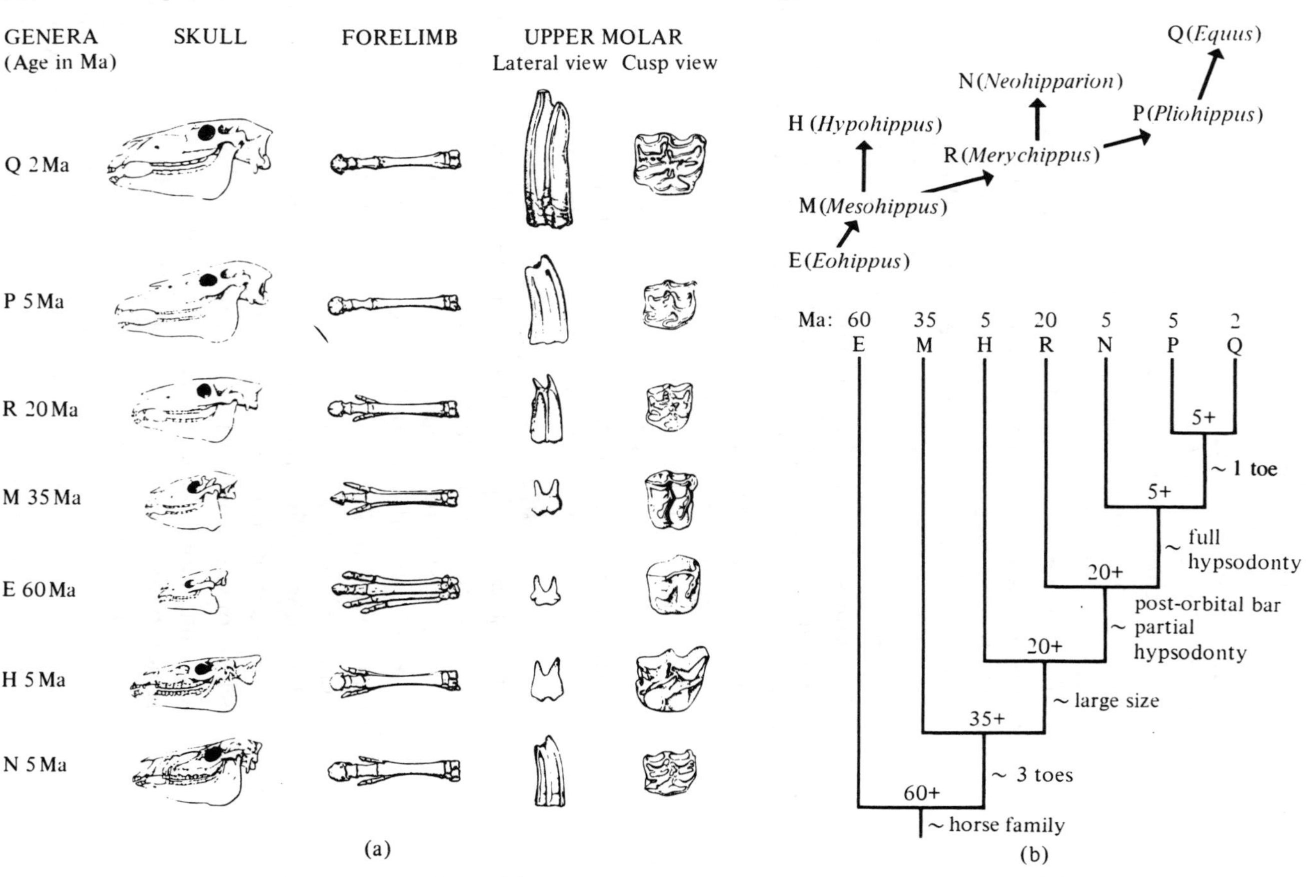

this case) form a sequence that diminishes regularly towards the most advanced taxon (Q), then this will be strong *prima facie* evidence for evolution. This is because the morphocline of cumulative advanced characters (E → Q) will correspond to a chronocline of fossils displaying those characters. In other words, E to Q will represent a *Stufenreihe,* while the sequence of dated nodes will represent an *Ahnenreihe* of hypothetical ancestors – in fact, a strongly corroborated if hypothetical ancestor-descendant sequence (Chapter 3, Section III).

It will have been noticed, however, that there is an objection to this proposal, which is illustrated by our horse cladogram. As it stands, that cladogram looks relatively poor on my criteria. The fossil dates do not form a regularly diminishing series because of the recent dates of H and N. As a result, two pairs of nodes have to be assigned the same minimum age. If, however, H and N were removed from the cladogram, then fossil and node dates would both form a perfect sequence. It is reasonable to doubt evidence for evolution (or anything else!) that is in part dependent on the selective removal of data.

It is not to be expected, however, that evolution consists entirely of *Ahnenreihen,* evidenced by *Stufenreihen.* This would be an unwarranted dogma equivalent to the progressionism of the nineteenth century. Nevertheless, if evolution has occurred, there must have been a phyletic line from the most primitive horse, to which *Hyracotherium* is the nearest known approximation, through to the extant *Equus.* It is improbable that one would be able to find such a line as an ancestor-descendant series, or *Ahnenreihe.* A well-established *Stufenreihe,* albeit complicated by "unsuccessful side branches", would nevertheless constitute strong evidence of evolution, even if it were necessary to eliminate some of these "branches" to make morphocline and chronocline coincide.

The morphocline referred to in this case is not of changing features but of cumulative apomorph characters. There are many such reconstructions of morphoclines from the fossil record, latterly using strict cladistic techniques as in the "Hennigian comb" method pioneered by Patterson and Rosen (1977) (see Chapter 3, Section III). Another well-known example is the cumulative origin of mammal characters seen in the fossil history of "mammal-like reptiles" (Kemp 1982, Chaps. 12 and 13). Gauthier, Kluge, and Rowe (1988) consider the case of the "mammal-like reptiles" in a review in which they are concerned with the effect of data from fossils on the pattern of amniote classification and thus phylogeny. They assigned each of twenty-nine amniote taxa a date based on its first appearance in the fossil record, and all were ranked according to that criterion. They then assigned a cladistic rank to each taxon, defined as the number of nodes separating that taxon from the most recent common ancestor of all amniotes – that is, from the root of their whole cladogram. Given these two-rank series of the same taxa, they were then able to calculate a correlation coefficient. This came out at 0.679, which is highly significant. There were, however, two major clades in their cladogram: Synapsida, or mammal-like reptiles plus mammals; and "Reptilia" (fossil Sauropsida less birds). The congruence between the chronocline of synapsids and the morphocline of cumulative mammal characters (expressed by the cladogram) was much greater than for those of the whole. I suggest that their technique might be used for assessing the strength of support for evolution of any cladogram of fossil taxa.

One case, still concerning the evolution of mammals, brings us to what I referred to as the second aspect of the search for evidence of transformation in the fossil record. We saw previously (Chapter 4, Section II) how the quadrate and articular bones, which form the jaw joint in most gnathostomes, were homologised by Reichert (1837) with two of the ear ossicles, the incus and malleus, respectively, of mammals. The homology is corroborated in considerable detail – including the use of embryological evidence. In primitive placental mammals, the incus is seen to arise from the back of the palato-quadrate cartilage (the primary upper jaw) and the malleus from Meckel's cartilage (lower jaw) (Goodrich 1930). In the pouch young of marsupials, such as the kangaroo, the quadrate-articular joint is still operative, and therefore, change-over to the mammalian condition occurs after birth. There has been enormous development of all the details of this homology since it was first proposed, and particularly the transformation is represented by the very rich record of mammal-like reptiles (Allin 1975, 1986). A morphocline extends in detail from early pelycosaurs, differing little from the most primitive amniotes, through to early mammals, which are usually characterised, somewhat arbitrarily, by the attainment of the squamosal–dentary jaw joint and the possession of all three ear ossicles – stapes, incus, and malleus.

Changes include the enlargement of the dentary bone until it alone forms the skeleton of the lower jaw; reduction and loss of the other dermal bones ("post-dentary bones") of the jaw, with the exception of the angular bone which becomes the mammalian tympanic; and thus the closer approach of the articular process of the dentary to the squamosal, a dermal bone of the main skull.

This time the morphocline is in changing arrangements and proportions of homologous bones, rather than the accumulation of mammalian characters, but there is again a high degree of congruence of the fossils arranged as a morphocline and the same specimens arranged as a chronocline. Once again this high degree of congruence represents strong evidence for evolution, particularly when coupled with the embryological evidence.

One final category of claimed fossil evidence for evolution remains to be considered: the existence in the fossil record of species or higher taxa said to be intermediate between major extant taxa. The most quoted cases are fossil vertebrates said to link an ancestral and a descendant taxon. Examples are *Ichthyostega* sp., said to be transitional between fish and amphibia; the anthracosaur amphibia, said to be ancestral to reptiles; and *Archaeopteryx,* said to link birds and dinosaurs. Patterson (1982) quotes Gingerich (1979, pp. 63–6):

> As is usual in well-documented transitions *Ichthyostega* shows a mosaic of primitive, intermediate, and advanced features linking it phenetically to earlier rhipidistian crossopterygians, on one hand, and to later labyrinthodont amphibians, on the other. . . . The important point is that, looking backwards in time, primitive reptiles converge with anthracosaurs in the lower Carboniferous. . . . The evidence indicates unequivocally that *Archaeopteryx* evolved from a small coelurosaurian dinosaur and that modern birds are surviving dinosaurian descendants.

The claim in each case is that an ancestral group has given rise to a descendant group via the fossils cited, a special case of the ancestor-descendant sequence. Patterson goes on to remark, not without sarcasm, that "This is the stuff of phylogeny. What all those statements have in common is the names of extinct paraphyletic (ancestral) groups."

It is reasonably certain that both rhipidistians and anthracosaurs, cited by Gingerich and scorned as paraphyletic by Patterson, are in fact monophyletic (Panchen 1980, 1985; Panchen and Smithson 1987, 1988), but if that is the case, neither would occupy the ancestral position claimed by Gingerich. In fact, Patterson's point is well taken. With *Archaeopteryx*, probably the most famous of all "intermediate" fossils, the claim is that it stands between coelurosaurs and birds, or more broadly between reptiles and birds. But if that is the case, both the Coelurosauria and the Reptilia are paraphyletic groups, and birds *are* nested (pun unintentional!) within coelurosaurs and within reptiles as a taxon of lower rank. (For a recent series of discussions on *Archaeopteryx*, see Hecht *et al*. 1985.)

That being the case, *Archaeopteryx* becomes an example of a special case within our category of congruence between a taxonomic morphocline and chronocline, to be ranged in a series perhaps extending from primitive therapod dinosaurs through to extant birds.

III. Biogeography

Nelson and Platnick (1981) give a brief account of the history of biogeography before the publication of *On the Origin of Species*. They draw attention to the importance of de Candolle's (1820) essay, "géographie botanique". In that essay, de Candolle considers what would now be termed "limiting factors" and then goes on to consider what he calls "stations" and "habitations" (in Nelson and Platnick's translation):

> By the term *station* I mean the special nature of the locality in which each species naturally grows; and by the term *habitation*, a general indication of the country wherein the plant is native. The term station relates essentially to climate, to the terrain of a given place; the term habitation relates to geographical, and even geological circumstances. The station of *Salicornia* is in salt marshes; that of the aquatic *Ranunculus*, in stagnant freshwater. The habitation of both these plants is in Europe; that of the tulip tree in North America. The study of station is, so to speak, botanical topography, the study of habitations, botanical geography...
>
> ... Stations are determined uniquely by physical causes actually in operation and... habitations are probably determined in part by geological causes that no longer exist today.

Thus, as Nelson and Platnick point out, de Candolle made explicit the distinction between ecological biogeography – the study of stations – and historical biogeography – the study of habitations. It is from historical biogeography that one might expect evidence for evolution. The distinction was made against the background of the knowledge and views of Linnaeus (who believed in the dispersion of all organisms as created from a literal Garden of Eden), Buffon, Latreille, and Humboldt who contrasted the mammalian faunas of Africa and South America. de Candolle then extended the Old World–New World contrast to suggest the division of the terrestrial globe into a series of regions, of which he lists twenty. These regions corresponded to his "habitations" and the differences between their floras demanded historical explanation. Later (de Candolle 1838), he was to extend his number of regions to forty: all are still recognisable areas of endemism. It is also of some importance that de Candolle recognised a small number of cosmopolitan species and discussed their probable modes of dispersal.

Soon after de Candolle's seminal work, zoologists started to delimit regions based on the distribution of animals. Prichard (1826) defined seven regions, based on the distribution of mammals. Later Sclater (1858) defined six regions based on birds. These six regions were eventually adopted by Wallace in *The Geographical Distribution of Animals* (1876). Sclater also considered the possibility of interrelationships between areas based on their faunas, so that a hierarchical classification of areas was envisaged. Wallace (1863), in the paper in which he proposed "Wallace's Line", presented an exercise in such an area classification within the Oriental region (Regio Indica). Unfortunately, as Nelson and Platnick point out, Wallace's technique does not utilise any information about species shared between the land-masses, and therefore, no area cladogram (see below) can be drawn, nor for that matter can any hypotheses be suggested of speciation giving rise to sister species on adjacent islands.

Before his 1863 paper, however, Wallace had already published the paper in which he enunciated the "law which has regulated the introduction of new species" (1855) and his paper on natural selection (1858). In his 1855 paper, Wallace set out a series of propositions that led him to enunciate the law that *"Every species has come into existence coincident both in space and time with a pre-existing closely allied species"*, and to claim that his law led to a "true system of classification". "The pre-existing allied species" is, in fact, the ancestor (or "antitype") so that Wallace's nine propositions that imply the law are in fact geographical and geological evidence for evolution. They are as follows:

Geography

1 Large groups, such as classes and orders, are generally spread over the whole earth, while smaller ones, such as families and genera, are frequently confined to one portion, often to a very limited district.
2 In widely distributed families the genera are often limited in range; in widely distributed genera, well-marked groups of species are peculiar to each geographical district.
3 When a group is confined to one district, and is rich in species, it is almost invariably the case that the most closely allied species are found in the same locality or in closely adjoining localities, and that therefore the natural sequence of the species by affinity is also geographical.
4 In countries of a similar climate, but separated by a wide sea or lofty mountains, the families, genera, and species of the one are often represented by closely allied families, genera and species peculiar to the other.

Geology

5 The distribution of the organic world in time is very similar to its present distribution in space.
6 Most of the larger and some small groups extend through several geological periods.
7 In each period, however, there are peculiar groups, found nowhere else, and extending through one or several formations.
8 Species of one genus, or genera of one family occurring in the same geological time are more closely allied than those separated in time.

9 As generally in geography no species or genus occurs in two very distant localities without being also found in intermediate places, so in geology the life of a species or genus has not been interrupted. In other words, no group or species has come into existence twice.

Having enunciated the law Wallace then goes on to develop his various propositions with examples, and to show that the geographical and geological distribution of organisms corresponds to the pattern of their classification into "group subordinate to group" (to use Darwin's phrase). Thus if Wallace's first proposition is correct, it should be possible literally to map the classification of smaller groups, "such as families and genera", on to the Earth's surface as internested areas, effectively Venn diagrams, which could of course then also be represented by a dendrogram (Chapter 2, Sections II and IV). Thus he attempted to establish the close correlation between patterns of classification and patterns of distribution. But Wallace was also aware of the correlation between taxonomic rank and geological time. The recent origin of geographical barriers ("vicariance events" as they are now known) is correlated with differences between related taxa on either side at a low level, endemic subspecies, or perhaps species. More ancient barriers separate higher taxa:

> When a range of mountains has attained a great elevation and has so remained during a long geological period, the species of the two sides at and near their bases will be often very different, representative species of some genera occurring, and even whole genera being peculiar to one side only. . . . A similar phaenomenon occurs when an island has been separated from a continent at a very early period.
>
> In all those cases in which an island has been separated from a continent, or raised by volcanic or coralline action from the sea, or in which a mountain-chain has been elevated, in a recent geological epoch, the phaenomena of peculiar groups or even of single representative species will not exist. Our own island is an example of this . . . while the Alpine range, one of the most recent mountain elevations, separates faunas and floras which scarcely differ more than may be due to climate and latitude alone.
>
> (*Wallace 1855*)

Thus, in this remarkable paper of Wallace's, some of the presuppositions of the most recent branch of systematic biology – vicariance biogeography – are already implicit. Wallace clearly had the concepts of the correlation of the distribution of animals (and plants) with the pattern of natural classification and of the correlation of the age (and effectiveness) of natural barriers with the ranks of the taxa which they separated. Later, as we have seen, Wallace (1863) attempted to outline an example in his Malaysian exercise, but, while he was clear on the delimitation of areas of endemism, he did not develop his ideas on the establishment of a pattern of relationship among areas. Patterson (1983) suggests that by the time of the publication of *The Geographical Distribution of Animals* (1876), Wallace had two incompatible aims. The first was that zoogeography should "reveal to us, in a manner which no other evidence can, which are the oldest and most permanent features of the earth's surface, and which the newest" – in other words, a development of his 1855 proposal of biogeography as corroborative of Earth history and as evidence for evolution. The second aim was to assume the permanence of extant geography and geology and to use zoogeography "as an application of the general theory of Evolution, to solve the problem of the distribution of animals"; in other words, assuming evolution, to explain the current distribution of animals by dispersal over the present-day map of the world from centres of origin for each taxon. Most historical biogeography, until the general acceptance of continental drift after the Second World War, was conducted within this second paradigm. Because it accepts evolution *a priori,* such dispersal biogeography cannot be evidence for evolution.

In the *Origin* (1859), Darwin, endorsed the aim of explaining the current distribution of animals and plants in terms of dispersal of each species from a single centre over the extant geography of the globe, with evolution to account for geographical relationships of taxa. Darwin allowed for the rising and sinking of land surfaces relative to sea level as a source of vicariance events. This included his own theory of coral reefs (Darwin 1842). More importantly, Darwin also set out to enlist biogeography as evidence for evolution in both chapters 11 and 12 of the *Origin* by showing that geographical distribution of organisms "cannot be accounted for by differences in physical conditions":

> In the southern hemisphere, if we compare large tracts of land in Australia, South Africa, and Western South America, between latitudes 25° and 35°, we shall find parts extremely similar in all their conditions, yet it would not be possible to point out three faunas and floras more utterly dissimilar. Or again we may compare the productions of South America south of lat. 35° with those of north of 25°, which consequently inhabit a considerably different climate, and they will be found incomparably more closely related to each other, than they are to the productions of Australia or Africa under nearly the same climate.

This leads to a discussion of physical barriers to dispersal and then to the "affinity of the productions of the same continent", looking at the replacement of one species of Rhea on the plains near the Straits of Magellan, by another northward of there on the plains of La Plata. Similarly, the rodents of South America, although of different species, are all of the South American type, whatever their environment. He then goes on to assert, as had Wallace, that there was no irrefutable evidence of any species having come into being at more than one site and that the wider distribution of many terrestrial plants was explicable by their greater powers of dispersal. This leads to a description of his own experiments on the viability of seeds in salt water.

Chapter 12 of the *Origin* is largely a discussion of dispersal – part evidence, part scenario. After talking about freshwater faunas, Darwin proceeds to a long and important discussion of oceanic islands. He makes five principal points, noting:

1. The paucity of species compared to a comparable area of continental mainland with, however, a high proportion of endemic species. He notes that nearly all the land birds of the Galapagos Islands are endemic, but only two of eleven sea birds are.
2. The absence of amphibia and terrestrial mammals, which cannot easily disperse over oceans.
3. The contrast between continental islands, such as Great Britain, and oceanic ones (with the West Indies as an intermediate case). Britain shares most of its fauna at the species level, including mammals, with the adjacent part of continental Europe.

4 The affinity of the fauna and flora with that of the nearest continental mainland. He cites the birds and the plants of the Galapagos. The Galapagos are then contrasted with the Cape Verde Islands which, although similar in volcanic origin, climate, size, and height, have a totally different fauna derived from that of Africa.
5 That within an archipelago, closely related but different species may inhabit different islands. When this phenomenon occurs, two explanations are possible: that the creatures are such that movement from one island to another is impossible, and/or that competition with the endemic related species would eliminate the invader. He cites the three endemic species of mocking thrush, each on its own island, in the Galapagos (but not "Darwin's finches"; see Lack 1947; Grant 1986) and land molluscs of Madeira and Porto Santo.

Historical biogeography for nearly a hundred years after the *Origin* was in the tradition of incorporating an assumption of "descent-with-modification" origin of taxa (at least terrestrial ones) as species at a particular focus and more or less probable scenarios of dispersal from centres of origin. Unhappily there was no consensus as to how "centres of origin" were to be pin-pointed. Cain (1944), in a critical review, listed thirteen criteria, not necessarily compatible with one another, for determining centres of origin. Other criteria involving fossil members of the group in question could be added to these (Patterson 1981a). The tradition continued after the acceptance of plate tectonics as exemplified by McKenna's (1973) splendidly titled paper "Sweepstakes, Filters, Corridors, Noah's Arks, and Beached Viking Funeral Ships in Palaeogeography". The first three items of that title are categories of dispersal route named by Simpson (1936, 1940, 1965); the others were added by McKenna to accommodate plate tectonics.

If one takes the evidence from traditional biogeography, it is less easy to construct a sequential logical argument for evolution than it is in the case of the evidence from the fossil record. Nevertheless, I think such an argument can be constructed. The first proposition is that any systematic description of the world's biota must include some element of historical reconstruction. Thus, like the fossil record, the distribution of organisms is a historical document. This was implicit even in Linnaeus's dispersal scenarios, was known to Buffon and Humboldt, and was made explicit by de Candolle's distinction between "stations" and "habitations". The second and third propositions are conflated in Darwin's account of the evidence from geographical distribution in the *Origin*.

The second proposition is that it is improbable that the distribution of organisms can be explained by the separate creation of each species in its present environment. All the items of evidence marshalled by Darwin and discussed above corroborate this. He also made a related point in presenting the theory of Natural Selection in Chapter IV of the *Origin:*

> No country can be named in which all the native inhabitants are now so perfectly adapted to each other and to the physical conditions under which they live, that none of them could anyhow be improved; for in all countries the natives have been so far conquered by naturalised productions, that they have allowed foreigners to take possession of the land.

Thus ecological adaptation in any environment is demonstrably imperfect.

The third proposition also derives from Darwin's evidence. It is that the resemblance between any two extant faunas (and to a lesser extent floras) is inversely related to the width of the barrier between them. Thus continental islands, such as Great Britain, have a fauna more closely related to that of the nearest continental mass than do oceanic islands such as the Galapagos, and similarly, islands within an archipelago have similar but not identical faunas. Nevertheless, the Galapagos fauna still resembles that of South America, while that of the Cape Verdes has African affinities.

The fourth proposition was made explicit by Wallace (1855), as we saw above. It is that it is possible to erect a hierarchical classification of faunal regions ("regions subordinate to regions", as one might say) based on degrees of resemblance between their faunas. Thus we have the concept of a classification of faunal regions suggesting, as an analogy with the relationship of organisms explained by common descent, the "relationship" of areas explained by the past history of their biota.

It will be noticed that in this fourth proposition we have perhaps moved from geographical data as evidence for evolution to geographical data as part of the *explanandum* of evolution. It could be argued that if the classification of faunal or floral areas simply reflects the sum of all the classifications of their faunas and floras, then we are dealing with, as it were, a series of superimposed taxonomic classifications.

The full development of this taxonomic approach to biogeography had to await a rigorous technique for the natural classification of organisms. That rigour is claimed by the exponents of cladistics, and it is no coincidence that vicariance biogeography has been developed by cladists, well within the past two decades; so we must say a little about vicariance biogeography as evidence for evolution.

Humphries and Parenti (1986) give an account of this developing discipline (see also Myers and Giller 1988), and we have seen that in outline it can be derived from Wallace's concept of the interrelationship of areas. In essence the method consists of producing a cladogram of geographical areas with known distinct faunas. All the areas under study must share members of two or more taxa that are not closely related, with different species of each taxon in each area. Thus cladistic classifications of species could be mapped onto present continents, giving an *area cladogram*. Each single area cladogram corresponds to a taxonomic cladogram using states of a single character. The hope is that a large number of area cladograms will be found to be congruent. One would then have a well-corroborated aggregate area cladogram in which species correspond to the characters of normal classification, and the areas to be classified correspond to species.

Area cladograms give the pattern, the *explanandum,* of vicariance biogeography; the *explanans* is a series of "vicariance events". Thus when a taxonomic cladogram is interpreted as the pattern of phylogeny (i.e., taken as a "tree"), each node represents a speciation event and the pattern of nodes represents a sequence of speciation events. Similarly, in an area cladogram each node represents a vicariance event in which a barrier arose, splitting what was once a continuous population of each taxon. So the pattern of nodes represents the sequence of vicariance events, produced by plate tectonics, orogenies, changes in sea level, glaciation, or whatever. Thus area cladograms make predictions (or more correctly retrodictions) about the geological history of the earth, which can be corroborated by historical geology.

So the *explanans* of cladistic classifications of species (or of any other classification of species claiming to be natural) is phylogeny, but the *explanans* of biogeography is not phylogeny but the sequence of vicariance events in Earth history. Furthermore, the technique of vicariance biogeography will work only if vicariance events are a necessary cause of speciation in a significant number of cases. Otherwise vicariance events will not be characterised by area cladograms based on taxonomy. Each independent taxon, which together with others makes up the area cladogram, must have its speciation events determined by the same vicariance events as all the other taxa.

Thus if the enterprise of vicariance biogeography, which is yet in its infancy, proves successful in producing many well-corroborated area cladograms, it will constitute almost irrefutable evidence of a pattern of speciation events, and thus be perhaps the best of all evidence for evolution.

IV. Conclusions

In discussions of the pattern of evolution, it is usual to distinguish two "modes" of evolution, as they were termed by Simpson (1944). The first, to use Simpson's original term, is *phyletic evolution*. This mode of evolutionary change in lineages of organisms is usually characterised as change in "morphology", but it also includes change in any taxonomic character, be it anatomical, behavioural, physiological, genetic, or biochemical. The second mode is *speciation* – the origin of two or more species from a single parent species.

In these two chapters reviewing the evidence, I have taken the traditional categories of "evidence of evolution" and have not discussed the cases of speciation or incipient speciation of the type considered in works such as Mayr (1963), White (1978) and Otte and Endler (1989). Nevertheless, in summarising our findings, we should try to make the distinction between evidence of phyletic evolution and evidence of speciation.

We saw in Chapter 4 that the traditional "evidence of evolution" included data that were not direct evidence of evolution but were used to construct classifications. Thus all well-characterised cases of taxic homology corroborate particular classifications – the *explanandum* of which phylogeny is the *explanans*. Special transformational homology, however, if singular (Chapter 4, Section II), can constitute one component of a case corroborating the occurrence of phyletic evolution. This is particularly the case if a series of more than two transformational homologies can be arranged as a morphocline. Alone such a morphocline does not constitute evidence of evolution. If, however, it is paralleled by ontogeny and, better still, by a corresponding chronocline from the fossil record, as with the evolution of the mammalian jaw and middle ear, it constitutes very persuasive evidence.

We also saw that there can be two types of parallel morphocline and chronocline – that of changing structures, as in the jaw–ear story, and that of the transmutation of whole organisms, as in the evolution of the horse family. Although the attempt to trace ancestor-descendant sequences in the fossil record may not be legitimate, a morphocline that is also a chronocline of whole animals (or at least their fossil remains), preferably culminating in extant species, may be interpreted as *Stufenreihe* and thus as evidence of an *Ahnenreihe*. The latter is then good evidence of evolution.

To use fossil evidence, it is necessary to establish its status in the series of steps I suggest in this chapter. They are summarised as follows:

1. that fossils are the remains of once-living organisms;
2. that fossils, in their stratigraphic setting, constitute a historical record;
3. that extinction is a real phenomenon;
4. that there is progression in the fossil record – that is, the record shows a succession of faunas (and floras) through geological time, such that the individual species within a fauna are taxonomically related to, but different from, species in faunas that precede and succeed them;
5. that the fossil record, despite major events, is not marked by any catastrophic event that appears to have eliminated the whole biota; and further that the first appearance of fossil taxa, and their disappearance, are not confined to periods of apparent mass extinction; and thus
6. that progression in the fossil record is more plausibly interpreted as phyletic evolution than as "catastrophism".

Turning to the evidence from the biogeography, we have a similar series of evidential propositions, although they do not form quite such a satisfying logical sequence:

1. Aspects of the present distribution of animals and plants can be explained only by their previous history (de Candolle).

2. Comparison of geographically distant but similar environments makes it improbable that each species was separately created in the environment to which it is most highly adapted (Darwin).
3. *Ceteris paribus,* the resemblance between any two faunas is inversely related to the width and/or effectiveness of the barrier between them.
4. Regions of geographical endemism can be arranged in a hierarchy of regions within regions corresponding to taxonomy of their contained species (Wallace).

The last two propositions might be taken as corroboration of the objective reality of a natural classification, rather than as evidence of evolution, the *explanandum* again. Now, however, with the development of vicariance biogeography using cladistic methods, that appears not to be the case. If area cladograms are the *explanandum,* then the *explanans* of the hierarchy they represent is a series of vicariance events – events that can be corroborated by purely geological data. Each node on an area cladogram represents taxa shared by the areas united by that node. The *explanans* of the divergence of daughter taxa from a node in a taxonomic cladogram is speciation; that from a node in an area cladogram is vicariance. But vicariance would not explain the coincidence of many taxonomic nodes unless speciation occurred. The more successful vicariance biogeography proves to be, the more strongly it will corroborate the hypothesis that speciation has occurred.

There is an irony in this fact. Vicariance biogeography was and is being developed by those who wish to emphasise the logical independence of taxonomy from any hypothesis of phylogeny, but the success of their new, and exciting, enterprise depends on evolution having occurred.

6

Methods of classification: The development of taxonomy

In this and the following chapters, we look at methods of classification. This chapter is historical; Chapter 7 takes up the story with phenetics and cladistics and concerns controversy as well as method. Chapters 8–10 are also much concerned with taxonomic controversy. Throughout we shall be looking at the concept of a "Natural Classification". As we saw in Chapter 2, naturalness up to the time of Linnaeus was often taken to inhere in the correct use of method. In order to show how this came about, I shall spend some time discussing the method of logical division as developed by Plato and Aristotle and eventually transformed into the taxonomic method used by Linnaeus. Another characterisation of naturalness was as the "plan of creation" which developed with "rational morphology" and the homology concept (Chapter 4). Finally, with the acceptance of evolution, some sort of relationship to phylogeny was taken to be the criterion, but, as we shall see in this chapter, it did not lead to any standardised taxonomic method for nearly a hundred years.

I. The method of logical division

In Chapter 2 (Section II), we noticed an early example of the so-called method of division, which appears in *The Sophist,* a later work of Plato. In *The Sophist,* Plato takes the art of angling and demonstrates its properties by making it one of the "taxa" in the lowest rank of a dichotomous dendrogram of the Arts (Fig.

2.3). This was an early version of the type of classificatory pattern that later appears as the tree of Porphyry, the *Stufenreihe,* and the "Hennigian comb". Plato's characterisation of angling may have been produced "somewhat playfully" (Oldroyd 1986), but in the spirit of the aphorism that all European philosophy is "footnotes to Plato" (Whitehead 1929, p. 53), it is possible to describe taxonomic method as a developing modification of Plato's ideas. I shall do this as a heuristic device rather than as an exercise in intellectual history.

The important features of Plato's method of division, as embodied in this example, are that, first, as we have seen, it is dichotomous. However, at each dichotomy only one of the two "taxa" produced is further divided, so that the whole is asymmetrical, with an axis leading from the highest ranking taxon (Arts) to the activity to be classified (angling), which thus occupies the lowest rank. It seems, therefore, that Plato's motive was to give a characterisation of angling by placing it in a hierarchy of human activities and not to classify all human activities, or at least "Arts", with angling as one of their number.

Plato's greatest pupil, Aristotle, was highly critical of Plato's method of classification (Pellegrin 1986). Aristotle, the greatest naturalist of the Hellenic world, wrote a series of treatises on animals in which he was much concerned, not with taxonomy in the modern sense, but with the classification of knowledge about living things. He proposed rejection of Plato's (*a priori*) dichotomous division for a number of reasons. Firstly, in a pre-echo of early battles over cladistics, Aristotle objected to dichotomous division on the grounds that it was not necessarily exhaustive. He cited the division of terrestrial animals by their mode of locomotion; they either walk, crawl (as in snakes), or undulate (as in worms). As Pellegrin points out, Aristotle seems to want to reject dichotomy *a priori* here; he could have divided terrestrial animals into those without and those with feet, and subsequently divided the former into crawling and undulating forms. Aristotle then pointed out that the classes produced by dichotomy frequently overlap. If animals were divided into gregarious and solitary, man would be in both categories, and an individual could change from one to another. Similarly, two classes produced by one mode of division would clash with those produced by another; grouping animals as terrestrial or aquatic cuts across a dichotomy into birds and quadrupeds. Aristotle further objected to division by "privation" (e.g., winged *versus* wingless), again paralleling the preoccupation of contemporary taxonomists – in this case, about "loss" characters (cf. Patterson's 1982 discussion of secondarily wingless insects; see Chapter 4, Section II). Aristotle was then able, using the same example, to set up a paradox. Subsequent division of the "winged/wingless" dichotomy could then, in the case of "winged", proceed in two different ways: If winged organisms were further divided by reference to the types of wings, an *artificial* classification based on the status of one character would be produced. On the other hand, if the second division were moved to a different *domain* (e.g., "domestic/wild"), the "domestic/wild" dichotomy would not necessarily be contained exclusively within the class "winged", and the arrangement would not be hierarchical.

In his discussion of Aristotle's classification of animals, Pellegrin argues convincingly that Aristotle was not engaged in a taxonomic enterprise at all: to Aristotle the method of logical division was a method of achieving knowledge (Oldroyd 1986). A class of objects is recognised intuitively by shared characteristics; mankind was one of Aristotle's examples. Man is then defined as a "rational animal". The definition embodies the "essence" of man: without the properties made explicit in the definition, man would not be man. The definition is then divisible into the feature or features unique to the class of mankind (i.e., "rational") and that or those shared with other classes of objects (i.e., "animal"; or rather, the definition of "animal"). In the original Greek, "animal" is the γένος (*genos*), mankind the εἰδος (*eidos*), and "rational" the διαφορά (*diafora*). Apart from the essence of man, expressed as a definition, any individual man may have two other sorts of characteristics: properties and accidents. *Properties* were those features that Aristotle thought could be inferred from the definition. In our example he believed that "being capable of learning grammar" was implied by "rational animal" (a doubtful proposition). Thus definitions and properties were "convertible predicates", delimiting the same class as the subject ("man") to which they referred. *Accidents* were not convertible predicates; they were neither part of the definition nor deducible from it. Characteristics that were not applicable to the whole class ("tall men", "short men") or that might or might not apply to an individual ("in a sitting position" was an example of Aristotle's) were accidents.

Thus Aristotle's method of division was presented as a method of investigating the nature of things, but there are five very important points to be made about it. First, there is Pellegrin's point that it was not a technique for producing a taxonomic classification, despite Aristotle's interest in animals. The system did not yield a hierarchy in his writings but merely the division of *genos* at any level into two or more *eide*. Second, and related to this, *genos* and *eidos* did not denote any particular level, as in the ranks of a classification. They are frequently translated into Latin as *genus* and *species* and then confused with the categories as Linnaeus used them. For example, Aristotle divided all animals (the *genos*) into those with blood and those without (two *eide!*), but at a much lower rank divided serpents (the *genos*) into oviparous and viviparous forms. Third, he divided the same *genos* (e.g., animals) in different ways, depending on his purpose at the time. Thus, in addition to the blood/bloodless dichotomy, he divided animals according to their mode of reproduction: viviparous / with perfect eggs / with imperfect eggs (which alter their size after being laid!) / with "scoleces" / with generative slime, buds, or spontaneous generation only. The fourth point to be made about Aristotle's method of division is that the essence of a *genos* and an *eidos* can be known *a priori* only in a *taxonomy of analysed entities* (Cain 1958): a triangle (*eidos*) is a plane figure (*genos*) bounded by three straight lines (*diafora;* or in Latin, *differentia*). The classification of organisms, and Aristotle's attempts at logical division with *genē* of animals constitute a taxonomy of *unanalysed* entities. It is not *a priori* that every *eidos* of Aristotle's that is based on observation has an essence. *A fortiori* there is no *a priori* reason to assume that any taxon in a classification of organisms has an essence in Aristotle's sense. Thus the fifth and perhaps most important point to be made about Aristotle's method of logical division as applied to taxonomy is that it requires an act of faith to assert that taxa have essences. That act of faith is a case of the philosophical stance of *essentialism*.

More important still is the relationship of essentialism to the concept of a natural classification. If the truth of phylogeny is not taken as axiomatic for classification, then the concept of a natural classification to which taxonomists can aspire requires that the taxa at every rank are real, that is, correspond to phenomena in nature. Furthermore, the diagnoses of these taxa, if correct, must embody their natures as essences. This is essentialism. I shall further argue that philosophical *idealism* may also be a necessary presupposition for the concept of a natural classification if phylogeny is not taken as axiomatic. For the origins of idealism we must return to Plato.

Plato's idealism is expressed in his development of the theory of "Ideas" (ἰδέαι) or "Forms". His doctrine is often introduced by noting that a geometrical figure, such as a triangle, may be imperfectly drawn so that the sides are not straight lines and thus the angles do not add up to precisely 180°, but it is easy to have a concept of a corresponding perfect triangle of which all drawn congruent triangles are imperfect representations. Similarly, one can have the concept of an ideal table of which all existing tables are embodiments. Plato pursued his idealism to the belief that concepts such as "beauty" or "justice" were also real and to be arrived at by finding what all (e.g.) beautiful objects had in common. Furthermore, to him true knowledge consisted of acquaintance with these ideal forms: data about the mere mundane world were opinions (*doxa:* δόξα). In his early writings Plato supposed that *Ideas* were immanent in the objects that represented them; beauty was immanent in beautiful objects and did not exist outside them. Later, however, he began to invest his *Ideas* with an objective reality of their own, until in the *Timaeus* they are regarded not only as transcendent, existing independently of mundane objects, but also as belonging to a higher level, or rather category of reality (Oldroyd 1986).

Thus there is an obvious parallel between Platonic *idealism* and Aristotelian *essentialism.* As applied to taxonomy, taxa can be compared to Plato's *Ideas,* having an existence in reality which transcends that of the individual organisms which the taxonomic names denote. The individual organisms are imperfect representations of the *ideas* recognised as the connotation of the taxon name. Presumably, higher taxa can be equated with *ideas* representing the perfect form, of which either all the individual organisms, or its subordinate taxa, are imperfect representations. If it is the subordinate taxa that are collectively represented by the form of the higher taxon, then there is an obvious parallel with the *Baupläne* of early nineteenth-century morphology.

Essentialism in taxonomy is thus subtly different from idealism. I shall take taxonomic *essentialism* to mean that the criterion of "naturalness" of every taxon of organisms at every rank is that a taxon can be characterised by a unique combination of features that together comprise its *essence* (the apomorph characters of transformed cladistics). *Idealism* assigns objective reality to taxa and to the *Baupläne* which represent them.

II. Linnean taxonomy

As we saw in Chapter 2, the Tree of Porphyry (Fig. 2.6) was a formalisation of the method of logical division in which, in contrast to Aristotle's usage of the method, a series of ranks with corresponding categories is standardised. In Porphyry the category of highest rank (now in Latin) was the *Summum genus,* divided by alternative *Differentia* into two *Subaltern genera* (although as we saw in Chapter 2, only one of these was named). The category of the lowest rank was *Infima species,* further divided only into *individua.* However, in the original tree the category *subaltern genus* was used at a series of ranks as required to construct the hierarchy between *summum genus* and *infima species.*

Cain (1958) is an excellent account of Linnaeus's use of logical division, giving a summary of Aristotle's doctrine of the five predicables, the only five ways in which a predicate B might be related to subject A. These are those we saw above: thus B might be a *Definition,* a *Genus,* a *Differentia,* a *Property,* or an *Accident.* Linnaeus took over his whole system from Aristotle and, in addition, attempted to incorporate the restrictions of the so-called *fundamentum divisionis.* As Cain puts it:

> In the genus triangle for example, if one species is isosceles, the others will be equilateral and scalene, the *fundamentum divisionis* being here the proportions of the sides: but to have one species isosceles and another right-angled means that two *fundamenta* are being used simultaneously, and this is illogical.

This is the same as the point which we saw that Aristotle made, that winged creatures should be divided according to the nature of their wings. However, Aristotelian taxonomists soon discovered that consistent use of one *fundamentum divisionis* was impossible. This is related to Cain's point that the Aristotelian system was valid only for a taxonomy of analysed entities.

Linnaeus used the categories that we have seen in the tree of Porphyry, except that he differentiated *subaltern genus* into two consecutive ranks so that his series of categories for logical division were *Genus summum, Genus intermedium, Genus proximum, Species,* which he explicitly paralleled with *Classis, Ordo, Genus, Species,* respectively (Linnaeus 1758, p. 7). Cain points out that Ray and other previous taxonomists used the same hierarchy. However, Linnaeus paralleled *Individuum* and *Varietas,* which is irrational: strict logical division cannot proceed to the individual; both *genos* and *eidos* are universals, or classes in the logical sense, despite the fact that the tree of Porphyry claims division to the individual. Variety is therefore just the introduction of a category at a lower rank than species, and, whether it is valid taxonomically or not, it has a different logical status from the individual.

The received opinion amongst most biologists is that Linnaeus inaugurated modern biological classification. This view is supported by the fact that the year of publication of the tenth edition of the *Systema Naturae* is taken as a datum by the *International Code of Zoological Nomenclature*. Article 3 of the latest (third) edition of the *Code* (Ride *et al.* 1985) reads:

> **Starting point.** – The date 1 January 1758 is arbitrarily fixed in this Code as the date of the starting point of zoological nomenclature. Two works are deemed to have been published on that date:
> 1. Linnaeus's Systema Naturae, 10th Edition; and
> 2. Clerck's Aranei Svecici.
>
> Names in the latter have priority over names in former. Any other work published in 1758 is deemed to have been published after the 10th Edition of Systema Naturae.

Received opinion correctly attributes four contributions to Linnaeus: (1) the binomial system in which every organism is identified by a generic and specific (formerly 'trivial') name; (2) that every taxon should be characterised by a diagnosis of those characters required for membership; (3) establishment of the standard hierarchy of categories – (Empire), Kingdom, Class, Order, Genus, Species (Variety); (4) the production, in volume 1 of the tenth edition of the *Systema,* of a much more comprehensive classification of animals than had been attempted before.

However, Linnaeus's work was much more a culmination of the classical "Aristotelian" approach to taxonomy than the inauguration of a new era as the International Codes of Zoology (and Botany) would suggest. Both the indispensable system of binomial nomenclature and the formalisation of brief diagnoses are in the Aristotelian tradition of logical division *per genus et differentiam (sive differentias).* The hierarchy derives directly from Platonic division and the tree of Porphyry, and Linnaeus's animal classification has more virtue as a catalogue than as a hierarchical arrangement.

Linnaeus regarded the genus as the fundamental unit of classification, partly in deference to the concept of *genos* and partly because he did not give high priority to resolving the problems of the higher-rank classification of animals. He recognised only six classes of animals, namely: I. Mammalia; II. Aves; III. Amphibia [including reptiles, cartilaginous fish, lampreys (and some bony fish in error)]; IV. Pisces; V. Insecta (including myriapods and spiders); VI. Vermes (all remaining known invertebrates). A rational classification of invertebrates, many groups of which had been recognised by Aristotle, was recommenced by Cuvier in his "Memoir on the Classification of the Animals Named Worms" (1795), and further improved by Cuvier himself and by Lamarck.

We have already noticed in Chapter 2 that Linnaeus developed a consciously artificial classification of flowering plants with his "sexual system" (details in Nelson and Platnick 1981, pp. 81–9). This was first developed in the first edition of the *Systema Naturae* (1735) and pre-occupied Linnaeus throughout the rest of his life. Linnaeus also continued to struggle with a natural system throughout the time that he was developing his artificial sexual one. The tenets of logical division suggested that he should use a single *fundamentum divisionis* and thus base the whole on the fructification only, including stamens and carpels, but also the fruit, seeds, and receptacle and the calyx and corolla. He also stated, however, in the *Philosophia Botanica* (1751), that a natural system should be based on aggregate affinity.

As with vertebrates in his animal classification, Linnaeus's classification of plants is strongly biased towards flowering plants, and for this reason the rules of the International Code of Botanical

Nomenclature (Greuter *et al.* 1988) use the first edition of Linnaeus's *Species Plantarum* (May 1, 1753) as the "start date" for the nomenclature of *all* Spermatophyta (flowering plants and gymnosperms) and Pteridophyta (ferns, clubmosses, etc.) but only *some* "lower" plants. Fossil plants all have a different start date.

III. Post-Linnean taxonomy

I have dwelt on Linnaeus's taxonomy, and its philosophical background in the writings of Plato and Aristotle, because of my contention that without a presumption of phylogeny, a "natural classification" presupposes idealism and/or essentialism. This conclusion, however, is at odds with the claim that classification is logically prior to phylogeny, as we have already noted. Phylogeny explains the pattern of classification, but if the taxa of classification represent Platonic *Ideas,* and can be characterised by a statement of their Aristotelian essence, then they must be immutable and evolution cannot have occurred, the point made by Mayr (1988). The parallel dilemma, which we have emphasised before, is that there is no *a priori* reason why the natural ordering of taxa should be a divergent hierarchy, but if phylogeny is taken to be a sufficient reason for the existence of hierarchy, the independent status of taxonomy is lost and phylogeny has no *explanandum*. The problem has been discussed, *inter alia,* by Scott-Ram (1990), but it has not been resolved.

In the period between the general acceptance of Linnean taxonomy in the latter part of the eighteenth century and that of evolution following the publication of the *Origin* in 1859, a series of studies, recorded by Mayr (1982), occurred which moved taxonomy away from the classical formalism of Linnaeus towards the state when the pattern of classification could reasonably be interpreted as phylogeny by Darwin and Wallace. The more important of these studies have been discussed in our Chapters 2, 3, and 4. At the beginning of Chapter 2, I introduced Mayr's (1982) distinction between an exclusive and an inclusive hierarchy in which he says that "the modern hierarchy of taxonomic categories is a typical example of an inclusive hierarchy".

The late eighteenth-century or early nineteenth-century taxonomist who accepted the natural ordering of organisms as a *scala naturae* would, if he produced a hierarchical classification, be turning an exclusive hierarchy into an inclusive one, which inevitably he would have to regard as an artificial arrangement. This corresponds to the distinction made by Lamarck between (the natural) "*distribution générale*" – or ordering the *scala* – and the (artificial) "*classification*" – or dividing the continuum of the *scala* to yield a hierarchy (see Chapter 2, Section I). Nelson and Platnick (1981, 1984) and Rieppel (1988) have developed the relationship between the *scala* as an exclusive hierarchy and what I here call the "Hennigian comb" (Chapter 2, Section II) as a corresponding inclusive hierarchy. Members of the *scala* are defined only by the characters that determine their position in the sequence. Then, as taxa, they have the cumulative advanced characters of all those members below them and lack those of all the members above. Only the advanced terminal member (usually *Homo sapiens!*) has any unique advanced characters; all the other members are defined in part by *privation* (Rieppel 1988, p. 132), the term used by Aristotle. The corresponding Hennigian comb converts the sequence of cumulative characters into an inter-nested series of taxa descending in rank towards the most advanced taxon (*Homo sapiens*) but has the inherent assumption that each individual member of the series as well as the terminal one has at least one unique character of its own (Rieppel 1988, p. 83ff.).

As we saw in Chapter 2, part of the history of taxonomy between Linnaeus and Darwin could be regarded as the rejection, but not the total abolition, of the concept of the natural order as a *scala*. Lamarck started with a taxonomic and later a phylogenetic picture of at least animals as a *scala* but later saw a more complicated arrangement in which his three grades (*animaux apathiques, sensibles,* and *intelligents*) cut across his major taxa. Similarly, MacLeay, in his invention of quinarian classification (Chapter 2, Section III), arranged his major insect taxa in two parallel *scalae* united by cross-connections representing grades. But MacLeay went further in at first distinguishing affinity (recognised by what we describe as homology) among members of the same *scala* and "relations of analogy" (recognised by what we describe as homoplasy) among pairs of taxa at the same grade. He then went on to arrange each *scala* as a circle, abolishing the concepts of degenerate and advanced, and to draw a systematic parallel between corresponding members of the two circles, thus making analogy part of the structural arrangement of the natural order. In his fully developed quinarian system he attempted a classification that is an inclusive hierarchy in which each taxon at every rank is divided into just five subordinate taxa. He thus started with parallel exclusive hierarchies and developed from them a single inclusive hierarchy.

Von Baer also evolved an *inclusive* hierarchy, essentially a Hennigian comb, from the *exclusive* hierarchy of the tradition of *Naturphilosophen,* with the additional important factor that both hierarchies featured the parallel between ontogeny and the pattern of classification (Chapter 2, Section II). Finally, Cuvier, followed by Agassiz, explicity rejected any *scala* arrangement, or exclusive hierarchy, with his four *embranchements* (Chapter 2, Section I).

Two other contributions of Cuvier's must be noted here. First was his use of characters derived from his extensive studies of internal comparative anatomy; second was his *a priori* weighting of characters. In all his studies of palaeontology and comparative anatomy, Cuvier attached great importance to the *correlation of parts.* Living animals were functional integrated machines and could not be atomised into a series of independent "characters". This led to the further conclusion that new and incomplete fossil specimens could have missing skeletal and other features restored by analogy with those correlations seen in their living relatives. From the concept of the correlation of parts, Cuvier was led to infer the principle of the *subordination of characters.* This led him to propose the hierarchy of characters in establishing and subdividing his four *embranchements,* as we saw in Chapter 2. Mayr (1982, p. 188) notes that at first Cuvier attached the highest importance to organs of nutrition and to the circulatory system but that by 1807 the nervous system was regarded by him as the most important character complex. Thus the condition of the nervous system was used to characterise his four major taxa, with the circulatory system subordinate to it and so on. As a result, Cuvier established a system of *a priori* weighting of characters based on the concept of a hierarchy of characters. The rank of a character in that hierarchy is the same as Riedl's (1979) concept of "Burden" (Chapter 8, Section III). Mayr (1982, p. 186) suggests that:

> Taking the entire period from Cesalpino to the present, taxonomic characters have provoked three major controversies: (1) Should one use only a single key character (*fundamentum divisionis*) or multiple ("all possible") characters? (2) Should only

> morphological characters be allowed, or also ecological, physiological and behavioural characters? (3) Should characters be "weighted" or not – and if weighted, by what criteria?

The use of a single key character for reasons other than convenience, as in dichotomous identification keys, could be considered good practice only by proponents of formal logical division, and Mayr suggests that Linnaeus and his forerunners such as Ray, Rivinus (Bachmann), and Tournefort maintained a partial fiction that their favourite characters gave the essence of taxa, but in fact chose the characters for convenience in identification. *A posteriori* weighting, favoured by Mayr, consists of the method whereby "organisms are first ordered into seemingly natural groups . . . and then those characters are given the greatest weight which seem to be correlated with the most natural groups" (Mayr 1982, p. 224).

Another important development in the years before the *Origin* was the formalisation of the concept of homology as opposed to homoplasy; developed as a non-phylogenetic concept by MacLeay, Geoffroy Saint Hilaire, and the German *Naturphilosophen,* it was codified by Richard Owen (Chapter 4). Finally, before the *Origin* two other controversies were more or less resolved in favour of a new-found pragmatism which further discredited classical logical division. These were the recognition of (1) *polythetic taxa* and (2) the move to "upward classification by empirical grouping" (Mayr 1982).

(1) The assumption of logical division is that any taxon will have a series of one or more characters that comprise its essence and characterise all its members. It is, however, often the case that individual species within the taxon, or even higher ranking subordinate taxa, lack one or more of those diagnostic characters. Thus fleas are pterygote insects lacking wings, snakes are tetrapods lacking legs, and whales are mammals lacking hair. So higher taxa may have a series of diagnostic characters, none of which taken alone is either sufficient or necessary for membership. Such taxa were termed *polythetic* by Sneath (1962), in contrast to *monothetic* taxa which "are formed by rigid and successive logical divisions so that the possession of a unique set of features is both sufficient and necessary for membership of the group thus defined" (Sneath and Sokal 1973, p. 20). Although the terminology is recent, the concept of a polythetic taxon appears to date back to Vicq-d'Azyr

(1792) and even to Ray. The acceptance of polythetic taxa negates essentialism and makes it necessary to characterise a "natural classification" in some other way, assuming that the concept is acceptable at all.

(2) Logical division is downward classification; alternative terms are *divisive* (Sneath and Sokal 1973) or *analytic* (Panchen 1982) classification. Clustering of species is upward, *agglomerative,* or *synthetic* classification. The sheer impracticality of logical division, which according to Mayr, caused even Linnaeus to "cheat", plus the recognition of polythetic groups, led to the acceptance of "upward" classification by the time of Darwin and Wallace. Notable in this changeover were Adanson's insistence on natural classification involving many characters (Chapter 7, Section I); Strickland's method, which so impressed Wallace (Chapter 2, Section III); and Darwin's own work on barnacles.

Thus, as we saw in Chapter 2, by the time of the publication of the *Origin,* the form of classification as a hierarchical clustering was probably fairly generally accepted. The brief enthusiasm for quinarian classification, particularly in Britain, was over, and no other numerological system had been suggested to replace it. Empirical clustering gave irregular patterns. As far as method was concerned, logical division was out of favour (though essentialism was to return), and the use of many characters to diagnose taxa was regarded as good practice, which made the acceptance of polythetic taxa almost inevitable (but see Chapter 10, Section II). It is probable that those who accepted the concept of a "natural classification", but not that of phylogeny as a criterion, regarded the pattern of classification as "the plan of creation". They would almost certainly have embraced a form of idealism referring to archetypes (Chapter 4) as their criterion of naturalness. With the publication of the *Origin of Species,* "evolutionary classification" was born, but what this implied in terms of methodology is remarkably unclear.

IV. "Evolutionary classification"

de Queiroz (1988) suggests that the reason for the lack of a coherent methodology in post-Darwinian traditional taxonomy is that "the Darwinian Revolution has only just begun in biological taxonomy". He attributes this late start to a failure to distinguish between two profoundly different ways of constructing classifications, a distinction credited to Griffiths (1974). These are, in Griffith's terminology (which differs from that of Simpson 1961a; see below), (1) classification and (2) systematisation. Classification in this sense is the ordering of elements or individuals into classes. A class in this (logical) sense is a group defined by a property or properties shared by its members. Systematisation is the ordering of entities into more inclusive entities. The more inclusive entity (a whole) is a system, whose existence depends on "some natural process through which its elements (component parts) are related" (de Queiroz).

Thus "classification" in Griffith's sense is based on essentialism. Classes as such cannot evolve or be related to one another in any literal sense, so that the ordering of classes in a classification cannot be based on phylogeny. Systematisation, on the other hand, is the reconstruction of a whole, a "chunk of the genealogical nexus" (Ghiselin 1974) or "unique, *individual* portion of the phylogenetic tree" (Patterson 1978). We consider the vexed problem of whether species and higher taxa are individuals or classes in logic in Chapter 14, but meanwhile we note de Queiroz's point that the Darwinian revolution in taxonomy is only just beginning, not because taxonomy until recently was entirely essentialist, but because the distinction between "classification" and "systematisation" (*sensu* Griffiths) was not made. Thus until the broad acceptance of cladistics in its original Hennigian form (see Chapter 7, Section III), there was no consistent methodology for producing a distinctively phylogenetic classification. It might well be asked in what way the work of taxonomists in the hundred years between (roughly) 1860 and 1960 was affected by their acceptance of evolution and thus the reality of phylogeny. The answer seems to be "very little". As many recent critics (e.g., Patterson 1981b) have noted, systematic zoologists (less so botanists) turned their attention to the tracing of ancestor-descendant relationships (see Chapters 3 and 5) and the elucidation of transformational homologies (Chapter 4). Both fields of endeavour were practised both to corroborate the truth of evolution and to decipher the pattern of phylogeny: they were not on the whole reflected in the written classifications of animals, as distinct from various sorts of diagrams representing phylogeny.

These preoccupations can be seen vividly in the classification of vertebrates. Linnaeus had four classes in his classification which comprised the vertebrates: Mammalia, Aves, Amphibia, Pisces. Reptilia and Amphibia were finally recognised as separate vertebrate classes, and thus of equal rank, by Gegenbaur in 1859 (Gadow 1901).

Aves have been recognised as a valid vertebrate taxon from Linnaeus onwards, although cladists question the category (class) to which they are usually assigned. The situation of Mammalia is similar. The class Pisces has had a more chequered history. The important division into bony and cartilaginous fishes was pre-evolutionary, made by Cuvier and Valenciennes in their *Histoire Naturelle des Poissons* (1828–48), but their Chondropterygii included the cyclostomes (lampreys and hagfish), primitive jawless vertebrates. Johannes Müller (1844) separated the cyclostomes as one of six sub-classes of the class Pisces and they were made one of the two "branches" into which all vertebrates were divided by Goodrich (1909). The history of fish classification up to about 1950 is reviewed by Berg (1958).

For our present purposes, I want to consider how the general acceptance of phylogeny and the enormous wealth of fossil material discovered and described affected the traditional *classification* of vertebrates. Patterns of phylogeny were freely reconstructed (Chapter 3), and some authors, notably Goodrich (1909, 1930) and Säve-Söderbergh (1934), attempted to base their classifications on them; however, fossil discoveries and phylogenies based on their use generally made little difference to the classification and ranking of living groups, most of which had been known before phylogeny was generally accepted.

Thus the birds, while a valid taxon, were known to be nested within the reptiles and more specifically within the archosaurs but were retained at the same rank as the Reptilia. A clade including mammals and "mammal-like reptiles" had been recognised as separate from other amniotes since Goodrich (1916) demonstrated two distinct morphoclines for the structure of the heart and aortic arches diverging from the amphibian grade (see Chapter 4, Section II), and since work on the mammal-like reptiles, from Cope onwards, demonstrated a series of reptilian forms spanning the gap between a primitive reptile grade in the Carboniferous, and recognised mammals in the Late Triassic (Kemp 1982, 1988b). Nevertheless, mammals have been retained as a class separate from the mammal-like reptiles in spite of the occasional suggestion (e.g., Reed 1960) that the whole clade should be embraced within the class Mammalia. Thus like Aves within the Archosauria, Mammalia are nested within the "reptilian" clade Synapsida, constituting with them the Theropsida of Goodrich. In the case of mammals, however, there was a period when, as with almost all the traditional vertebrate classes, mammals were regarded as polyphyletic (for a contemporary review, see Hopson and Crompton 1969), or perhaps at least diphyletic (e.g., Hopson 1969; rejected by Patterson 1981a). Nevertheless, this view did not, as de Queiroz notes, "lead to their [i.e., the advocates of polyphyly] rejection of the taxon Mammalia". Nor did it lead to the inclusion in the Mammalia of all those synapsids which those advocates regarded as descendants of the nearest common ancestor of all those creatures which they regarded as mammals. Had they been so included, Mammalia would have become miraculously monophyletic again!

Class Reptilia, on the other hand, was irretrievably a grade group, paraphyletic in cladist terminology because incomplete without the mammals and birds. The grade status is emphasised by the division of the theropsid clade into mammals and mammal-like reptiles. Similar arguments apply to the class Amphibia. Numbers of fossil primitive tetrapods were discovered from the middle of the last century onwards, and as knowledge of them increased, they were regarded as non-amniote and thus not to be referred to reptiles. Almost universally they were placed within the class Amphibia. Early patterns of classification are reviewed in Panchen (1970). Thus like the Reptilia, Amphibia was regarded as a grade group, so that "amphibian" to most authorities meant "non-amniote tetrapod" as it does to many today (e.g., Carroll 1988). The only dissenters were those (Holmgren 1933, 1939, 1949; Säve-Söderbergh 1934, 1945; Jarvik 1980, 1986) who regarded the "Amphibia" as diphyletic. They were certainly wrong (Panchen and Smithson 1987, 1988) but, unlike the mammalian workers, adjusted their classifications accordingly. When a strong case was made for the monophyly of the extant Amphibia by Parsons and Williams (1963: see Milner 1988), they renamed the extant group "Lissamphibia", with the same denotation as the original class Amphibia, rather than disturb the grade group with its vastly expanded and invalid membership.

The history of "Pisces" is, as I have said, more complex, but again, in most cases, there was no one-to-one correspondence between classification and reconstructed phylogeny. We have seen that the separation of Chondrichthyes (cartilaginous fishes) and Osteichthyes (bony fishes) was made in the first half of the nineteenth century, with the cyclostomes recognised as a group outside both at about the same time. By the time of Goodrich's (1909) contribution to the Lankaster *Treatise on Zoology,* cyclostomes were recognised as a group outside and equal in rank to all other vertebrates, the latter named "Gnathostomata". Goodrich retained "Pisces" for the fishlike gnathostomes both in 1909 and as an avowedly grade group (with the same denotation as "Grade Ichthyopterygii") in 1930. Even in the 1960s, "class" (or "superclass") "Pisces" can still be found (Jarvik 1960; Camp, Allison, and Nichols 1964; Weichart 1965 who includes the Agnatha). These classifications are reviewed by Olson (1971).

Thus while there may have been great efforts to use fossils to establish phylogenies since the general acceptance of evolution, classifications have been based largely on traditional taxa, often at the same rank. We have seen this for the vertebrate classes, and I suspect that it could be demonstrated for other vertebrate categories and for the taxa of other phyla. It therefore seems to me that the growing inconsistencies between reconstructed phylogenies, whatever the validity of the methods used, and the classifications purporting to represent them, were due to taxonomic inertia. Any text attempting to explain the methods of traditional taxonomy thus as evolutionary was sure to be to some extent a work of *post hoc* rationalisation.

In the 1960s, when criticism of evolutionary classification was gathering force, two books, both by major contributors to the Synthetic Theory of evolution (Chapter 12), were seen as summaries of, and justification for, traditional methods. These were Simpson's *Principles of Animal Taxonomy* (1961), developed in part from an earlier essay (Simpson 1945), and Mayr's *Principles of Systematic Zoology* (1969), effectively a revised edition of Mayr, Linsley, and Usinger (1953).

Simpson (1961a) elevates the inertia I have spoken of above into a principle in the name of stability:

> A published classification in current use should be changed when it is definitely inconsistent with known facts and accepted principles, but only as far as necessary to bring it into consistency.
> (*Simpson 1961a, p. 112*)

Throughout the book he emphasises that however data are arranged or phylogenies reconstructed, the production of classification is an art and not a science:

> Taxonomy is a science, but its application to classification involves a great deal of human contrivance and ingenuity, in short, of art. In this art there is a leeway for personal taste, even foibles, but there are also canons that help to make some classification better, more meaningful, more useful than others.
>
> That the best basis for classification is evolutionary has already been decided. That means in the first place that a classification should be *consistent* with all than can be learned of the phylogeny of the group classified. *That classification can or should express phylogeny is an evident error.*
> (*p. 107; italics in last sentence mine*).

Simpson's book includes a series of useful definitions of all the terms he uses. His distinction between *taxonomy* and *classification* (as in this quotation) and their distinction from *systematics* are expressed thus (his pp. 7–11):

> *Taxonomy is the theoretical study of classification, including its bases, principles, procedures and rules.*
>
> *Zoological classification is the ordering of animals into groups (or sets) on the basis of their relationships, that is, of course by contiguity, similarity or both.*
>
> *Systematics is the scientific study of the kinds and diversity of organisms and of any and all relationships among them.*

The implicit distinctions have been adopted in this book unless some other source is stated specifically. In zoological (and presumably botanical) classification, association into taxa, is, according to Simpson, to be by contiguity and/or similarity. By "contiguity" Simpson meant membership of a single system, as in members of a single clade; "similarity" was used in the sense of characters in common. The use of contiguity, similarity, or both is symptomatic of Simpson's attitude to the relationship between reconstructed phylogeny and classification, which should be *consistent* with each other but not necessarily more (see above).

When one turns to the reconstruction of phylogeny, Simpson was emphatic that fossils provided the best evidence despite deficiencies and bias in the record – "fossils provide the soundest basis for evolutionary classification" (p. 83) – citing, at this point, confirmation of the close relationship of horses and rhinoceroses by the convergence backwards in time of the horse family to the primitive rhinoceros condition (see our Chapter 5, Section II). Phylogeny was to be reconstructed largely by the tracing of ancestor-descendant sequences (see Chapter 3), but, this done, classification had to be consistent with it to the degree that taxa must be monophyletic according to Simpson's very relaxed use of that term:

> *Monophyly is the derivation of a taxon through one or more lineages* (temporal successions of ancestral-descendant populations) *from one immediately ancestral taxon of the same or lower rank.*
>
> (*p. 124; italics in original*)

On this criterion all the theories of the polyphyletic origin of tetrapod classes argued over during this century would be cases of Simpson's monophyly: the proposed diphyletic origin of tetrapods (? super-class) from osteolepiform plus porolepiform or dipnoan fishes (Jarvik and Holmgren, respectively), of reptiles from two groups of "labyrinthodont amphibia" (Olson 1947; Watson 1951), and of mammals from different advanced theropsid groups (see above).

One more feature of Simpson's view of the relationship of classification and phylogeny must be noted. In his *Tempo and Mode in Evolution* (1944), he had proposed the existence of *quantum evolution* as a mode distinct from speciation and phyletic evolution, which consisted of the rapid evolution of presumed small populations across a metaphorical non-adaptive zone from one adaptive type to another. This appears rather like the speciation events which Eldredge and Gould envisage in their theory of punctuated equilibria (Chapter 12, Section IV). Generally, however, Simpson saw evolution as a gradualistic process, with speciation resulting in the daughter species diverging gradually from each other, and phyletic gradualism ("horotelic evolution") prevailing between speciation events. Thus to him a single lineage or interbreeding population could evolve without any speciation event until the descendant population differed morphologically from the ancestral one sufficiently to be regarded as a different "successional [and taxonomic] species" – that is, speciation by anagenesis. He (Simpson 1961a, pp. 165–8) gives instructions for chopping up such a lineage found in the fossil record. He does not give examples here but develops one (erroneous as it turns out; see Maglio 1973) in elephants (see Simpson 1953, pp. 387–8).

Simpson's *Principles of Animal Taxonomy* (1961a) was published just before the two principal challenges to traditional taxonomy (phenetics and cladistics) had really got under way. He refers to Hennig's (1950) work in German with generous praise (Simpson 1961a, p. 71, footnote), referring to partial but substantial agreement (!), but does little to expound Hennig's system. By the time that Mayr had published *Principles of Systematic Zoology* (1969), Sokal and Sneath (1963) had started to codify phenetics, Hennig (1966) had been translated into English, and cladistics was beginning to attract a following. As a result, Mayr spends some time in the criticism of phenetics and cladistics. According to Mayr (1969, pp. 65ff.) there are five, and only five, theories of classification:

1. Essentialism (Aristotle to Linnaeus)
2. Nominalism
3. Empiricism
4. Cladism
5. Evolutionary classification

"Essentialism" we have considered above (Section I, this chapter). Nominalism is the stance that "only individuals exist. All groupings, all classes, all universals, are artifacts of the human mind". Mayr considers phenetics to be nominalist (but see next section). Empiricism requires no theory of classification. "Provided enough characters are intelligently evaluated, a natural system (the meaning of 'natural' being very different from the Aristotelian one) will emerge automatically". This approach is perhaps represented by Strickland's classification of birds (see Chapter 2, Section III, this volume) and is taken by Mayr to characterise a developing theme in post-Linnean, pre-Darwinian taxonomy. "Cladism" we consider more fully in Chapter 7, but mention briefly in the next section. Evolutionary classification is, needless to say, the approach Mayr favours. It is in many respects similar to that of Simpson (1961a). Mayr finds Simpson's definition of monophyly acceptable, but goes further than Simpson in attempting to characterise a "natural classification". Simpson (1961a, pp. 54–7) presents a rather inconclusive discussion. He agrees that populations and species can often be apprehended as natural entities and that "species exist because they evolved". Higher taxa are less intuitively obvious as natural entities, "but evolution does also produce higher taxa". "In short, if such a thing as natural classification can meaningfully be achieved, it must be by evolutionary classification".

Mayr attempts greater precision. He emphasises the various desirable "operational" aspects of a classification: that it must have *explanatory* and *predictive* value (see Chapter 7, Section I, for the phenetic use of these criteria), and it must have a "strong *heuristic* aspect". It must, as a theory, also be provisional and subject to amendment. He then goes on to develop his concept of "genes-in-common" as a criterion of naturalness in classification, a concept that he develops further in his subsequent attacks on cladistics (Mayr 1974, 1981). "If we knew the entire genotype [i.e., genome] of each organism, it would be possible to undertake a grouping of species that would accurately reflect their 'natural affinity' " (Mayr 1969, p. 81). This was the justification that Mayr (1974) gave for the separation of birds from other sauropsids, or *Homo sapiens* from other living hominoids (Chapter 2, Section IV). Both speciation and anagenesis result in genetic change. Development of the avian or hominid grade produces a degree of genetic difference that overrides that due to cladogenesis. Thus, while man may be closer cladistically to the African apes than either is to the orang-utan, he has fewer "genes-in-common". As we saw in Chapter 2, the assumption in this case about genes-in-common is false (see also Chapter 9).

Advocacy of "evolutionary classification" continued into the 1970s and up to the present day, notably by Mayr himself, but to some extent this must be seen as a rearguard action against the increasing dominance of phenetics and cladistics. Nevertheless, some valid points have been made by the evolutionary taxonomists. These will be considered in context in our next chapter. Meanwhile in this last section we must summarise various definitions and ideas that have arisen from the foregoing historical account.

V. Conclusions: The historical basis of taxonomy

The purpose of this chapter, taken together with previous ones, has been to show how the "structure" of taxonomy, including the patterns of classification and the shared assumptions about method, have evolved. We saw how the method of logical division arose within Greek philosophy as a method of achieving knowledge. This eventually became formalized in Linnean taxonomy in the binomial system (generic and specific or trivial name), still used in nomenclature today. Also borrowed from Greek philosophy was the convention (breached for polythetic taxa) that every taxon could be characterised by a diagnosis, a list of features necessary and sufficient for membership of that taxon. Linnaeus also introduced a hierarchy of categories, whose ranks corresponded to those of the tree of Porphyry. Thus a Linnean taxonomist, whatever his views as to the true arrangement of organisms, could represent his *classification* as an inclusive hierarchy. Since the eclipse of quinarian classification, no other arrangement has received wide support despite our knowledge of the hybrid origin of some plant species. With the acceptance of this consensus all taxonomic controversy (as distinct from controversies about specific classifications) has in some way concerned the "naturalness" of classifications.

Mayr's five theories, listed in the previous section, form a convenient framework for considering these taxonomic controversies. Of the five, "nominalism" can be separated from all the others. A true nominalist would reject the phrase "natural classification" as meaningless: "all groupings, all classes, all universals, are artifacts of the human mind". Mayr, as we saw, regarded phenetics as nominalist, but pheneticists have a criterion of naturalness (Chapter 7), whatever one may think of its validity. Nominalism is a possible stance, but one that probably has no extant practitioners. Buffon seems to have been a nominalist but a non-practising one (Chapter 2, Section I). It could be suggested that phenetics is nearer to empiricism, the polar opposite of nominalism, and pheneticists themselves make this claim (Sneath and Sokal 1973) – given an appropriate technique a "natural" system emerges – but this is to anticipate the discussion in Chapter 7. Here we must summarise the meaning of "natural" in essentialism, evolutionary classification, and (to anticipate again) cladism.

Under "essentialism" Mayr conflates essentialism and idealism as I have used those terms (see this chapter, Section I). An essentialist believes that taxa are real *because* each can be characterised by the diagnosis, and that a natural classification is an inclusive hierarchy of real taxa. An idealist believes in the objective reality of the *Baupläne* themselves. As in Plato's theory of forms, an idealised bird is characterised by the diagnosis of the class Aves, but has an objective reality of its own. In rejecting the use of phylogeny to validate their classifications, but retaining a concept of naturalness, transformed cladists must be essentialists, idealists, or empiricists, or perhaps a combination of two or all three of these.

"Cladism", as it was known to Mayr (1969), was "phylogenetic": to an original (Hennigian) cladist a natural classification is one that reflects the pattern of cladogenesis of the group classified. "Evolutionary" classification, on the other hand, attempts to reflect all the results of evolution: cladogenesis, anagenesis, grade (as in birds or *Homo sapiens*), and, in Mayr's view, genes-in-common. Such a classification may have more informational content than a cladistic one, but unhappily when all its proposed aims are conflated, they can not be disentangled. None of the constituent features can be recovered from the final classification, or the pattern it represents, without extrinsic information such as an independently reconstructed phylogeny.

7

Methods of classification: Phenetics and cladistics

We have seen in Chapter 6 and the previous chapters that dissatisfaction with traditional taxonomy gave rise, after the Second World War, to two distinct attempts at a remedy – phenetics and cladistics. In this chapter, I review the methodology of these two schools, as they were originally proposed and developed. In each case I describe the principles and methods, which I hope will be sufficiently clear to non-taxonomic readers to enable them to follow the subsequent discussion. In each case there has been an important development of the original philosophy and method. In the case of phenetics, various biochemical techniques have used phenetic methods, but also, and importantly, "numerical cladistics" has diverged from phenetics and converged on cladistics. In the case of cladistics, "transformed cladistics" is regarded by some as an offshoot and rival to mainstream cladistics and by others as a logical development of the mainstream. These will be dealt with in Chapters 8 and 9. In Chapter 10, I shall attempt to sum up the present state of taxonomy and to draw conclusions that I regard as pivotal to the theme of this book.

I. Phenetics

Pheneticists used to trace their history back to Adanson, a contemporary of Linnaeus (Sokal and Sneath 1963, p. 50). In *Familles des Plantes* (1763), Adanson produced sixty-five rival classifications, each based on the states of a different character. Superimposed, as it were, all these classifications might be regarded as a single comprehensive phenetic classification, but this was not Adanson's aim (Nelson 1979). He had already arrived at a division of plants into fifty-eight families, which he regarded as natural. Each of his sixty-five "artificial" classifications was then tested against his natural families to see how many of the latter were dismembered by species within them being assigned to more than one group. His aim, in fact, was to show the futility of artificial classifications based on one or a few characters (like Linnaeus's sexual system for plants).

Vernon (1988) gives an interesting and useful account of what he describes as the "founding of numerical taxonomy", and a personality rather than subject-based history is given by Hull (1988). Although some of the principles of phenetics can be traced back to the last century, it was a series of papers published by three groups of people in 1957 and 1958 that led to the origin of phenetics as a conscious school of taxonomy. The three groups, with their publications, were Michener and Sokal (1957) at the University of Kansas, Cain and Harrison (1958) at the University of Oxford, and Sneath (1957a, 1957b) at the University of Leicester. Development of a methodological consensus was very rapid: other papers of importance were Rogers and Tanimoto (1960) and particularly Sneath and Sokal (1962). By 1963, it was possible for Sokal and Sneath to have produced a textbook on the subject, *Principles of Numerical Taxonomy*.

We have noted the distinction between "agglomerative" (synthetic) and "divisive" (analytic) clustering methods (see Chapter 6, Section III). Clustering in phenetics is agglomerative (Sokal and Sneath's term), starting with the items to be classified and clustering them into successively larger groups. The items to be clustered, usually species, are *Operational Taxonomic Units* (OTUs). For clustering to be possible by mechanical means there must be some measure of "affinity" between individual OTUs; this is the reciprocal of "phenetic distance". Affinity bears no necessary relationship to genealogical relationship. Pheneticists recognise "patristic similarity" (due to homology) and homoplastic similarity as both contributing to "phenetic similarity".

We saw in Chapter 2 (Section IV) that the results of phenetic clustering are usually represented by a phenogram, which is a minimal Steiner tree (Page 1987) – that is, one in which the original points (in this case OTUs) are connected so that they are all terminal (i.e., none appears at an interior node). It is also normally a dichotomous dendrogram, so that there is a bifurcation at each node. In these respects a phenogram is similar to a cladogram as produced by cladists, but, as the affinity between OTUs represents both homology and homoplasy in unknown proportions, it does not represent the pattern of phylogeny. Therefore, pheneticists are obliged to have a different procedure for the reconstruction of phylogeny (Chapter 9). We saw in Chapter 2 that phenograms differ from cladists' diagrams in that phenograms record the measure of affinity of any two branches that diverge from a node. At the lowest rank, the length of the line from the node to either OTU is a measure of the *taxonomic distance* between the two OTUs. One final point is that pheneticists do have criteria by which they judge the naturalness of a classification (and so are not "nominalist" as Mayr asserts; see previous chapter, Section V), which derive from the work of Gilmour and others in the late 1930s onwards, but can be traced back to John Stuart Mill in the last century.

Theories and methods in phenetics are described in Sneath and Sokal (1973), effectively a second edition of Sokal and Sneath (1963). Rohlf and Sokal (1982) present "a flow chart of taxonomic procedures", and Sokal (1986) gives a more recent summary review. Rohlf and Sokal list a series of stages in a taxonomic study, designating each by a letter symbol (Fig. 7.1).

The **O** stage consists of the original study of the OTUs. The specimens are described in terms of a series of "characters". To a pheneticist a character is a variable shared by all the OTUs. Each OTU will then be characterised by a "state" of that character. Thus "number of digits in the manus" (hand) might be a character in classifying tetrapod vertebrates. States of that character would be "one digit" (or finger), "two digits", and so on. Not only morphological characters are acceptable: physiological, chemical, behavioural, and even ecological characters ("habitats, food, hosts, parasites, population dynamics, geographical distribution"; Sneath and Sokal 1973, p. 91) are also admitted. Not acceptable *together* are two characters that are logically correlated, because one would be redundant. Sneath and Sokal give the example of redness of the blood as a state of one character, and presence of haemoglobin as a state of another. Only one, presence of haemoglobin (as more precise), should be recorded. Similarly, two characters involving different measurements of the same phenomenon are to be avoided, such as the radius and circumference of the same circular structure.

Much of the potential trouble arising from use of redundant characters could be said to arise from the pheneticists' insistence on using the maximum possible number of characters. The original justification for this requirement was embodied in four hypotheses (Sneath and Sokal 1962). All these are based on ideas of the correlation of characters and the genome of organisms, and so come very close to Mayr's criterion of genes-in-common. First is the "*nexus hypothesis*" (1), which is that every taxonomic character is likely to be affected by more than one genetic factor and also that every active gene affects more than one character. Thus there is a complex nexus of cause and effect. The larger the number of characters sampled, the more representative of the genome that sample is likely to be. Next is the "*hypothesis of non-specificity*" (2), which is the assumption "that there are no large and distinct classes of genes affecting exclusively one class of characters (such as morphological, physiological, or ethological) or restricted regions of organisms (such as the head, skeleton, or leaves, etc)". Thus sam-

Fig. 7.1. "Scheme illustrating the customary flow of procedures in a numerical taxonomic study". O = description of OTUs; X = data matrix; S = similarity matrix. Information in S can be summarized as ordinations (M), (unrooted) trees (T), or dendrograms (i.e., phenograms) (D). Only the latter are considered in the text. All can be expressed as classifications (C). (From Rohlf and Sokal 1981. Published by permission of the Society of Systematic Zoology, National Museum of Natural History, Washington, D.C.)

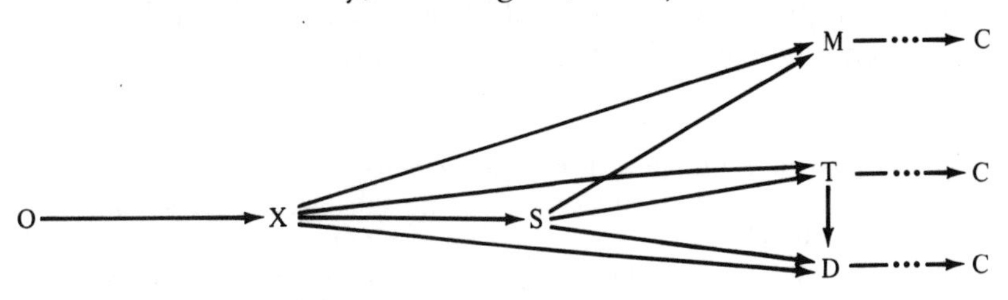

pling one body region or one type of character would not restrict information to a particular class of genes. Therefore, different samples of characters, if numerous enough, should yield similar classifications. A classification of Lepidoptera based on their caterpillars *should* be similar to one based on the adult butterflies and/or moths. Thus "there are no *a priori* grounds for favouring one character over another".

The last two hypotheses suggest that, as more and more characters are added, the rate of gain of new taxonomic information and change in similarity measures will diminish, eventually stabilising (or reaching an asymptote). They are the "*hypothesis of the factor asymptote*" (3), stating that the information content will reach an asymptote; and the "*hypothesis of the matches asymptote*" (4), stating that the similarity coefficient (below) between any given pair of OTUs will reach an asymptote. Sneath and Sokal conclude, however, that one "would need a very large number of characters to approach the asymptote".

Unhappily, these hypotheses, which Sneath and Sokal (1962) saw as giving empirical validity to their techniques, soon came to be doubted by the authors themselves (Sneath and Sokal 1973, pp. 96ff.). The nexus hypothesis, while "undoubtedly true in a general way . . . has lost some of its relevance for phenetic taxonomy". Impressed (or depressed!) by the complexities of the relationship of genotype to phenotype, Sokal and Sneath no longer felt that a large sample of character states necessarily gives a fair representation of the genome.*

Similarly, the nonspecificity hypothesis had now "been shown to be only partially correct. Identical classifications are not produced from different sets of characters for the same OTU's". [Subsequently, Adey *et al.* (1983) have reported a direct refutation of the hypothesis within a plant family.] There are various methods of producing measures of congruence between classifications based on different character sets which we shall mention below. The factor asymptote hypothesis had "been shown to be of little utility", even when attention was transferred from the genome to the

*To allay possible confusion, current genetic practice is to use the word "genotype" to refer to some part under study of the total genetic programme of an individual, with "phenotype" as its physical manifestation, whereas "genome" refers to the total genetic complement of an individual, represented by all the chromosomes with their included genes in a somatic cell.

"phenome" (its phenetic manifestation), and, as far as the matches asymptote hypothesis was concerned, "It seems doubtful to us at this time that there is a single parametric measure of similarity between OTU's" (Sneath and Sokal 1973, pp. 106–7).

Returning to phenetic procedure, the next stage is the production of a *data matrix* (**X**). Cain and Harrison (1960) described phenetic relationships as "arrangement by overall similarity, based on all available characters without any weighting". Subsequently, it was agreed (e.g., Sokal 1986) that this meant without *a priori* weighting, a position, as we have seen, also held by Mayr. The data matrix, therefore, consists of a table in which *t* OTUs are ranged along the top, and the *n* characters studied down the side. States of all the characters are then recorded in the cells of the matrix (Fig. 7.2). Coding for the character matrix is simple if each character is binary – that is, occurs in only two states. The simplest is the case where the states are present or absent, but all binary characters can be coded 0 or 1, using any arbitrary criterion, for computer manipulation. Coding for characters in which there are more than two states for the OTUs under study, or in which the states form a continuum, is more difficult.

Sneath and Sokal (1973, pp. 147ff.) sub-divide characters for coding purposes as follows:

> (1) Two-state characters (all-or-none characters, binary characters, presence–absence characters) [as noted above]; (2) quanti-

Fig. 7.2. A phenetic binary character matrix. The states of each character (e.g., present = 1; absent = 0) are recorded for each OTU.

Characters	OTUs
	1 2 3 4 . . . t
1	1 0 1 1 0
2	0 1 0 0 0
3	1 0 1 0 1
4	0 0 0 0 1
n	0 1 1 1 0

tative multistate characters (sometimes called just quantitative characters) including both continuous and meristic ones; and (3) qualitative multistate characters.

Continuous quantitative multistate characters are those where direct measurement of a continuum is possible. Sneath and Sokal cite "the amount of a chemical produced by a bacterial strain, length of an animal, or amount of pubescence on a leaf". In this case, it is possible to enter the actual measurement on the data matrix. Frequently, however, the computer program to be used demands binary coding. The continuum then has to be broken up into two or more lengths. Thus coding may be as two states, for example: (1) "leaf slightly pubescent", (2) "leaf very pubescent"; or the character may be scored as a series of unit characters, each with two states, thus:

	OTUs			
Characters	1	2	3	4
Slightly pubescent	0	1	1	1
Moderately pubescent	0	0	1	1
Highly pubescent	0	0	0	1

Note that four original states, represented by the four different OTUs in the table, require only three binary characters for complete coding. Zero coding in all three characters, as in OTU number 1, represents the least development of pubescence. Meristic quantitative characters are those in which the states do not form a continuum but already form a series of discrete numbers, as with (e.g.) the number of trunk vertebrae of a vertebrate. Qualitative multistate characters are, *ceteris paribus,* better recorded as a series of discrete states A, B, C, *et cetera* but if binary coding is required, it is done in a similar fashion, for example:

	OTUs		
Colour	1	2	3
Red	1	0	0
Yellow	0	1	0
Blue	0	0	1

Thus OTU1 is red, OTU2 is yellow, OTU3 is blue. There are, however, difficulties in principle with this sort of coding:

> This is not an easy task inasmuch as the recording has to be done in such a way that a positive score on one of the new characters does not automatically bring about negative scores on all other such characters derived from the same qualitative character. The reader recognises this as the problem of avoiding logically correlated characters. . . . This type of recording is therefore best reserved for situations where it is evident that the new characters are logically independent: for example, the states "stem spiny", "stem hairy" and "stem red", can appropriately be coded as three separate characters, because any combination of the three can theoretically occur.
>
> (*Sneath and Sokal 1973, p. 149*)

Having completed coding for the data matrix, the next stage is production of a similarity matrix (**S**). This is done by comparing every OTU pairwise with every other OTU. If the similarity matrix is actually drawn, its triangular pattern resembles the distance tables published at the back of motoring atlases, in which the cells in the matrix record the distance between any town and any other town. The axes of such a table – ordinate and abscissa – both list all the towns in the same sequence, usually alphabetical. In a similarity matrix, the OTUs are listed usually by arbitrary number or letter along each axis, and the cells record either a measure of similarity, Sjk, where j and k are OTUs, or dissimilarity (taxonomic distance), Ujk. The former gives a similarity matrix, the latter a dissimilarity matrix. In computer phenetics the program may demand the input of merely the original data matrix, so that no similarity matrix is actually drawn but will be inherent in the program.

There are an enormous number of possible pairwise functions expressing similarity or dissimilarity, and many pheneticists have allowed their enthusiasm to run away with them in divising new ones: "Numerous association coefficients appropriate for binary data are described in various reviews. . . . Many of these have been applied only rarely (often just once in the original paper proposing their use)" (Sokal 1986, p. 428).

The two most important categories of pairwise functions for entry into the similarity matrix are the *association coefficients,* referred to by Sokal, and distance coefficients. Association coefficients have merely to record matches and mismatches between the binary states of all the characters for a pair of OTUs. These matches and mismatches can be summarised in a 2 × 2 frequency table for the two OTUs thus:

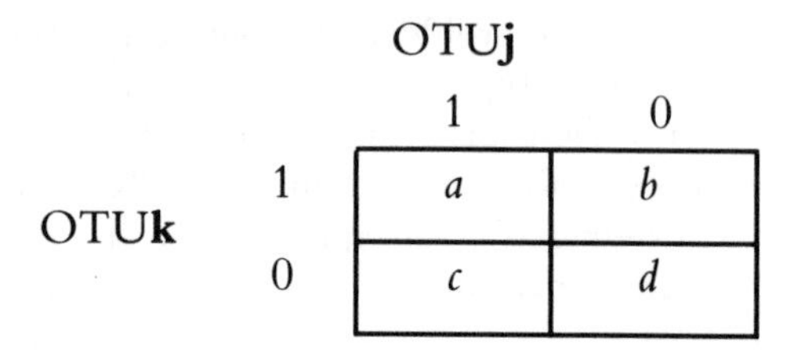

a represents 1-state characters shared by **j** and **k**; *d* represents 0-state characters shared by **j** and **k**; *b* and *c* represent the mismatches. The the total number of characters coded in each OTU is *n*. Thus $n = a + b + c + d$. Matches are symbolised by m ($= a + d$) and mismatches by u ($= b + c$). The *simple matching coefficient,* first used in phenetics by Sokal and Michener (1958), is

$$S_{sm} = \frac{m}{m + u} = \frac{m}{n} = \frac{a + d}{a + b + c + d}$$

This is the normal coefficient to use for binary characters, whether the 1 and 0 states represent different positive features or represent presence and absence, respectively. If, however, for presence or absence, the presence (1) state is rare when all characters are considered [i.e., $(a + b)/n$ and $(a + c)/n$ are very small fractions], then the simple matching coefficient will indicate a spurious high similarity based on the absence of features. Under those circumstances the *coefficient of Jaccard* (Sneath 1957a) can be used:

$$S_j = \frac{a}{a + u} = \frac{a}{a + b + c}$$

This thus omits negative (zero) matches.

Association coefficients will serve for simple binary characters, but where the character states are coded as numerical measures, *distance coefficients* are used. What these have in common is that they represent the average state-difference between two OTUs by summing all the differences. Because only the difference is of interest and because the relationship of any two OTUs should be a symmetrical one, all state differences must be positive. This is signified by $|X_i\mathbf{j} - X_i\mathbf{k}|$ for states X of the *i*th character of the OTUs **j** and **k.** A simple coefficient is then

$$\text{M.C.D.} = \frac{1}{n} \sum_{i=1}^{n} \left| X_i\mathbf{j} - X_i\mathbf{k} \right|$$

where *n* is the number of characters as before. This expression is the *Mean Character Difference* (M.C.D.), originally proposed for taxonomy by Cain and Harrison (1958).

From consideration of this type of coefficient, it is possible to picture the nature of the clustering procedure in phenetics. A simple case can be illustrated directly (Fig. 7.3). If the value of *n* was 2 – that is, if only two characters were being coded – then the OTUs could be represented by points on a normal two-dimensional graph, in which the abscissa represented measurable states of character X_1, and the ordinate, of character X_2. For three characters, it is still possible to represent "character space" by a three-dimensional graph (Fig. 7.4). If *n* is larger than 3, it is no longer possible to represent the graph, or character space, on the printed page, or, probably for anyone to picture it. Nevertheless, an *n*-dimensional graph is inherent in phenetic computer programs, with each axis representing a character and the points on the graph representing OTUs. The distances between a pair of OTUs in this space is a measure of their character difference. Inevitably the OTUs will form clusters. The clustering procedure (to which we shall return in the next section) produces a hierarchical classification from these clusters.

In two dimensions – that is, for two characters – as well as for more dimensions than two, the distance along a straight line between any two OTUs is the *Euclidean distance*. This can be measured directly from a two-dimensional graph, but it can also be calculated: the reason is that the Euclidean distance forms the hypotenuse of a right-angle triangle of which the other two sides are each parallel to an axis and each of these represents the state difference between the OTUs for a character. Thus in Figure 7.3, the distance between the OTUs **b** and **d** is given by Pythagoras's Theorem:

$$\Delta^2\mathbf{bd} = (X_1\mathbf{b} - X_1\mathbf{d})^2 + (X_2\mathbf{b} - X_2\mathbf{d})^2$$

In order to represent Euclidean distances in two or more dimensions, the *mean character difference* referred to above can be modified accordingly; otherwise it will under-estimate the Euclidean distance. A further refinement is to divide the summed differences by R_i, the range of the *i*th variable in the data table. This ensures that the scales of different measurements do not affect

Fig. 7.3. Four OTUs – **a, b, c, d** – plotted according to their states in two-variable characters X_1 and X_2. The solid line joining (e.g.) **a** and **d** represents the Euclidean distance between **a** and **d** and is the hypotenuse of the triangle whose other two sides represent the state differences between those two OTUs. (After Sneath and Sokal, *Numeral Taxonomy*. W.H. Freeman & Co. 1973. Used by permission of the publisher.)

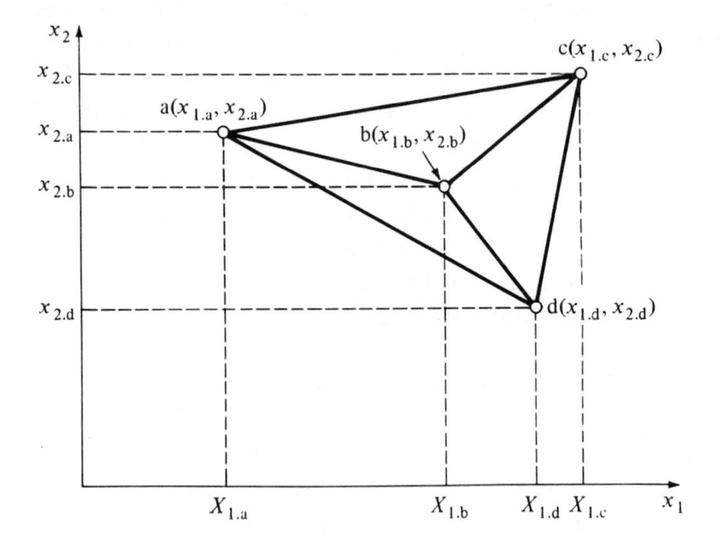

the result. The whole can then be turned into the *similarity coefficient* by subtracting the distance coefficient from one. We then arrive at a *coefficient of "overall" similarity* [as presented by Farris (1977) in a clear, concise, but critical review of phenetic methods] of

$$S_{jk} = 1 - \frac{\sum_{i=1}^{n} \frac{|X_i\mathbf{j} - X_i\mathbf{k}|}{R_i}}{n}$$

Fig. 7.4. Four OTUs in three-dimensional space, plotted accordingly to their states in three characters X_1, X_2, X_3. (After Sneath and Sokal, *Numeral Taxonomy*. W. H. Freeman & Co. 1973. Used by permission of the publisher.)

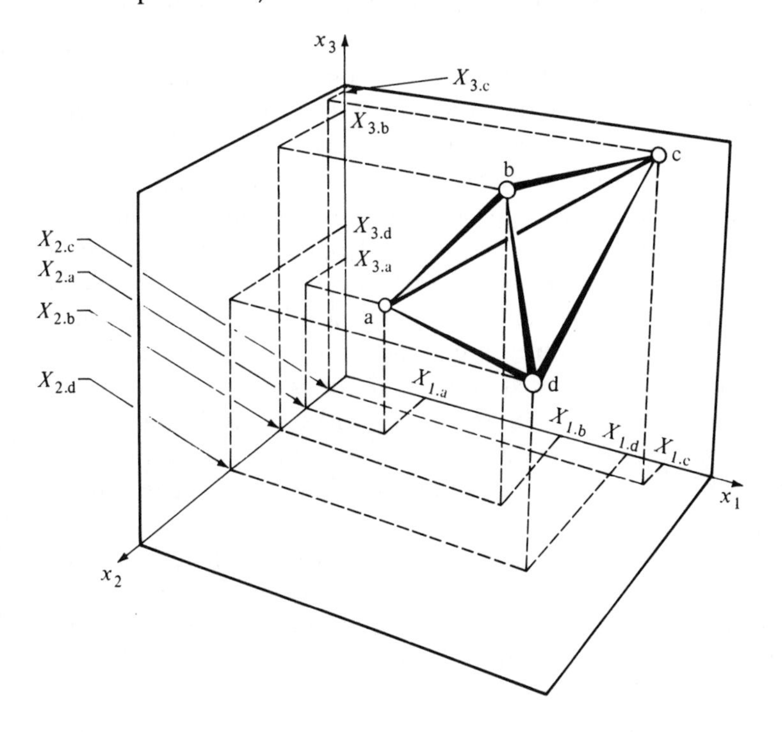

based on the M.C.D. Or, for Euclidean distances,

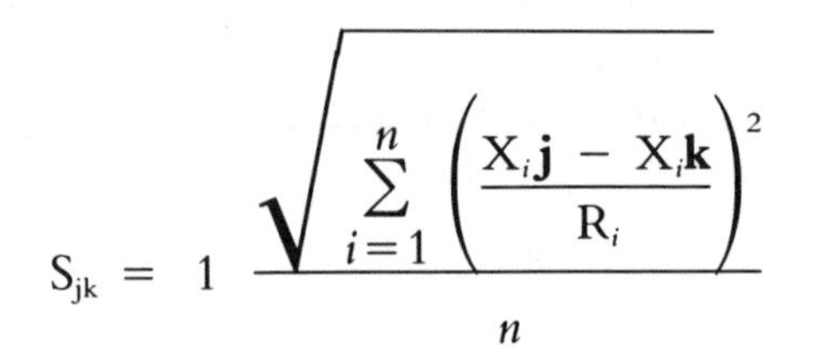

$$S_{jk} = 1 \quad \frac{\sqrt{\sum_{i=1}^{n} \left(\frac{X_i\mathbf{j} - X_i\mathbf{k}}{R_i} \right)^2}}{n}$$

Through the use of an association coefficient or a coefficent of similarity, or some other measure as discussed by Sneath and Sokal (1973, pp. 114–47), values can be entered for each possible combination in pairs of the OTUs in the similarity matrix.

II. Phenetic clustering

Sneath and Sokal (1973, pp. 201–45) review a selection of the enormous numbers of clustering methods that have been devised. We shall consider only one phenetic method in any detail, but must note a few preliminaries. Most methods used in taxonomy are *s*equential, *a*gglomerative, *h*ierarchic, *n*onoverlapping methods (SAHN: Sneath and Sokal 1973, pp. 214ff.). They are *sequential,* as opposed to simultaneous, in that the algorithm used produces a sequence of operations rather than a single one; *agglomerative,* in starting the whole operation with distinct OTUs; *hierarchic,* in producing a normal Linnean hierarchy (or rather a dendrogram which can be converted to it); and *non-overlapping,* in that the hierarchy is divergent with no reticulations (see Chapter 3, Section IV). Most SAHN algorithms also represent pair-group methods, rather than variable-group; only one OTU or cluster is admitted at a time, and a dichotomous phenogram results.

Within SAHN pair-group methods, however, there is an important choice to be made in methods of associating clusters with one another to produce clusters of a higher rank. If one pictures the clusters in hyperspace referred to above, there are three obvious ways of measuring the distance between one cluster and a neighbouring cluster and thus assessing whether any cluster or OTU is nearer to a given cluster than to any other. *Single-linkage clustering* (Sneath 1957b), or nearest-neighbour technique, is described by Sneath and Sokal (1973, p. 218): "An OTU that is a candidate for an extant cluster has similarity to that cluster equal to its similarity to the closest member within the cluster. Thus connections between OTU's and clusters and between two clusters are established by single links between pairs of OTU's". They note that this method often leads to "long straggly clusters". At the other extreme is *complete-linkage clustering,* or furthest-neighbour clustering, where measurement for the OTU or candidate cluster is to the most distant OTU in the extant cluster. This method yields very compact clusters, but clusters join other clusters with difficulty and with relatively low overall similarity values. The compromise is *average-linkage clustering* in which the extant and any candidate clusters are in effect measured from their centres. There are, needless to say, various ways of doing this, but I shall describe only one involving what is the most frequently used method of phenetic clustering.

This is the "*unweighted pair-group method using arithmetic averages*" (UPGMA), developed by Sokal and Michener (1958) but first used by Rohlf (1963). It is *unweighted* because it gives equal weight to each OTU within a cluster, rather than giving whole clusters equal weight, however many members they have. It is a *pair-group* method admitting only one cluster at a time, as noted above, and thus gives a dichotomous phenogram. The phrase "*arithmetic averages*" denotes the fact that it is average-linkage clustering, but also that the distance of the candidate OTU to the extant cluster is taken as the average of the distances between that OTU and the members of the extant cluster. (The alternative is *centroid* clustering, whereby the distance is measured from the OTU to the geometrical centre of the cluster in hyperspace.)

The process of clustering commences by uniting the pair of OTUs, from the whole set, that have the greatest similarity coefficient. That pair can then be represented by a single taxon which replaces them in the similarity matrix. Again, as presented clearly by Farris (1977), we can refer to this taxon, combining OTUs **j** and **k,** as **p.** The formula for computing the similarity between **p** and any other OTU **m** can then be computed from the coefficients for **j** and **m** and **k** and **m,** using the following general formula:

$$S_{pm} = \frac{\mathbf{NjSjm} + \mathbf{NkSkm}}{\mathbf{Nj} + \mathbf{Nk}}$$

This formula is used throughout the clustering procedure. **Nj** is therefore the number of OTUs in the taxon **j,** and **Nk** the number in **k.** If **j** represents one of the original OTUs, then **Nj** = 1; similarly for **Nk.**

Thus the whole clustering operation is a matter of uniting the two closest OTUs, representing them as a single taxon (**p**), looking for the next closest pair (which may or may not include **p**), and continuing the procedure until all the OTUs have been included. The completed operation may be represented by a phenogram in which vertical distances between node and node, or between node and OTU, are measures of *dis*similarity. Two OTUs each separated by a short (identical) distance from the node that uniquely unites them have a high similarity coefficient; those with a greater distance have a lower one.

The final stage is the production of a written classification from the phenogram. In looking at the form of phenograms in Chapter 2 (Section IV), we noted Sneath and Sokal's (1973, pp. 294–6) proposal of "phenons" – horizontal lines drawn across the phenogram at various levels of percentage similarity (Fig. 2.11). The branches cut by a line at any percentage phenon line then represent separate taxa at that level. Thus in the example in Figure 2.11, the 75 per cent line defines four taxa two of which are further branched before reaching the OTU (e.g., species) level. Thus the phenon lines are equated with rank, and the phenons can be equated with taxa, although Sneath and Sokal are at pains to distinguish phenons (phena?) from taxa: "Since they are groups formed by numerical taxonomy, they are not fully synonymous with taxa; the term 'taxon' is retained for its proper function, to indicate any sort of taxonomic group" (their p. 295). I take this to mean that phenons are a subset of taxa, but it is not entirely clear. The whole idea of using phenons has a number of disadvantages:

1. The choice of percentage level is quite arbitrary, introducing a completely subjective element into what is claimed to be as nearly as possible an objective process.
2. Because the nodes of a phenogram are all at different levels, more than one phenon line may cut through a single vertical branch. Thus in Figure 2.15, the 55, 65, and 75 per cent lines all cut the same branch (terminating in OTUs 1, 2, 5, and 9). If all three lines are used to define ranks, then three nested monotypic taxa result.
3. There is no compatibility between different phenograms: different clustering programs will produce different percentage similarities among OTUs and possibly different patterns of phenogram, and likewise the same method will produce non-comparable percentage similarity levels in two phenograms representing different groups of organisms.

McNeill (1980) discusses the conversion of phenograms into classifications. His method of classification is to assign a rank level to every node, and then to extend this rank to the same dissimilarity level on branches below a node where there is no node at that level. Thus a line of rank corresponds to a node at most branches but not all. He goes on to perform the exercise for a group of plants and shows that the resulting dendrogram of classification can be simplified by conflating ranks to eliminate monotypic taxa and thus to give either of two previously accepted classifications (with a "certain amount of adjustment"!).

Pheneticists have given considerable attention to criteria for judging the effectiveness of classification. These optimality criteria are reviewed by Sokal (1986). He suggests four sets of criteria: (1) closeness to the original resemblance matrix, (2) maximum similarity within taxa, (3) maximum predictive value, and (4) stability of classification. Closeness to the original resemblance (similarity) matrix is measured by comparing the distance between all the OTUs taken in pairs as measured from the phenogram and comparing those distances with the distances represented in the original similarity matrix. The distance between any two OTUs on the phenogram is the distance to either of them from the nearest node or branching point that connects them. That distance is the *cophenetic value* or distance. It is thus possible to set up a *cophenetic matrix* from the phenogram, which can be compared directly with the original similarity matrix. Details are given in Sneath and Sokal (1973, pp. 278–80). The resulting measure is the *cophenetic correlation coefficient*. Of phenetic methods UPGMA scores highly, but is rivalled by some numerical cladistic techniques.

Maximum similarity within the taxa and maximum predictive value are closely related. Similarity can be measured if there is a valid criterion of "compactness or of dispersion of the OTUs in the attribute space" (Sokal 1986, p. 434). The measure is the variance or sum of squares of the clusters, based on all the character state differences recorded in the similarity matrix. The smaller the variance, the greater the resemblance between the OTUs in the cluster. The greater that resemblance, the greater is the possibility of making true general statements about the members of the cluster. Prediction is used in this sense. Various mathematical measures of predictive value have been suggested, and are noted by Sokal, but there is also an important philosophical point involved. The pheneticists' characterisation of a *Natural Classification* is based on the notions of maximum similarity and predictive value. This characterisation is attributed by pheneticists to the work of Gilmour (1937, 1940, 1961; Gilmour and Walters 1963) and used to be attributed, erroneously, to Adanson (see previous section). Its clearest expression is in John Stuart Mill's *A System of Logic*... (1843). Quoting from the 1974 edition (vol. 8, p. 714):

> The ends of scientific classification are best answered, when the objects are formed into groups respecting which a greater number of general propositions can be made, and those propositions more important, than could be made respecting any other groups into which the same things could be distributed.... A classification thus formed is properly scientific or philosophical, and is commonly called Natural in contradistinction to a Technical or Artificial, classification or arrangement.

Concerning similarity, Sneath (1961) makes the point that Mammalia is a natural taxon because it is characterised by a large number of correlated features and its denotation corresponds to their distribution, but a taxon confined to horses, mice, and rabbits is not a natural taxon because most of the features that would characterise such a group are not exclusive to it but are common to other mammals, particularly placental mammals. Predictive power of classifications is summed up in a sentence by Fitch (1979): "The essence of predictivity . . . is the degree to which a specific classification agrees with characters not used in the formulation of that classification". Both Ruse (1980a) and Mayr (1981; references and notes) quote Mill's arch-rival William Whewell (see Chapter 13), saying something similar:

> Thus the correspondence of the indications of different functions is the criterion of Natural Classes. . . . And the Maxim by which all Systems professing to be natural must be tested is this: – that the *arrangement obtained from one set of characters coincides with the arrangement obtained from another set.*
>
> (*Whewell 1840, vol. 1, p. 521; his italics*)

This, as Ruse (1976, 1980a) points out, corresponds to Whewell's concept of *consilience* applied to taxonomy. It relates to both predictive power and stability. Apart from Sokal's two types of prediction quoted above (homogeneity and heterogeneity), we can cite another pair, consilience using different characters, as above, and predictions about organisms. In the latter case, the most natural classification is the one that yields the most predictions about a newly discovered organism assigned to a taxon within it. A newly discovered small hairy tetrapod with a heterodont dentition will be recognised, without dissection, as a mammal. Immediately, still without dissection, one can make numerous predictions (or more correctly *retro*dictions) about its internal anatomy. It will have a false palate, three ear ossicles, a diaphragm, a four-chambered heart, and so forth. If it is a male, one can even infer the characters of the female, such as suckling the young, without even seeing one. Furthermore, one can make discoveries about the character states of one species or individual within a taxon, even when these states were not considered in the original classification, and then predict their distribution among other members of the taxon.

The features of a natural classification of organisms may be contrasted with, say, the classification of library books. University and public libraries normally classify their books by subject matter. The classification is hierarchical. A book on animal taxonomy will be classified successively in the "taxa" natural science, biology, zoology, general zoology, taxonomy. But its position will give no information about characters not employed in the original classification such as size, shape, number of pages, binding, price, language (unless programmed in the original classification), and so forth.

Returning to Sokal's (1986) optimality criteria, the last was stability. As he says, three types of stability are commonly discussed: (1) character stability, the "robustness of a classification to the addition of new characters or to different selections of characters"; (2) OTU stability, the robustness of a classification to the addition or subtraction of OTUs; and (3) the consistency of the classification when different phenetic methods, such as differences in character coding or different coefficients of similarity, are used. I do not wish to pursue these criteria any further. It is, however, necessary to make the point that all the optimality criteria, together with the concept of "Gilmour-naturalness" as Farris (1977) calls it, are related to the pheneticists' four hypotheses used to justify their method, and particularly to the second one – the hypothesis of non-specificity. To the extent that pheneticists themselves regard these hypotheses as discredited, they must regard the optimality criteria as discredited as well.

Two further features of phenetics must be emphasised. The first concerns the distribution of points, representing OTUs, in hyperspace. If they are randomly distributed as with the distribution of points (such as individual organisms) in real space, the points will occur in clusters. If they did not, that is, if the points were uniformly separated, that would be over-distribution. Thus in phenetics random distribution of organisms in character space will yield a phenogram. There must, therefore, be a null hypothesis for phenetic taxonomy, that the data have "no structure at all"; for "clustering will impose structure even on random data" (Sokal 1986, p. 432). This theoretical possibility seems not to worry Sokal too deeply, however, since it is "unlikely that the organisms representing a taxonomic group have no taxonomic structure".

The second feature of phenetics constitutes its principal difference from cladistics. Phenetics can handle *distance data* only. All differences between organisms in phenetic classification are coded as distances – summed measures of similarity or dissimilarity in the pairwise comparisons of the similarity matrix. We have already seen that the distances between OTUs are compounded of primitive and derived "patristic similarity" corresponding to the symplesiomorphy and synapomorphy of cladistics (Chapter 4, Section II) and homoplasy, which cannot be disentangled. It is also the case, however, that there is no necessary structural relationship between the pattern of clusters in hyperspace and the inclusive Linnean hierarchy into which they are converted. Hence two operations are necessary to convert that pattern into a classification: First is the production of a phenogram, and then of a classification. The classification is presented as a hierarchy, not because the natural ordering of organisms is necessarily hierarchical, but because convention demands it.

III. Cladistics

Phenetics as a school of taxonomy originated from the coming together of several widely scattered workers; the origin of cladistics, on the other hand, is always associated with one man, Willi Hennig. His preliminary work was published in German in 1950 as *Grundzüge einer Theorie der phylogenetischen Systematik* and did not attract wide attention. That work was then revised by 1961, but the revision was not published until 1966, in a translation into English, as *Phylogenetic Systematics.* This translation had its critics (Dupuis 1984), and two other versions, one in Spanish (Hennig 1968) and the other in the original German (Hennig 1982), have been published subsequently. A clearer account of Hennig's views is presented in his book on insect phylogeny (in German 1969, in English 1981), and there is also a short review with the emphasis on insect taxonomy in Hennig (1965). Dupuis (1979) charts reactions to the development of Hennig's ideas, and Hull (1988) presents an account of controversies surrounding cladistics, with emphasis on the personalities involved.

We have seen many features of cladistics in previous chapters, particularly Chapter 2 (Section IV), Chapter 3 (Section III), and Chapter 4 (Section II), so to some extent this section will be a summary, but it is necessary to look at facets of cladistic procedure in some detail.

In principle, "Phylogenetic Systematics", as Hennig termed his method, is based on a simple but brilliant idea. In sexually reproducing organisms diversification in evolution is due to speciation. Unlike anagenesis or phyletic evolution, speciation or cladogenesis is a quantum event that either happens or doesn't. Thus that part of phylogeny that has the potential for unambiguous representation is the pattern of cladogenesis. There need be no argument about the measurement of phenetic distance, grade, or genes-in-common: a simple cladogram could represent the pattern of speciation events culminating in the species to be classified. If speciation were always divergent, that pattern would represent a divergent, inclusive hierarchy. The pattern of Linnean classification is also a divergent, inclusive hierarchy. In phylogenetic systematics the two patterns should be isomorphic.

The system will work, according to Hennig, only if each speciation event in the past is detectable in the present complement of characters of each organism classified. A species, to be valid taxonomically, must have one or more unique characters that distinguish it from all closely related species. These characters are *autapomorphies* (Chapter 4, Section II). But that species will also have characters that it shares uniquely with the species (one or more) from which it most recently split. Those characters were autapomorphies of their common parent species, but are also the *synapomorphies* that unite them as sister-groups. Hennig appears to have believed that most speciation was dichotomous, so that sister-species occurred naturally in pairs, but this was by no means a dogma of his. Methodologically, however, the assumption is made that every species has a single sister-species, and cladistic analysis proceeds on this assumption. Some early cladists, however, did accept dichotomous speciation as dogma (e.g., Brundin 1968; Eldredge and Tattersall 1975), and critics of Hennig's methods in general pounced on this on the assumption that it was a universal belief.

Thus the essence of cladistics is to cluster the "*terminal taxa*" (corresponding to the pheneticist's OTUs) by discovering for each a sister-group with which it is united by one or (much preferably) a number of synapomorphies. As in phenetics, the sister-group of a terminal taxon (probably species) may or may not be another terminal taxon. The pair of sister-taxa so recognised then becomes a taxon in its own right: its distinguishing features are the synapomorphies used to characterise it. The synapomorphies uniting a pair of sister-taxa become the autapomorphies of the taxon that uniquely includes them.

Symplesiomorphies, as we saw in Chapter 4, are characters defining a group at a higher rank than that under consideration. This distinction results in the most important difference between phenetics and cladistics. Phenetic groups are defined by all the characters ("states") they hold in common, cladistic groups only by those unique to them. Pheneticists also use homoplastic character states, without necessarily recognising them as such, to characterise groups, but in cladistics homoplastic characters have to be detected and rejected. It is therefore an important part of cladistic procedure to detect symplesiomorphy and homoplasy, whereas in phenetics "overall similarity" does not distinguish synapomorph, symplesiomorph, and homoplastic characters: all are aggregated as a single distance measure.

Wiley (1981) provides an excellent account of cladistic methods from the viewpoint of a phylogenetic or Hennigian cladist – that is, one who believes that the purpose of cladistic method is to reconstruct phylogeny. Following Hennig, he points out that the basic operation in cladistics is embodied in the statement "A shares a common ancestor with B not shared by C" (Hennig 1966). The whole of reconstructed phylogeny is embodied in conjunction of such "three taxon statements". In the simplest case, representing an ideal world, the truth of that three-taxon statement would be evident because A would share one or more synapomorphies with B while, in each case, C would have an alternative character (or a different state of the same character, as a pheneticist would say). But homoplasy occurs, and there are likely to be apparent synapomorphies uniting B and C and/or A and C. In the closed system of the three taxa alone, these contradictory apparent synapomorphies may be either symplesiomorphies or homoplasies. They can be identified as one or the other only by information extrinsic to the data from the three taxa.

Symplesiomorphies will, by definition, be characters occurring in a group larger than that confined to the three taxa. Thus any character occurring in a taxon outside the three (ABC) may be suspected to be a symplesiomorphy if it also occurs within A + B + C. This pattern of character distribution is the basis of the technique of *out-group comparison* for polarising characters. Watrous and Wheeler (1981) propose an "operational rule" for out-group comparison:

> Operational rule – For a given character with 2 or more states within a group, the state occurring in related groups is assumed to be the plesiomorphic state. If the character contains only 2 states, the alternative state is assumed to be apomorphic.

Thus the ideal for out-group comparison would be that a particular state of a character (a') occurs in some but not all of A+B+C while an alternative state (a) also occurs; a also occurs in the sister-group of A+B+C (say D). Then a is plesiomorph and a' apomorph within A+B+C. As an example we may refer to an example given in Chapter 4 (Section II) in the discussion of homology. If A, B, and C represent birds, crocodiles, and other extant reptiles, respectively, then A+B+C is the taxon Sauropsida. Birds (A) lack a left aorta, whereas crocodiles (B) and "other reptiles" (C) possess one. We shall assume (*pace* Gardiner) that mammals (or Theropsida) is the sister-group of Sauropsida, forming the Amniota; and further that the amphibia are the sister-group of the amniotes. Of all these, birds are unique in lacking a left aorta. Crocodiles and other reptiles have one, but so do mammals and amphibians. Thus within A+B+C the presence of a left aorta is plesiomorph: its absence is an autapomorphy of birds.

Continuing with the same example, birds (A) and crocodiles (B) share the character of a completely divided ventricle to the heart, whereas in "other reptiles" (C: viz. snakes, lizards, *Sphenodon,* and turtles) the ventricle is partially divided. Here I introduce, quite deliberately, a further complication. Is the shared character of a completely divided ventricle in A and B a synapomorphy? Mammals also have a completely divided ventricle, and, obviously, this forms part of Gardiner's claim for a sister-group relationship of mammals and birds. We can now ask two questions: (1) Is the divided ventricle a synapomorphy of birds and crocodiles? (2) Are mammals more closely related to birds + crocodiles than they are to "other reptiles"? As mammals are in contention, we cannot take them as the out-group to A+B+C, but it is reasonable to look to the amphibia, which have an undivided ventricle. Assuming the correctness of the taxon Sauropsida for the moment, "other reptiles" have a partially divided ventricle, amphibians have no division: it is therefore a reasonable hypothesis that a completely divided ventricle is a synapomorphy of crocodiles and birds, and this is supported by other characters; thus for Sauropsidae the classification is (A + B)C.

Turning now to Gardiner's (1982) challenge that mammals are the sister-group of birds, the controversy here is not a matter of distinguishing synapomorphy and symplesiomorphy, but synapomorphy and homoplasy. It is a question of *taxic homology* (Chapter 4, Section II). Orthodox opinion (which I hold to be correct; see Kemp 1988a and Gauthier *et al.* 1988 and below, this section) would contend that a greater number of valid characters unite crocodiles and birds than unite mammals and birds. Thus any unique resemblances between mammals and birds, including the ventricular septum, are homoplastic. This is the principle of *parsimony* in action. All cladistic groupings are to be regarded as hypotheses of relationship. If two different hypotheses of relationship are in contention, then the one to be accepted is the one supported by the greater number of apparently valid characters. Cladistics depends absolutely on parsimony, yet the principle is difficult and contentious. Later, in Chapter 8, I shall devote a whole section to it.

Summing up basic cladistic procedure, terminal taxa (whether species or higher taxa) are clustered as pairs of sister-groups. Each terminal taxon is defined by one or more autapomorphies. Sister-groups are united by synapomorphies, which then comprise the autapomorphies of the higher taxon so formed. Synapomorphies are unique to the sister-group pair and, as taxic homologies, are distinguished from homoplastic characters by congruence with other characters. The principle invoked is the principle of parsimony: a hypothesis of a sister-group relationship must be supported by a greater number of apparent synapomorphies than that for any rival pairing.

A question about all of this may have occurred to the alert reader: if characters are polarised by out-group comparison, must there not have been a pre-existing classification before one can identify the out-group? How does one start cladistic classification from scratch? A partially satisfactory answer to this question is given by Watrous and Wheeler (1981):

> However, we agree with Eldredge (1979 [b]: 171, footnote) that some progress has been made during the last few centuries, and a practical starting point is to follow the existing classification.

Also the proposal of a hypothesis of arrangement can be refined by introduction of more data, particularly data representing different sorts of characters (corresponding perhaps to the pheneticist's "non-specificity hypothesis"!). This Hennig refers to as "reciprocal clarification". There are, however, other possible approaches to polarising characters which we shall consider in Chapter 8.

As is the case with phenetics, theoreticians of cladistics have spent much more time discussing how to produce dendrograms than they have considering how these should be turned into classifications. There is, however, one point on which all cladists, from Hennig onwards, are insistent: all taxonomic groups, to be valid, must be *monophyletic.* The term "monophyletic" is also used by cladists in a more rigorous sense than by other taxonomists. We have already noted (Chapter 6, Section IV) Simpson's very lax definition of monophyly. Other taxonomists, both "evolutionary" (Mayr 1969, pp. 75–6) and phenetic (Sneath and Sokal 1973, pp. 40, 46–50), are more restrictive, with the concept of a monophyletic taxon as one in which all the members are descended from a single species. Cladists, following Hennig, go one important step further. He introduced the concept of a *paraphyletic* group as one descended from a common ancestor, but whose membership is incomplete. The notorious cases, which we have referred to before (Chapter 2, Section IV; Chapter 6, Section IV), are the class Reptilia (Sauropsida *minus* Aves) and the family or sub-family Pongidae [Hominidae (*sensu lato*) *minus Homo sapiens*]. Monophyletic groups to Hennig are those that contain *all* the known descendants of a common ancestor. This led to protests from non-cladists, with the claim that Hennig had misused the term "monophyletic". Ashlock (1971, 1972, 1980) proposed that taxa should be divided into those that are polyphyletic and those that are monophyletic (i.e., descended from a common ancestor), and then the latter distinguished as paraphyletic and *holophyletic.* The latter term corresponds to Hennig's monophyletic. Differences about the definition of monophyly might perhaps be regarded as mere words about words.

There has, however, been a difference of some substance about the meaning of polyphyly and paraphyly within cladistics, with different definitions given by Hennig (1966), Nelson (1971), and Farris (1974). Farris's definitions, favoured by most phylogenetic cladists (e.g., Wiley 1981, pp. 82–92), appear clear and unambiguous. As phrased by Wiley, a *monophyletic group* is one "that includes a common ancestor and all of its descendents". A *paraphyletic* group "includes a common ancestor and some but not all of its descendents". A *polyphyletic* group is one "in which the most recent common ancestor is assigned to some other group and not to the group itself". In addition to these definitions, Farris gave a simple algorithm in each case to test the status of phyly of any given taxon. Unfortunately, these have given rise to difficulties (Ashlock 1980; Oosterbroek 1987).

If cladograms are regarded as representing phylogeny, there is a strong justification for rejecting paraphyletic groups: any study of the history of a taxon or any generalisation about taxa must refer to complete entities – to "a limb with *all* its dependent branches, twigs and terminal twiglets and leaves" (using Darwin and Wallace's metaphor; Chapter 3, Section II). Thus if birds are, cladistically, a sub-group of dinosaurs, as is now generally accepted (Gauthier 1986), then dinosaurs did not become extinct at the end of the Cretaceous, because birds *are* dinosaurs.

As we have seen in previous chapters, "evolutionary classification" might well be characterised as a technique that allows paraphyletic groups, when it deems it desirable knowingly to remove a grade group (such as Aves) from an otherwise monophyletic group (*sensu* Hennig) (such as Sauropsida). This stance has been maintained in a series of works by Mayr (1969, 1974, 1981).

Other early objections to cladistic classification concerned the matter of ranking. Hennig (1966) originally insisted that "sister groups be coordinate, and thus have the same absolute rank"; in a classification of extant sauropsids, Aves would be given the same rank as Crocodilia. If fossil sauropsids also appeared within the classification, Aves would be downgraded even further, probably as sister-group to some group within theropod dinosaurs (Gauthier 1986). The other principal ranking problem was that if each pair of sister-groups at every rank was to be regarded as a named taxon and all the ranks were to be kept separate in classification, then an enormous number of category names would be required, going far beyond the names normally used in the Linnean hierarchy. A much discussed case was McKenna's (1975) classification of mammals, which required an enormous number of ranks and thus categories, without (as Patterson and Rosen 1977 pointed out) the fossil "mammal-like reptiles" which would have completed the clade Theropsida. If n species are the terminal taxa to be classified, the number of ranks will lie between $(1 + \log_2 n)$ and n (McKenna 1975; Panchen 1982). The former, smaller, number represents a completely symmetrical cladogram, where $n = 2^x$ (where x is any whole number) and every branch has a dichotomy at every rank. The addition of one species to such a perfect cladogram requires an additional rank! The other extreme, where the number of ranks is equal to n, represents our old friend the "Hennigian comb" (Chapter 2, Section II; Chapter 3, Section III).

IV. Cladistics and fossils

The classification of fossils has long been a vexed topic, not only for cladists but also for taxonomists of every persuasion. As far as cladistics is concerned, we can identify four controversial areas, strongly interrelated, in the use of fossils in classification and the reconstruction of phylogeny. These are (1) the use of age as a ranking criterion, (2) the problem of ancestral forms in cladograms, (3) the treatment of morphoclines from the fossil record, and (4) the question, Can data from fossils upset a classification based on extant species? To these can be added the obvious fact, affecting any taxonomic procedure, that fossils are invariably incomplete, yielding little or no biochemical and "soft anatomy" data, and frequently lacking in "hard part" data as well.

Hennig (1966, pp. 154–93) devoted a long discussion to the question of ranking. He concluded not only that sister-groups should have the same rank, but also that rank should depend on position in the cladogram, that is, that the nearer to the root of the cladogram a dichotomy between two groups occurred, the higher their rank should be. Thus the cladogram and the classification resulting from it should have identical hierarchies. Hennig further postulated that rank would be related to absolute age, on the basis of an overall steady rate of cladogenesis. This view immediately leads to difficulties in classifying fossils. Are two sister-species from the Cambrian to be given the rank of (say) subphyla? Brundin (1966) discussed these difficulties, asking whether it is possible to classify fossils and recent organisms together. In 1968, he addressed the problem of placing fossils in a cladogram and suggested that the cladogram should have a time axis, so that fossil terminal taxa would appear at the appropriate level below the extant ones. Crowson (1970, pp. 249ff.) went so far as to suggest that fossils and extant groups should be classified separately. His system was actually attempted by Patterson and Rosen (1977), who saw some "minimal" advantages. They rejected it, however, because of the enormous number of categories necessary to place (in their case) a small number of Jurassic fishes and because of the loss of information content.

Schaeffer, Hecht, and Eldredge (1972) and most subsequent workers have agreed that fossils should be treated as terminal taxa, but there has been some disagreement as to how this should be done. Schaeffer *et al.* were quite clear that fossil and recent terminal taxa should be treated in exactly the same way:

> Extinct and living representatives must be compared on the basis of the same characters, regardless of whether or not additional (unpreservable) characters are used for reinforcing conclusions about the relationships of the living (and indirectly of the fossil) ones.

The majority opinion now is that extant organisms and fossils should be classified together but treated differently, but this involves our third controversial area (above), so I shall consider it after the problem of ancestors.

In a resolved cladogram of extant organisms the original phylogenetic assumption was that each node represented the hypothetical ancestor of the monophyletic clade that diverged from it. Thus the node uniting two sister-species represented their common ancestral species. When fossil species were introduced into the system, the principal theoretical problem was to ask, If a fossil representing the ancestral species of one or more extant (or later) species were found in the fossil record, how would it be recognised? As we have seen, the only diagnostic characters of the ancestor would be the synapomorphies uniting its descendant species. If it had any autapomorphies of its own, it would be debarred from ancestral status. If no autapomorphies were found in the fossil the "ancestral species" could only be regarded as *incertae sedis* within the group of which it was the putative ancestor. In practical terms, if it were fossil rather than a living relict, one could never be sure of its lack of autapomorphies compared to the characters available in extant forms. The dilemma has been discussed by Hennig (1966), Farris (1976), Engelmann and Wiley (1977), and Platnick (1977). Wiley (1981, pp. 222–5) reverses the opinion of Engelmann and Wiley (1977) by suggesting tentatively that "in the future some ancestor might be identified by some other criterion (for example, a biogeographic criterion . . .)".

Wiley goes on to suggest a convention for the classification of a fossil ancestral species in which it is given the rank of the whole descendant group. He considers the case where "one of the species of *Archaeopteryx*" was identified as the stem species of birds. It now seems almost certain that there is only one known species of *Archaeopteryx* (Hecht *et al.* 1985) – that is, *A. lithographica* Meyer – whereas Wiley assumed that there was more than one. Wiley's suggestion may be updated with only one species (the ancestor) and a recent classification of birds (Cracraft 1988):

Supercohort Aves (*Archaeopteryx lithographica*)
  Cohort Ornithurae [all other birds]
    Plesion Hesperornithiformes
    Subcohort Carinatae
    Plesion Ichthyornithiformes
    Infracohort Neornithes
    (? category) Palaeognathae
    (? category) Neognathae

There are several difficulties here. The first arises because there is only one species of *Archaeopteryx,* identified, for the sake of argument, as the ancestor. As the ancestor it has the autapomorphies of the "supercohort" (more usually class) Aves and is thus the sister-group of the Crocodilia, if the orthodox classification of living forms is correct, but has no apomorph features at any rank below that. Thus either *Archaeopteryx lithographica* is a paraphyletic species, genus, and so on (up to the category supercohort) or all birds are members of the species *A. lithographica!* The second difficulty is that, with only one species of *Archaeopteryx,* the well-based taxon Ornithurae, whose name refers to the abbreviated tail found in all other birds, lacks a sister-group. Thus despite the fact that Aves and Ornithurae do not have the same denotation, the former appears monotypic, containing only the latter. I conclude that cladistic classification cannot cope with ancestors. This conclusion was reached long ago by Nelson (1972a, 1972b, 1973).

A third difficulty is not endemic in Wiley's proposal but is illustrated by my attempt to express Cracraft's grouping using Wiley's convention. It is that, in expressing a part of a cladogram in Hennigian-comb form, one is faced with two alternatives, both unsatisfactory. In Cracraft's cladogram the first five nodes (Aves, Ornithurae, Carinatae, Neornithes, Neognathae) all represent named taxa with a well-defined series of autapomorphies. The choice is between (1) using these names to characterise taxa, which leads to the problem with numbers of ranks to which I have already referred, and (2) using a sequencing technique but supressing the taxon names and the information they represent in the classification. Thus this third difficulty brings us to the third controversial area in the use of fossils referred to on p. 158. That was "the treatment of morphoclines from the fossil record".

Because ancestors cannot be recognised, any morphocline represented by fossils must be a *Stufenreihe* expressed as a Hennigian comb. Nelson (1972a, 1974) suggested a convention for reducing the number of categories when a comb was presented as a classification. This is known as "sequencing". A series of taxa at the same rank in a cladistic classification indicates successive terminal taxa in the Hennigian comb from which the classification was derived. The sequencing convention can be applied whether the terminal taxa include fossils or not.

A convention that can be used in addition to Nelson's sequencing was proposed by Patterson and Rosen (1977). It is represented in the bird classification above but was developed for Patterson and Rosen's fish classification which we noted briefly in Chapter 3 (Section III). Fossil terminal taxa, species, or otherwise, are given the unranked category of "plesion":

> We therefore propose that fossil groups or species sequenced in a classification according to the convention that each such group is the (plesiomorph) sister-group of all those, living and fossil that succeed it, should be called "plesions". Plesions may be inserted anywhere (at any level) in a classification, without altering the rank or name of any other group.

> They may bear a categorical name representing any conventional rank . . . these ranks being those already existing in the literature.
>
> (*Patterson and Rosen 1977, p. 160*)

Thus sequencing and classifying fossil taxa as "plesions" is a fairly satisfactory practical way of turning a *Stufenreihe* cladogram or Hennigian comb into a Linnean classification, but the production of comb cladograms has considerable theoretical importance as well.

As we noticed briefly in Chapter 3 (Section III), Hennig (1969) suggested a way of applying his techniques of classification to fossils by describing the part of a Hennigian comb consisting entirely of fossil taxa (later plesions) as the "*Stammgruppe*" while the extant members that terminate the comb were referred to as the "**Gruppe*". Jefferies (1979) subsequently suggested the appropriate term "crown-group" for the latter. Thus in the example of actinopterygian classification set out by Patterson and Rosen (1977), extant teleosts are the crown-group, while the comb sequence of fossil forms with cumulative teleosts autapomorphies are the stem-group, which is paraphyletic by definition. Similarly in Cracraft's (1988) classification of birds (above), *Archaeopteryx,* Hesperornithiformes, and Ichthyornithiformes are stem-group birds; Neornithes constitutes the crown-group. The system is set out in a methodological preamble to Jefferies (1979). It acquires its theoretical importance because both Jefferies and Patterson (e.g., 1980b) take the view that construction of a Hennigian comb is the correct use of fossils in systematics. The latter work provides a historical account of speculation about the origin of tetrapod vertebrates:

> Dealing with another hotly disputed area of vertebrate phylogeny, the hominids, Tattersall and Eldredge (1977) . . . argue that the accepted method has been to write scenarios . . . and that the construction of trees, and of scenarios, is necessarily secondary to a more basic activity, constructing cladograms. . . . I suggest that we might make progress with the problem of tetrapods by taking that to heart, and going back to the problem that Owen and Bischoff argued about – what are the characters of tetrapods? With a comprehensive answer to that question, *we could make better use of the fossil record, looking in it not for ancestors, but for the sequence in which those tetrapod characters arose.* [my italics]

Thus the expectation is that when fossils and extant taxa are classified together, the fossils will be ranged in sequence as members of the stem group such that they possess the crown-group autapomorphies in the manner a, $a + b$, $a + b + c$, until the full complement ($a \ldots n$) is reached in the most primitive extant taxon, representing the crown-group.

I believe that accepting this expectation as dogma has a number of serious objections to it. These were noted briefly in Panchen (1982) and developed in detail in Panchen (1991). In the latter work, I cited my objections as (1) the expectation referred to above as, in effect, a self-fulfilling prophecy; (2) "a too-exclusive interest in placing the crown-group in a linear hierarchy, in the same way that the "tree of Porphyry" is more concerned with classifying the "Individua" than exploring the properties of the subaltern genera" (see Chapter 6). I suggested, however, a more serious objection, that the technique is liable to yield an invalid cladogram and thus classification. As an example, I looked in detail at the classification of fossil non-amniote tetrapods and the place in that classification of *Crassigyrinus scoticus* from the Carboniferous of Scotland. Skull and skeleton of *Crassigyrinus* both have a number of primitive, or possibly degenerate, features. Because of these it would be placed as a very primitive stem-group tetrapod, above *Ichthyostega* (*the* primitive tetrapod from the Devonian of Greenland), but below other Paleozoic non-amniotes. It is placed in this position by Milner *et al.* (1986). It is my opinion, however (Panchen 1985, 1991), that *Crassigyrinus* has at least four valid synapomorphies with the fossil Anthracosauroideae. But the anthracosauroids are usually associated with the amniotes (Panchen and Smithson 1988) and would appear much higher up on a Hennigian comb. Thus, with the *scala naturae* mentality that results in a Hennigian comb, the *Crassigyrinus*–anthracosauroid link would probably not even be considered. More importantly, however, neither would the possibility that early tetrapods split into two major clades, one containing *Ichthyostega,* extant Amphibia, and associated fossil groups; the other, *Crassigyrinus,* anthracosaurids, and amniotes. Dr. Smithson and I believe that this fundamental dichotomy is correct (Panchen and Smithson 1988).

Crassigyrinus can also be used to illustrate another weakness concerning traditional cladistic practice, that of out-group comparison, but we shall consider that in Chapter 8. Meanwhile our fourth and last question in regard to the classification of fossils (p. 158) was, Can data from fossils upset a classification based on extant species?

This question was reviewed by Patterson (1981b), who concluded that "instances of fossils overturning theories of relationship based on Recent organisms are very rare, and may be nonexistent". Subsequently, two important cases have been cited, and it is of interest to attempt to discover whether any general principles might be involved. The first involves the controversy to which I have already referred a number of times: the nature of the relationship of birds (Aves) and Mammals (Mammalia). Gardiner (1982) caused a considerable stir amongst vertebrate taxonomists by producing a cladogram (Fig. 7.5a) of the major groups of tetrapods in which he claimed to show that Aves and Mammalia were sister-groups comprising the taxon Haemothermia, a grouping cautiously accepted by Owen (1866) but discounted since. Gardiner claimed that seventeen synapomorphies united birds and mammals. Later more were added by Gardiner himself and by Løvtrup (1985), bringing the grand total to twenty-seven. There appears to be some support for Gardiner's hypothesis from trees constructed from molecular sequence data for myoglobin and haemoglobin but not most other proteins (Bishop and Friday 1988; see also Chapter 9). The orthodox position, as we have seen, is that Crocodylia is the sister-group of Aves and that "other reptiles" are more closely related to the group so formed (Archosauria) than are mammals (Fig. 7.5b). Following cladistic precepts, Gardiner based his initial cladogram on extant taxa although he did then consider the position of a few well-known fossils.

There was an enormous amount of criticism of Gardiner's ideas, but two works – Kemp (1988a) and Gauthier *et al.* (1988) – discussed them in detail. Kemp was able to reduce the list of putative mammal–bird synapomorphies to nine. The others were either not unique to mammals and birds, or not truly homologous, while in two cases the character state in crocodiles is not known. Kemp then listed twenty-four possible bird–crocodile synapomorphies but concluded that "confidence in the superiority of the Archosauria hypothesis is not overwhelming in pattern [i.e., "transformed"] cladist terms, and it behoves us to seek further agreements that might tend to corroborate or to refute it, relative to the alternative".

Fig. 7.5. (a) Cladogram showing the relationships of the major groups of tetrapod vertebrates according to Gardiner (1982). (b) The orthodox arrangement.

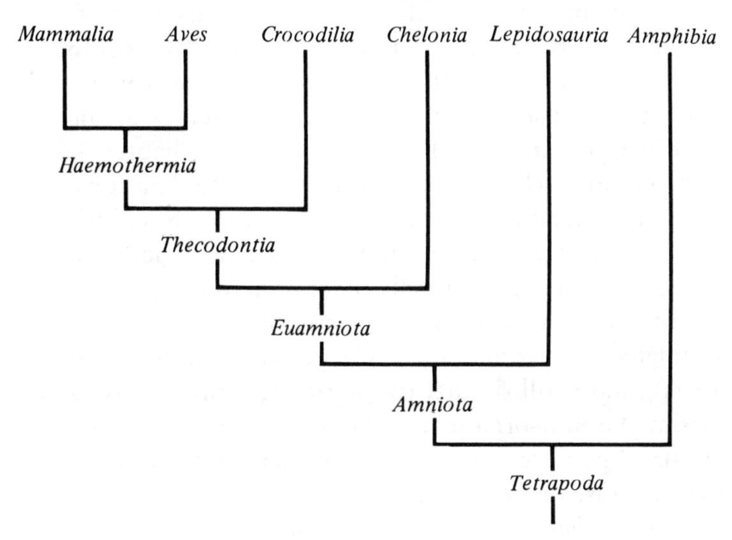

(a)

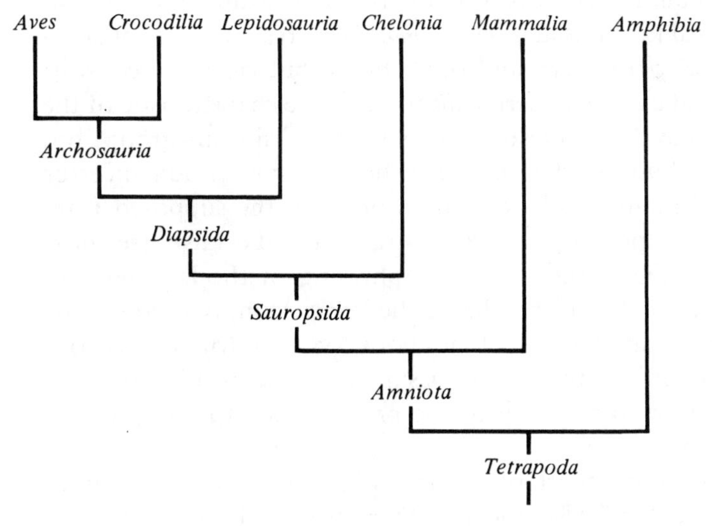

(b)

Kemp then discussed such further arguments based on data from fossils. We saw in Chapter 5 (Section II) that the "mammal-like reptiles" (i.e., stem-group mammals) show a morphocline representing the apparent change from the quadrate and articular bones of the jaw joint to their respective homologues, the incus and malleus of the mammalian middle ear, a homology first established by Reichert (1837) (see Chapter 4, Section II). Kemp claimed, *contra* Patterson (1982), that stages in that transformational homology can be used taxonomically. If the stages in the morphocline are congruent (*sensu* Patterson; see also Chapter 4, Section II) with the other characters, then they will be data for a Hennigian comb classification of stem-group mammals culminating in the crown group – that is, Mammalia (*s.s.*) The most primitive of these stem-group mammals are known to have lacked the fossilizable characters of Gardiner's Haemothermia – for example, a consolidated atlas vertebra, an enlarged cerebellum (from endocranial casts), and skull dermal roofing bones sunk below a soft dermis (from the non-ornamented bone surface). Thus in a proposed mammal–bird sister-grouping, these characters, thought to be homologies, must be homoplastic.

On the bird side, *Archaeopteryx* possesses several characters that unite it with crocodiles. It is also a stem-group bird, so that these characters would have to be homoplastic in Gardiner's view. But the small birdlike dinosaur *Compsognathus* shares a number of characters with *Archaeopteryx* and other birds, but lacks the consolidated axis and expanded cerebellum said to be characteristic of the Haemothermia. It thus also belongs to the bird stem-group, but further contributes to the view that the nearest common ancestor of birds and mammals lacked some or all of the supposed Haemothermia synapomorphies. As a "transformed cladist" (see next chapter), Gardiner might find the phrasing of the last sentence unacceptable, so I shall re-phrase the conclusion as follows: the lowest ranking taxon that includes both Aves and Mammalia contains fossil members that can be assigned to one or other of these groups but lack some or all of the proposed synapomorphies of Haemothermia.

I described one aspect of the investigation by Gauthier *et al.* (1988) in Chapter 5 (Section II). Their principal aim, however, was to "test the proposition that fossils cannot overturn a theory of relationships based only on the Recent biota". Once again they tested Gardiner's proposal of Haemothermia as supported by Løvtrup and once again showed that many of the proposed synapomorphies were invalid. They then produced a cladogram of Recent major amniote taxa, using a cladistic computer program (see Chapter 9). The cladogram agreed with orthodoxy in showing Aves and Crocodylia as sister-groups (the Archosauria) but disagreed in showing mammals as sister-group to extant archosaurs, and Chelonia (turtles) as nearer to this grouping than were Lepidosauria (lizards, snakes, *Sphenodon*). Gauthier *et al.* then presented a cladogram incorporating fossil amniotes, giving a total of twenty-nine terminal taxa. In that case all the extant taxa with their stem groups appeared within the cladogram in their orthodox relative positions (as in Fig. 7.5b). Notably the primary amniote dichotomy was between Synapsida, including mammals, and "Reptilia", including all extant reptiles and birds. In order to discover the fossil taxa responsible for the change in cladogram topology, they tried recomputing several times with various terminal taxa omitted. Collectively and severally the mammal-like reptiles were responsible, once again, for the switch in pattern (the relative position of birds and crocodiles, being orthodox, was unchanged).

Thus Kemp and Gauthier *et al.* effectively reach the same conclusion. If there are known fossils that share some but not all the autapomorphies of an extant group, those extant autapomorphies lacking in the fossils must be homoplastic rather than synapomorphous if they appear to be shared with some other extant group.

This principle need not imply acceptance of the stem-group/crown-group concept, of which I am suspicious (Panchen 1991) and which is rejected by Gauthier *et al.* The principle used in our second case, however, involves further inferences uncongenial to some cladists.

In a recent review (Panchen and Smithson 1987), Dr. Smithson and I set out to investigate the relationship of Tetrapoda to the various bony fish taxa, extant and fossil, reputed to be related to it. Following strict cladists precepts, we investigated the extant taxa first, concluding that Dipnoi (lungfishes) was the extant sister-group of Tetrapoda and that Actinistia (coelacanths) was the sister-group of Dipnoi plus Tetrapoda. As in the mammal–bird case we subsequently inferred that primitive Dipnoi probably lacked the dipnoan–tetrapod synapomorphies. These were (1) separation of pulmonary and systematic blood circulation in heart and aortic arches, (2) heart functionally divided, (3) ventral aorta (in front of heart) developed as a truncus arteriosus, and (4) presence of a glottis and epiglottis. All of these are "soft anatomy" characters not accessible in fossils, but we nevertheless asserted that they were probably absent in the soft anatomy of the earliest Dipnoi from the Devonian period. All four characters are related to pulmonary respiration in living dipnoans, allowing aerial respiration in deoxygenated water or in air-filled aestivation burrows. *Carboniferous* dipnoans were probably similarly adapted: they occurred in coal swamp faunas with tetrapods and, like the living dipnoans, had a small operculum (gill cover). Carboniferous aestivation burrows are also known. Early and Middle Devonian lungfishes, on the other hand, were mostly marine (Campbell and Barwick 1987) and, where known, had large opercula and a well-developed gill skeleton, unlike the extant species.

We inferred, therefore, that they were primarily or exclusively gill breathing, that they lived in an environment where lung breathing was unnecessary, and that their inferred anatomy showed the primitive dipnoan condition. Thus the supposed dipnoan–tetrapod synapomorphies were absent in primitive Dipnoi and are thus homoplastic between Dipnoi and Tetrapoda. The correctness of our inference, which would refute the sister-grouping, depends not on parsimony or any other cladistic procedure, but on palaeoecology and the reconstruction of functional anatomy. A scenario may refute a cladogram (see Chapter 10).

8

Methods of classification: The current debate

In Chapter 7, we saw that the characteristic features of phenetics were the production of a classification based on methods that were either based on distance data, or conflated (in cladist terms) plesiomorphic, synapomorphic, and homoplastic characters to yield distance measures. Cladistics, on the other hand, claimed to distinguish these features and was thus originally proposed as a technique for the simultaneous generation of classification and phylogeny. In the description of cladistics in the last chapter, the philosophy and methods reviewed applied to this original Hennigian "phylogenetic systematics"; however, there and in the previous chapters, I have also mentioned "transformed cladistics". This will be discussed in some detail here.

I also suggested in Chapter 7 that there was some dissatisfaction with the polarisation of characters by "out-group comparison". Coincident with the development of transformed cladistics has been the attempt to polarise characters by ontogeny, and we shall also consider this in relation to out-group comparison. The question of parsimony will, as promised, also arise in relation to the cladists' axiom that the natural order of organisms is an inclusive hierarchy. Further consideration of parsimony will, however, be presented in Chapter 9, where I describe the techniques of numerical cladistics.

I. "The transformation of cladistics"

The quotation marks enclosing the title of this section refer to the title of a paper by Platnick, "Philosophy and the Transformation of Cladistics" (1980). A general awareness of change in the background philosophy of cladistics began in the late 1970s and early 1980s, but its roots lie in Hennig's original system. We have already seen elsewhere (Chapter 7, Section III), that Hennig was not wedded to the dogma of dichotomous speciation. This is emphasised by Platnick with a quotation from Hennig (1966, p. 210): "If phylogenetic systematics starts out from a dichotomous differentiation of the phylogenetic tree, this is primarily no more than a methodological principle."

It follows from this quotation that Hennig accepted the possibility that the "correct" cladogram need not be isomorphic with the pattern of cladogenesis for any given group. Thus Hennig paved the way for the distinction between cladograms and trees which I discussed in Chapter 3 (Section III). The idea that a cladogram could represent several possible trees is stated explicitly by Schaeffer *et al.* (1972):

> If phylogenetic affinity is assessed on the basis of comparative morphology, it follows that a cladogram of the sort illustrated in Fig 2A [see our Fig. 8.1a] represents the best possible expression of relationships. This is because the cladogram is based on fewer assumptions (regarding ancestor–descendant relationships) than the usual phylogenetic tree, which is a highly specific statement, and is frequently but one of several that could be derived from the same morphologic and stratigraphic data (Fig 2B–D) [Fig. 8.1b–c, this volume]. In other words, a cladogram claims less than, but actually says as much as, a phylogenetic tree.

The trees that Schaeffer *et al.* illustrate (Fig.8.1b–c) differ from their "parent" cladogram in two important ways: Assuming palaeontological data, the geological range of all five species concerned is shown on the vertical axis, and it is assumed in two cases (b and c) that one species, a terminal taxon in the cladogram, may be ancestral to another which is also a terminal taxon. Whether the time ranges of the terminal taxa are represented or not, the multiple derivation of trees from a single cladogram always involves at least one of two presumptions: (1) that of a pair of cladistic sister-species, one may be ancestral to the other; (2) that species, as terminal taxa, may arise by hybridisation from two "parent" species that may or may not be sister-species. Platnick (1980) shows two possible ways, based on Hennig's views on speciation, in which an ancestral species and its daughter species would appear as sister-groups in a cladogram (Fig. 8.2). In the first (Fig. 8.2b), anagenesis only is supposed to have occurred. One terminal taxon represents the ancestral population; the other, its descendants after some detectable morphological or other change has occurred, but

Fig. 8.1. A cladogram (a) and three phylogenetic trees (b–d), all constructed from the same data. 1 and 2 are exant taxa, 3–5 are known only as fossils; the known stratigraphic ranges of all five taxa are shown as vertical solid lines in b–d. Broken lines are conjectural lines of descent. (After Schaeffer, Hecht, and Eldredge 1972, published by permission of the Plenum Publishing Corp.)

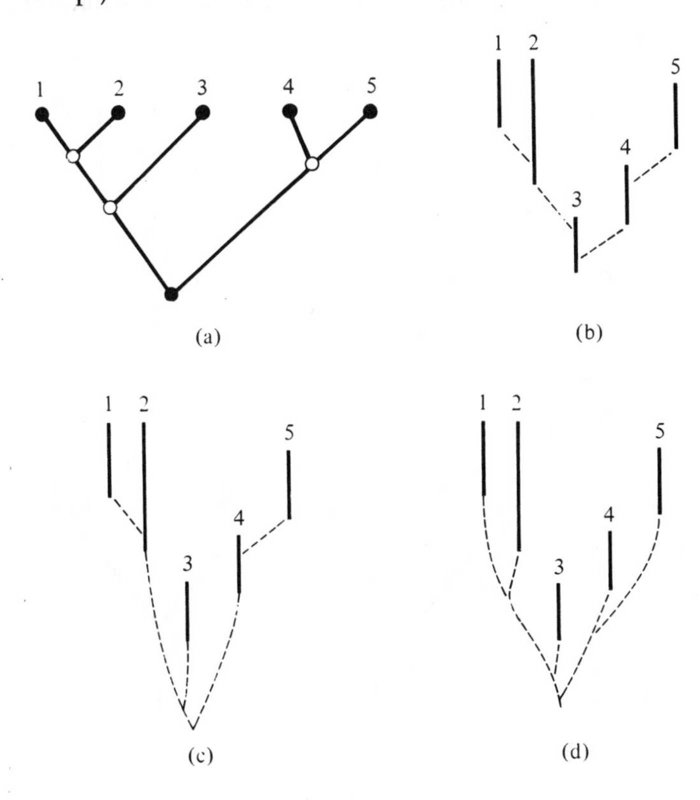

without any splitting of the ancestral population. But according to Platnick, the daughter population must show "at least one additional apomorphic character" – the detectable change. In this case, according to Hennig's views, speciation cannot be said to have taken place.

In the second case, speciation can be said to have occurred (Fig. 8.2c) because a daughter population has split off from the parent stock and is detectably different. The parent population, however, is taken to continue morphologically unchanged, so that samples

Fig. 8.2. Possible phylogenetic trees (top) and their resulting cladograms (bottom). In (b), according to Platnick, the ancestral population (1,2) will appear as sister-group to its descendants despite the fact that no speciation has occurred. (After Platnick 1980. Published by permission of the Society of Systematic Zoology, National Museum of Natural History, Washington, D.C.)

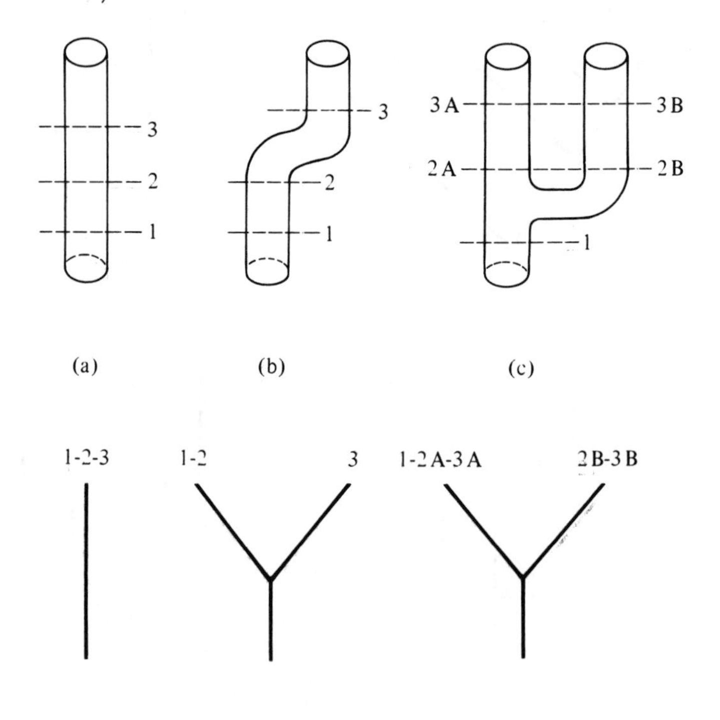

of that parent population from before and after the speciation event could not be distinguished taxonomically as different species, despite Hennig's (very controversial) view that they should be. Thus Platnick concludes, reasonably, that "Hennig's views on limiting species at branch points are irrelevant to cladistic practice". In both cases, the "daughter species" has acquired at least one autapomorphy of its own which distinguishes it from its "parent species".

There is, however, a difficulty that was referred to in Chapter 7. The daughter species has at least one autapomorphy, but what of the parent species? If a tree is to be regarded as derived from a cladogram, when the tree shows one terminal taxon as ancestral to another, then the "ancestral" taxon must lack autapomorphies. Platnick (1985) allows that this might be the case (see also Cracraft 1989). In Platnick's words:

> The concept that species taxa are individuals creates no difficulty for cladistics, as there is nothing in cladistic theory that requires species taxa to have defining characters (i.e. autapomorphies) . . . [A]s terminal taxa, species must have unique *sets* of apomorphic characters, but need not have any autapomorphies. [*my italics*]

Thus Platnick claims that of a pair of sister-species in a cladogram, one could be ancestral to the other. In the case of higher taxa, however, this could not be the case. If higher taxa are the terminal taxa in a cladistic analysis, then placing one taxon in a position ancestral to its sister-group in a tree derived from the cladogram *must* correspond to a statement that the "ancestral" taxon is paraphyletic. Stem-groups (Chapter 7, Section IV), paraphyletic by definition, are not terminal.

Thus trees implying ancestry of one terminal taxon to its sister-group are at best confined to those with species as terminal taxa, and their validity depends entirely on the acceptance of the taxonomic validity of species without unique defining characters. It follows that species must represent a different sort of entity from any taxon of higher rank. Platnick refers to the possibility that species are individuals, in the philosophical sense, while asserting that higher taxa are not (whatever they are). The idea that species are philosophical individuals (Ghiselin 1974) has generated an enormous literature, and the flow shows no sign of abating. We shall have to consider it in Chapter 14, but meanwhile I shall state my view that transformed cladistics is of necessity idealist and essentialist (Chapter 6, Section I) *in its methodology*. Whether it is in its epistemology is a question we shall reserve for Chapter 10. [The distinction between the methodology and the epistemology of cladistics is made very clearly by Hull (1980).] That being so, I believe that the species and other taxa ordered by cladistic methods are classes. Species must therefore have defining characteristics (i.e., autapomorphies), but implicit in the claim that cladistics produces natural classifications is the hypothesis that each taxon recognised, including species, is a philosophical individual. As I say, I shall postpone discussion of this point, but meanwhile note that any species recognised by cladistic taxonomy must have at least one autapomorphy and cannot, therefore, be depicted as ancestral to another species in a tree derived from a resolved cladogram. Wiley (1979a, 1979b) shows that for three species (X,Y,Z), the only valid trees not isomorphic with one of the possible cladograms are those derived from an unresolved trichotomy (Fig. 8.3)!

If this argument is accepted, then the only way in which more than one tree could be derived from a single resolved cladogram is if reticulate evolution is assumed to have occurred, that is, if some of the species recognised by cladistic methodology are in fact of hybrid origin. It is interesting that Platnick (1985), in defending transformed cladistics against the views of traditional Hennigian cladists, attempts to refute the one-cladogram–one-tree dogma only with a hypothetical example involving hybrid species. Depending on the inheritance of the sampled characters, the hybrid species would be expected to favour one parent species or the other; but if by chance a taxon of hybrid origin displays all the autapomorphies of its two parent taxa, and these are present in equal numbers, then it will be clustered as an unresolved trichotomy with its parent taxa and thus still not yield a cladogram isomorphic with its "true" phylogenetic tree. But while Platnick shows that a cladistic analysis can yield an inclusive divergent hierarchy even when the actual phylogeny is in part convergent, he does not explain how the phylogeny can be known to be reticulate if, as Farris (1983) – quoted with approval by Platnick – says, "Synapomorphies constitute the only available evidence on genealogies." If this were true, no species could ever be known to be hybrid: in fact, of course, as I noted in Chapter 3 (Section IV), hybrid species, at least in plants, are known to exist, and their hybrid nature can be checked by crossing their putative parents, if both parent species are extant.

Also, as I have noted previously, hybrid species constitute the only feature of the pattern of phylogeny that is open to direct

Fig. 8.3. The seven possible phylogenetic trees, showing the relationships of three species X, Y, and Z which form a monophyletic group. (a), (b), and (c) are isomorphic with their corresponding cladograms; (e), (f), and (g) would yield the same cladogram as (d), isomorphic with the latter, and no other. (After Wiley 1979a. Published by permission of the Society of Systematic Zoology, National Museum of Natural History, Washington, D.C.)

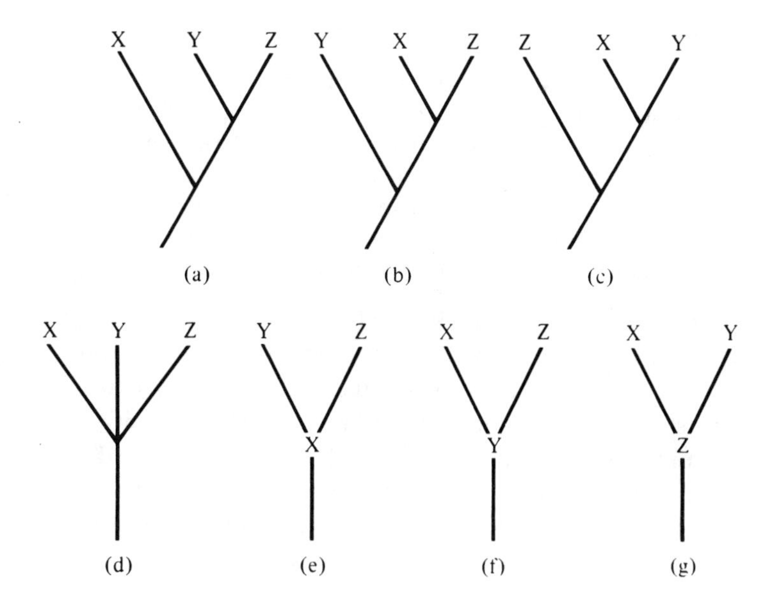

experimental testing. Their existence appears to refute the fundamental axioms of transformed cladistics. As Platnick (1980) formulated them, the requirements are

> [First,] that nature is ordered in a single specifiable pattern which can be represented by a branching [i.e. divergent] diagram or hierarchical classification; second, that the pattern can be estimated by sampling characters and finding replicated, internested sets of synapomorphies; and third, that our knowledge of evolutionary history, like our classifications, is derived from the hierarchic pattern thus hypothesized.

Thus, as expressed by Hull (1980):

> Cladistic classifications do not represent the order of branching of sister-groups, but the order of emergence of unique derived characters, whether or not the development of these characters happens to coincide with speciation events. It is not the emergence of new species which is primary but the emergence of new traits (Tattersall and Eldredge 1977: 207; Rosen 1978: 176). In general, cladists seem to be moving towards the position that the particulars of evolutionary development are not relevant to cladism. It does not matter whether speciation is sympatric or allopatric, saltative or gradual, Darwinian or Lamarkian, just as long as it occurs and is predominantly divergent.

Or, as expressed by Nelson (1979):

> In a cladogram, the branch point represents the generality of supposedly true statements ("synapomorphies") that can be made about the terminal taxa. In a phyletic tree, the branch point represents a supposed speciation event. A cladogram is an atemporal concept . . . a synapomorphy scheme . . . [M]any different trees may be derived from a single cladogram; but only one cladogram can be derived from any given tree.

Later Nelson and Platnick (1981) devoted a part of their book to what they referred to as "component analysis" (*not* to be confused with the established statistical technique of principal component analysis; e.g., see Thorpe 1976), in which they classify the patterns of cladograms and the trees they claim may be derived from them.

Perhaps one should not have been surprised that the reaction to "transformed cladistics" was as vehement in the early 1980s as that to Hennigian cladistics in the late sixties and early seventies.

In fact, it was often the original opponents of phylogenetic systematics in any form who transferred their anathema to transformed cladistics. Also, as before, the response from cladists yielded nothing in passion to their opponents.

Critiques of transformed cladistics, by those who accept to a greater or lesser extent the superiority of cladistic analysis, include those of Charig (1981, 1982), Beatty (1982), Brooks and Wiley (1985), Kemp (1985), and Ridley (1986). All object to the idea that cladistic classification can be pursued, in theory and in practice, without reference to evolutionary theory. Thus to Charig cladistic analysis ("character distribution analysis") has as its aim the production of "the branching pattern of phylogeny" expressed as a cladogram, but the resulting hypothesis of phylogeny may be modified by using evidence "from any other valid source, e.g. palaeontology". To that extent, he accepts Hennig's methodology but sees the creation of a classification from a cladogram as involving more freedom of action than any orthodox cladist would allow. Taxa may be derived from the cladogram in an arbitrary manner, provided that each corresponds "to a single continuous segment of the dendrogram". Paraphyletic groups are permissible and can be defined by sharing synapomorphies derived from their common ancestor and by lacking the autapomorphies of the groups excluded from them (i.e., definition by privation as rejected by Aristotle; see Chapter 6, Section I). But the main point of Charig's discussion, as with the other critics, is that classification must be "justified" by the acceptance of evolution.

Beatty (1982) makes the point about the essentialist nature of "pattern cladistics". All taxa, without assuming phylogeny, must have defining characters and are therefore classes. This applies equally to species, which thus "cannot evolve *with respect to their defining properties*" (his italics). Thus according to Beatty, "pattern cladistics" and evolutionary theory are actually contradictory of each other rather than pattern cladistics being neutral with respect to evolutionary theory.

Ridley (1986) repeats the idea that it is necessary to "justify" classifications by the acceptance of phylogeny. He also makes the more remarkable claim that it is necessary to accept the theory of natural selection as well as the "fact" of phylogeny in order to do justifiable cladistics:

> Each character has to be referred back to the theory of natural selection, which acts as a continual reminder of how classification uses the theory of evolution. Information here will obviously be flowing from the theory of evolution into the classification: from 'scenario' to 'cladogram'. The scenario will have to be understood in order to construct the cladogram (Fisher 1981).

The claim, however, becomes less remarkable when one realises that by "natural selection" Ridley means "adaptation"; the postulated mechanism of evolution is confused with its observed or reconstructed result. Ridley's reference to Fisher (1981) is to a worked-out case similar to that by Panchen and Smithson (1987) referred to in Chapter 7 (Section IV). But Fisher is careful to distinguish selection – "adaptation as a *process*" in evolution (presumably caused by selection) – from "adaptation as a *result* of evolution". It is adaptations as a result, in his case as in ours reconstructed from fossil evidence, that feed back to test the form of the cladogram.

Brooks and Wiley (1985) represent phylogenetic cladistics as practitioners. They see transformed cladistics as nested within phylogenetics, a methodology without its attendant theoretical structure ("*deep structure* or *measures of explanatory adequacy:* Chomsky 1965"). They complicate the picture somewhat by expressing their feelings about the inadequacy of Neo-Darwinism: they want a new evolutionary framework (Brooks and Wiley 1986; see also Chapter 12), whereas, according to them, the "pattern cladists" responded to the same dissatisfaction by jettisoning any theoretical framework whatsoever. Their substantive differences from the pattern cladists, however, concern the matter of polarisation by out-group rather than ontogeny [as is the case with Kluge (1985) in an immediately following paper]. There is also, however, the difference of opinion, as we saw above, regarding the equivalence of cladograms and trees (Wiley 1979a, 1981).

There is another criticism of transformed cladistics voiced by Charig (1982), Beatty (1982), and Ridley (1986). In Beatty's words:

> In a flurry of articles in the late 70s, cladists adopted Popperian ideals as their own, and further distinguished their brand of systematics from other brands in terms of this supposedly superior set of aims and methods (e.g. Wiley 1975; Platnick 1977,

> 1979 [i.e., 1980]; Platnick and Gaffney 1977, 1978a, 1978b; Cracraft 1978; Gaffney 1979; see also Cartmill, 1981 . . .). . . . An indication of current uneasy feelings about evolutionary biology is the decision on the part of some cladists to shuck whatever evolutionary aspirations and connotations were originally associated with their program.

We shall be looking at Popper's philosophy of science in Chapter 13, and its relevance to biological theory in Chapter 14. Meanwhile, however, there are two points to be made. The first, made by Platnick (1985), is that by no means all transformed cladists are followers of Popper; he cites particularly Patterson (1978, 1982). The second, and more important, is that whether various cladists still regard themselves as adherents of their version of Popper's philosophy of science or not is of psychological interest only, because (1) those taxonomists who have attempted to apply Popper's philosophy to classification have always misunderstood his original ideas; (2) the version of Popper's ideas which they misunderstood were subsequently rejected, or at least strongly modified by Popper himself (Panchen 1982); and (3) as we shall see in Chapters 13 and 14, Popper's methods of theory testing cannot be applied to biological and, least of all, taxonomic hypotheses.

Kemp's (1985) critique of transformed cladistics is part of a review of methods of classification and of phylogeny reconstruction. His opening sentence is:

> The theory of evolution *predicts* that all organisms are related to one another by a unique, hierarchical genealogy; how can we hope to discover that genealogy, or more formally, how should we propose and test hypotheses of phylogenetic relationship? [*my italics*]

This is obviously at odds with the central point that I have been trying to make so far in this book. The theory of evolution does not *predict* the existence of a genealogical hierarchy; it is a theory to *explain* a previously recognised hierarchy as genealogy. Nevertheless, Kemp's paper is perhaps the most cogent critique of transformed cladistics. He makes the point about known phenomena of reticulation and convergence in phylogeny, however reconstructed; the first (as in plants) known from extrinsic evidence, the second appearing in any attempt to produce a taxonomic hierarchy. He makes a further point about the fact that maximum congruence in classifications does not necessarily yield the same result as minimum homoplasy (Farris 1983). But Kemp's most important point, also made here and by many other authors, however, is that transformed cladistics will yield a divergent hierarchy, whatever the true nature of the taxonomic organisation of organisms. The hierarchy is axiomatic. Kemp also goes on to mention likelihood methods as a rival to the use of parsimony (see Chapter 9, Section III), but the underlying assumption is that evolution has occurred. The axiom that the hierarchy of classification represents phylogeny is added to the axiom that the natural order is a hierarchy: both are taken by Kemp, as by phylogenetic cladists, as *a priori*.

Transformed cladists have been less than lucid in pointing out the reversal of *explanans* and *explanandum* in taking phylogeny as axiomatic in classification, although they have been clear and insistent that cladistics could proceed satisfactorily without the assumption of evolution. The fact that the reversal has occurred in all phylogeny-based taxonomic methods has been spelt out with admirable clarity by Brady (1985; see also Panchen 1982). He makes the distinction between *explanans* and *explanandum* quite clearly and talks about the presumption of hierarchy which must underlie any taxonomic theory that makes no assumptions about process:

> If we let the aggregate of characters determine our interpretation (parsimony), we can find a fairly clear hierarchy of groups *in many cases*. In these cases such classifications show a high degree of stability, and for this reason, and the fact that the characters were themselves homologies . . . they may be termed 'natural'. About the groups that we cannot resolve we may say nothing, for our knowledge is insufficient to make a judgement as to whether these groups are somehow different or simply have not been resolved *as yet*. Our claims to knowledge must rest upon the well resolved groups, and these are most interesting to the investigator . . .
>
> We do not, or should not, advance explanatory or process theories prior to the discovery of a particular order in appearances to which the theory is addressed . . . [I]f we lose the distinction between the detection of pattern and its explanation by a process hypothesis, we lose the reason for our enquiry, not merely historically but logically.

Thus "transformed cladistics" is cladistics stripped down to two axioms and three items of methodology. The axioms are (1) that organisms (or at least species) are ordered in a divergent inclusive hierarchy ("groups within groups"); (2) that a taxon at any rank is defined by one or more characters unique to it and recognisable in all its members as taxic homologies (apomorph characters) so that these characters also form a hierarchy isomorphic with that of the organisms.

The procedure, therefore, is (1) to recognise valid characters within taxa; (2) to identify the inter-nested pattern of characters ("polarise" them) by out-group comparison or ontogeny; and (3) to eliminate non-homologies (homoplasies), regarded by some cladists as "mistakes", by parsimony.

II. Out-groups or ontogeny

Correlated with the increasing emphasis on the distinction between cladograms and trees, which has characterised the development of cladistics, has been the concept of the cladogram as representation of a nested series of taxic homologies and thus the equation of synapomorphy and homology (Chapter 4). The concept, as we saw, was expounded by Patterson (1982) in his paper on homology. But if synapomorphy equals taxic homology, then synapomorph characters, the only characters used in cladistic clustering, must be homologues and must be nested in the natural hierarchy. The concepts of synapomorphy and symplesiomorphy could thus be disposed of as redundant. Cladistics seeks what it takes to be a natural hierarchy of characters. Within that hierarchy the more general character (at a more inclusive node on the cladogram) includes all those less general ones normally regarded as derived from it (at nodes that diverge from the more inclusive one). Nelson and Platnick (1981, p. 328) use the idea of the hierarchy of characters to counter the idea that transformed cladistics is anti-evolutionary because of its essentialist nature:

> The rationale for this depreciation [of essentialism] seems to be that if evolution occurs, the characters of species (and hence groups) may change in the future; therefore, species and groups cannot be permanently characterised by means of a single character or set of characters such that the character or set is necessary and sufficient for membership in the species or group. The ar-

> gument seems to rest on the misleading use of character states: it assumes that when a species is modified, and acquires a new apomorphic character (state), it no longer has, or is no longer recognisable as having the original plesiomorphic character (state). In other words, according to this argument, we cannot use characters (such as fins) to define groups (such as Vertebrata), because some members of those groups (such as tetrapods) may acquire apomorphies (such as limbs). If one accepts the validity of ontogeny or outgroup comparison (i.e. parsimony) or any other possible test of hypotheses about character transformation, the argument is obviated.

Thus, according to Nelson and Platnick, fins are the more general character, limbs the more specialised, lower ranking character. Thus "presence of paired fins" still characterises all gnathostome vertebrates including tetrapods. Patterson (1982) makes a similar point about "loss" characters. One way of testing this nesting of characters is to look at the ontogeny of the organisms concerned. This idea was developed by Nelson (1978). As an example, he took the case of the ontogeny of a flatfish. In the early stages of ontogeny, the eyes are symmetrically placed, on one side of the head; later, migration of one eye can be followed, so that eventually both eyes are situated on the same side, in the characteristic manner of flatfish. The question Nelson posed was: "Consider an example: a species of fish with an eye on each side of the head, and a second species with both eyes on one side. Which condition of the eyes is the more primitive?"

> Without further information, no rational answer would be possible. But suppose we study the ontogeny of both species, and find that embryos of both species have character x [symmetrical eyes], and that during subsequent development of species B [of A and B] character x is transformed to character y. With this ontogenetic character transformation we may answer the original question: character x is more primitive than y.

From this apparently naive example, Nelson attempts to generalise a method of polarising characters in taxonomy. The transmutation in ontogeny is observable and involves no speculation about phylogeny, and no previous knowledge of taxonomic grouping as required by out-group comparison. Thus cladistics could be related empirically to a real series of phenomena, the ontogeny of the organisms being classified.

What are, or should be, classified in cladistic methodology are species, but what the term "species" represents in this context is the ontogeny of all the individual organisms making up the species. Hennig (1966) introduced the term *character-bearing semaphoront* for the individual organism to be compared with other semaphoronts in erecting a classification. Semaphoronts must be at a comparable stage of development if their characters are to be compared. A classification of holometabolous insects, in which some species are represented by larvae and others by adults, would be invalid, although one could use a classification based solely on larvae to corroborate one based on adults. The stage used, larval or adult, for each terminal taxon, is the semaphoront. But ideally it should be total ontogenies that are classified. Patterson (1982) points out that the ontogenetic argument, as developed by Nelson (1978),

> also explains Wiley's (1979[c], p. 315) conundrum, how a cranial capacity of 1200 cc can be characteristic of *Homo sapiens*, when no newborn individual has the character. The character of *H. sapiens* is not to have that cranial capacity, but both to lack it (general condition) and to have it (special condition), whereas all other organisms exhibit only the general condition.

Nelson's (1978) method of polarising characters by ontogeny is related obviously to Von Baer's laws (Chapter 2, Section II), and Nelson attempts a new formulation of the "biogenetic law" (confusingly Haeckel's term; see Chapter 3, Section II) which he, Nelson, regards as "falsifiable":

> *[G]iven an ontogenetic character transformation, from a character observed to be more general to a character observed to be less general, the more general character is primitive and the less general character advanced.*

Criticisms of Nelson's method frequently invoked *paedomorphosis* (Rieppel 1979; Stevens 1980; Arnold 1981), cases where the derived condition of the adult is taken to have arisen from the more general one of related species by loss of some adult characters in ontogeny, so that some juvenile characters are retained to the adult stage (as in the Axolotl). But several authors (de Queiroz 1985; Weston 1988) have pointed out that this will cause trouble only when species are being classified using their adults as semaphoronts, and also that the essence of Nelson's proposal is the transformation from a more general to a less general character in one organism compared to the persistence of the more general character in another. This rejoinder also applies to cases, such as the development of foetal membranes (amnion, chorion, allantois, and yolk sac), in the embryos of amniote vertebrates. The derived condition may appear early in ontogeny, but it is still not the more general character amongst vertebrates as a whole. Lundberg (1973) and de Queiroz (1985), however, suggest that in looking at whole ontogenies rather than semaphoronts (usually adult organism) in taxonomy, the characters should not be "instantaneous morphologies" (de Queiroz's phrase) but the actual transformations themselves. Unhappily this destroys Nelson's criterion of generality. Whereas Nelson would say that if, in the ontogeny of two related organisms, one showed the transformation from x to y (as in the eyes of flatfish; see above) while the other showed x to x (the "normal" condition), then x is the more general, plesiomorph character and y the derived or apomorph one. If, however, $(x \rightarrow y)$ is a character and $(x \rightarrow x)$ is a character, there is nothing to choose between them in terms of generality, and the ontogenetic method of polarising characters fails; one is thrown back on out-group comparison. From this, de Queiroz draws the further conclusion that transformed cladistics arose in part for the following reason:

> When instantaneous morphologies are treated as characters, both ontogenetic and evolutionary polarities exist at the same hierarchical level (i.e. both exist between characters), and it is easy to confuse them.

However,

> [w]hen ontogenetic transformations are treated as characters, ontogenetic and evolutionary polarities are clearly distinguishable since they exist at different hierarchical levels: ontogenetic polarities exist within characters, evolutionary polarities between them.

De Queiroz's views are criticised by Weston (1988) and Kluge (1988). Weston, despite being a phylogenetic cladist, defends Nelson's methodology. Kluge accepts both what he describes as "horizontal" and "vertical" characters. *Horizontal* characters are those normally accepted – de Queiroz's "instantaneous morphologies";

vertical characters are transformations. Kluge accepts de Queiroz's point about the latter and recommends the out-group criterion for polarising them. Polarisation by out-group comparison, as we saw in the last chapter, must depend on some pre-existing hypothesis of grouping. It is also frequently taken to depend on the assumption of phylogeny. Maddison *et al.* (1984) made that assumption in a detailed review of the method, although they do say at one point:

> Our results hold whether one views a cladogram as a diagram that indicates common ancestry (Nelson, 1974) or as a diagram that indicates only the patterns of character distribution (Nelson & Platnick, 1981). If the cladogram is interpreted as indicating character distributions, then the assignment of states to nodes can be interpreted as merely a book-keeping procedure to keep track of character distributions. Under this interpretation outgroup analysis finds the assignment to the outgroup node that requires the fewest hypotheses of characters in the outgroup portion of the cladogram, and the results presented below hold *mutatis mutandis*.

In the system presented by Maddison *et al.*, the "outgroup node" is that node on the cladogram (assuming ancestry) that represents the common ancestor of the in-group (that being classified) and the sister-group of the in-group (Fig. 8.4A). Given only that sister-group as the "out-group", any character within it that also occurs in the in-group is taken to be plesiomorph in the latter. This is simple out-group comparison (see Chapter 7, Section III), and, assuming the correct choice of out-group, the result is "decisive". However, if there is a series of out-groups in a Hennigian comb arrangement – that is, forming the stem-group of the in-group – then these may alter opinion as to the primitive condition by having different states of any given character from those in the first out-group (Fig. 8.4B, C). The results then, can be either "decisive" or "equivocal".

There are, however, cases where there will be uncertainty in assessment of the out-group node because of doubt about the arrangement of a series of out-groups. Maddison *et al.* are careful to distinguish this case from an equivocal result. The latter (see Fig. 8.4) occurs when two or more interpretations of the primitive state are equally probable, using the criterion of parsimony:

the former occurs "when there are several plausible outgroup resolutions which yield different ancestral state assessments (some of which may be equivocal, others decisive)". Obviously, if the arrangement of out-groups as a stem group is resolved, the uncertainty will disappear, and the authors note that a resolved classification of the out-groups will do the trick. But this again will depend on out-group comparison, and an infinite regress is

Fig. 8.4. Out-group comparison: a and b are alternative states of the character being assessed; the triangle represents the in-group; the spot with indicated state, the out-group node. In A the left assessment is more parsimonious than the right, which requires a postulated state change (bar). In B and C, left and right are equally parsimonious; the result is equivocal. (After Maddison *et al.* 1984. Published by permission of the Society of Systematic Zoology, National Museum of Natural History, Washington, D.C.)

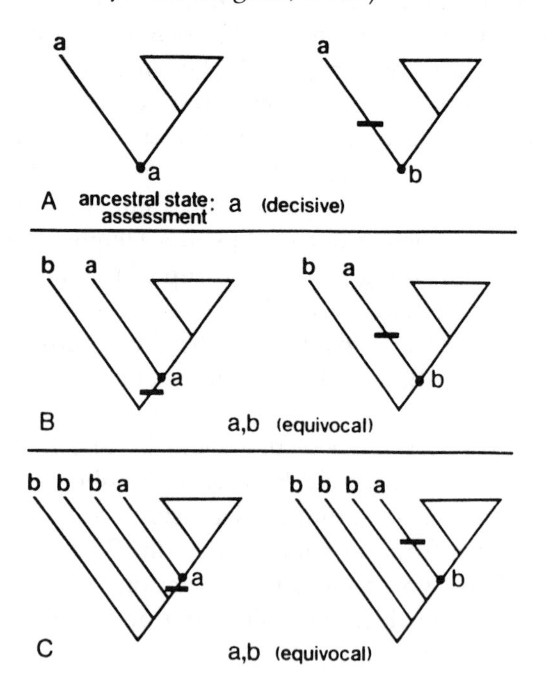

set in train. This is the principal objection to out-group comparison, and no doubt it seems more cogent when phylogeny is not assumed, as in transformed cladistics.

In the polarisation of characters, fossils, as always, pose special problems. Obviously ontogeny cannot be used in the classification of an exclusively fossil group, so that out-group comparison is the only resort. As I pointed out in the Chapter 7 (Section IV), the primitive Carboniferous tetrapod *Crassigyrinus* illustrates the difficulties involved. We saw how *Crassigyrinus* had a number of characteristics that appeared to be synapomorphies shared with a more advanced group of early tetrapods, the anthracosaurs, which in their turn may be related to reptiles and thus to all amniotes (Panchen and Smithson 1988). But *Crassigyrinus* is also a very primitive tetrapod, little above the Devonian form *Ichthyostega* in many of its skeletal features. *Ichthyostega* and *Crassigyrinus* differ from each other in having (to use phenetic terminology) a number of contrasting states of the same character. It is therefore important to hypotheses of the relationship of tetrapods to different groups of bony fish to decide which of the alternative states of each character are primitive, uniting tetrapods and their closest fish relatives, and which represent autapomorphies of the Tetrapoda.

We also saw in Chapter 7 that on the basis of living material, and exclusively "soft anatomy", the lungfishes (Dipnoi) appear to be the extant sister-group of Tetrapoda (Rosen *et al.* 1981; Panchen and Smithson 1987). Following cladistic practice, both Rosen *et al.* and Panchen and Smithson then attempted to classify a number of fossil fish groups in the light of the extant cladogram of (Dipnoi + Tetrapoda) + Actinistia. Rosen *et al.* concluded that the Dipnoi were still the sister-group of tetrapods, fossils notwithstanding, as did Forey (1987; one of the *alii* of Rosen *et al.*). We, on the other hand, arrived at the more orthodox conclusion that the Osteolepiformes, an exclusively fossil group, was the sister-group of tetrapods. As the (then) best known very primitive tetrapods, *Ichthyostega* and *Crassigyrinus* should show character states shared with the fish group that is the true sister-group of tetrapods. To polarise the character states in *Ichthyostega* and *Crassigyrinus,* one must resort to out-group comparison; but which is the out-group? In my description of *Crassigyrinus* (Panchen 1985), I drew up a table of the contrasting characters polarised on the alternative assumptions that Dipnoi and Osteolepiformes were the sister-group of Tetrapoda, reproduced here as Table 8.1. I summarised my conclusions as follows:

> In Table 2 each character state of a pair may be regarded as primitive, depending on whether Osteolepiformes or Dipnoi is the sister-group of tetrapods, and this is indicated in the table. However, the problem posed by these rival primitive characters may be stood on its head, so that each character state may be used as a synapomorphy uniting tetrapods and the tetrapod sister-group that the state favours. The vicious circle is thus closed: a character state is 'proved' primitive because of its occurrence in the 'outgroup', that is the sister-group of tetrapods; but the same character state is used as evidence . . . to establish the same sister-group relationship.

III. Parsimony

We saw in the previous section that Maddison *et al.* invoke the criterion of parsimony in attempting to determine the primitive state (that in the out-group node) in their review of out-group comparison. We also saw in the quotation in the last section that Nelson and Platnick (1981) appear to equate out-group comparison and parsimony. But parsimony is also used in cladistics to distinguish synapomorph from homoplastic shared characters, and is the basis of Patterson's congruence test of taxic homology (Chapter 4). The common feature in the cladistic use of parsimony, or economy of hyopothesis, is the reduction to a minimum of hypotheses of states or events which the cladist regards as undesirable. In the case of out-group comparison, it is character state changes that are minimised; in the case of homology, the number (or ratio) of homoplastic characters.

The simplest use of the criterion of parsimony in cladistics is in a "three-taxon test" (Chapter 7, Section III) in which the proposition is that (say) A is more clearly related to B than either is to C. The proposition is corroborated if A and B share more characters than either does with C. However, as pointed out by Engelmann and Wiley (1977; also Wiley 1981, pp. 110ff.), one is dealing with a closed system, and synapomorphy, homoplasy, and symplesiomorphy cannot be distinguished. It is quite likely, therefore, that some, even a majority, of the A + B shared characters are symple-

Table 8.1. *Contrasting characters of* Ichthyostega *and* Crassigyrinus *of controversial polarity*

	Ichthyostega		*Crassigyrinus*	
(ix)	braincase divided by suture	(O)	braincase consolidated	(D)
(x)	parasphenoid underlying ethmoid region only	(O)	parasphenoid underlying whole braincase	(D)
(xi)	course of infraorbital lateral line canal cf. dipnoans	(D)	course of infraorbital canal; cf. osteolepiforms (Panchen 1973)	(O)
(xii)	external nostril marginal	(D)	external nostril not near jaw margin	(O)
(xiii)	pars facialis of lateral rostral reduced	(D)	pars facialis not reduced	(O)
(xiv)	"fused" postparietals	(D)	separate postparietals	(O)

Key: (D), pleisomorph assuming Dipnoi to be the sister-group of tetrapods; (O), pleisiomorph assuming Osteolepiformes to be the sister-group of tetrapods.
Source: After Panchen 1985.

siomorphies as demonstrated by out-group comparison (in the terms of Maddison *et al.*, they are shared by the out-group node). That being the case, when these characters are removed from consideration, A + C characters or B + C characters may be found to outnumber the non-plesiomorph A + B characters, and the (A + B)C pattern must be rejected. Thus out-group comparison must precede analysis of the in-group, whether the latter contains three or more terminal taxa. Maddison *et al.* describe this as "the two-step procedure". Parsimony is used in the first step, in the sense of minimising (in phylogenetic terms) convergence and reversals in evolution, by minimising the number of evolutionary steps needing to be postulated in the evolution of the stem-group (Fig. 8.4). In terms of transformed cladistics, this means minimising the number of contradictions of the nested pattern of characters arrived at. The characters of the in-group can then be polarised and all characters appearing at the out-group node rejected as plesiomorph. Parsimony in the in-group then consists of a simple vote. If the number of apparently apomorph characters favouring a particular sister-group relationship (say, A + B) outnumber any rival pairing, then A + B is accepted, and so on until (if possible) the whole in-group is resolved. As we have seen before, this amounts to a hypothesis of a uniquely shared ancestor (of those taxa considered) in the case of phylogenetic cladistics, but simply one of uniquely shared taxic homologies, with a divergent hierarchy of characters, in the case of transformed cladistics.

"Global parsimony" (Maddison *et al.* 1984) is the desirable situation in which the in-group cladogram "is one which requires the fewest hypotheses of convergence and reversal (in all the characters examined) within the ingroup and among the outgroups". In terms of transformed cladistics, where hypotheses of evolutionary convergence and reversal are not entertained, the contradictions in the nested pattern of characters referred to above also means minimising homoplasy; reversals and convergences both result in homoplasy (Fig. 8.4). Thus the universal rule of cladistics is to "*minimise hypotheses of homoplasy*". Maddison *et al.* conclude that their "two-step procedure" achieves global parsimony only under certain circumstances:

> when: (a) outgroup relationships are sufficiently resolved beforehand; (b) outgroup analysis is taken to indicate the state not

in the most recent common ancestor of the ingroup, but in a more distant ancestor; and (c) ancestral states are considered while the ingroup is being resolved, not merely added afterward to root an unrooted network.

Considering parsimony in more general terms, a number of criticisms have been levelled at it as a criterion for resolving cladograms, particularly in phylogenetic cladistics. The first concerns the claim that by using parsimony, cladists were following the hypothetico-deductive method of Karl Popper. This claim, alluded to above, is simply false (Panchen 1982) – an assertion I shall return to and justify in Chapters 13 and 14. However, as I said in that earlier review (Panchen 1982), "The fact that cladists have misunderstood Popper does not necessarily invalidate their methodology". Nevertheless, other criticisms of the use of parsimony, and suggestions for alternative strategies, must be considered. My review was addressed explicitly to "the use of parsimony in testing *phylogenetic* hypotheses". I had suggested previously, in a running correspondence on cladistics in *Nature* (Panchen 1979), that the use of parsimony in a three-taxon test (assuming that out-group comparison had already taken place) would give valid results only if

> (a) the organisms comprising the three taxa could potentially be atomised into a series of discrete diagnostic characters of equal taxonomic weight, such that the state of each character was in no way correlated with the state of any other (or alternatively that only characters having these features were used); and also (b) if a competent cladistic taxonomist could be reasonably sure that in each taxon, the relatively tiny number of apparent synapomorphies that he uses in his test has a similar ratio of true to 'false' synapomorphies as does the class of all its characters.

The first criticism concerns the vexed problem of weighting characters – a problem for all methods of taxonomy, not just cladistics. It is worth noting three overlapping classes of reasons for weighting. Hecht (1976) and Hecht and Edwards (1977) were concerned principally with the reliability. On their scale of 1 to 5, their lowest category comprised "loss characters . . . which have no developmental information to indicate the pathway by which the loss occurred" and therefore give "zero information as to monophyly". Their highest category comprised characters that are "innovative and unique for the morphological series and . . . usually indicate new functional or adaptive trends". They cite the amniote egg, the artiodactyl astragalus, and the gekkonid (a lizard family) cochlea for the latter. The second reason for weighting characters is degree of independence. The lowest category here would be that containing characters whose presence is totally dependent on the presence of another character that is a synapomorphy *at the same rank.* An example would be the presence of the incus (quadrate) as an ear ossicle in mammals, correlated with the presence of the malleus (articular). One of these is always (as far as we know) present in association with the other (see Chapter 5, Section II). The *reductio ad absurdum* of this category is when two "characters" are different expressions of the same thing, as in the cases stigmatised by the pheneticists: red blood and the presence of haemoglobin, or diameter and circumference of the same structure. The highest category of independence is more difficult to exemplify. No character stands alone so that it neither affects nor is affected by any other, but hair in mammals is perhaps not part of a character complex in quite the way that any one of three ear ossicles is. The third class of reasons for weighting is related to Riedl's (1979) concept of "Burden" (Chapter 6, Section III). In this case, however, there is less of a problem for cladistics than for rival methods: characters of high burden tend to be apomorphous at high rank, those of low burden at low rank. They will, in Patterson's phrase, "weight themselves" in a way that is not the case in phenetics. Thus I now see the problems inherent in my 1979 criticism (a) above as problems for all taxonomists, with cladistics faring rather better than phenetics.

My criticism (b) (Panchen 1979) goes more to the root of the cladistic parsimony procedure. Here it is essential to distinguish its use in phylogenetics from its use in transformed cladistics. In phylogenetics a real phylogeny is taken as *a priori,* and the cladistic aim is that proposed by Hennig – to reconstruct the pattern of cladogenesis and to use this as the basis of a classification. In this case the use of parsimony is vulnerable. A real phylogeny, accessible or not, is "out there" and may have many constituent clades where characters resulting from parallel evolution, convergence, and even hybridization together outnumber divergent characters resulting from cladogenesis. In Panchen

(1982), I cited Butler's (1982) work on mammalian teeth. In this he showed a series of parallel lines of tooth evolution in mammals. The same morphoclines are also parallel chronoclines, and it is virtually impossible to interpret them as other than parallel evolution. On these tooth characters alone, parsimony would dictate arranging the roughly contemporary forms together in a single taxon and the whole as a Hennigian comb of stem-group mammals. The knowledge of more complete fossils shows that this is wrong; a series of orders extends the parallel through time. In this case an adequate sampling of characters, and the precept of classifying extant forms first in the case of the classes with living members, should show that the parallel characters are in a minority; but there is no good reason to assert that in evolution character divergence is always commoner than the phenomena resulting in homoplasy.

This problem is tackled by Farris (1983). He states that the use of parsimony does not imply that homoplasy is rare. If for the A,B,C three-taxon case, there are ten apparent synapomorphies uniting A+B, and only one uniting B+C, then most of the AB characters might be "homoplasious" [*sic!*]. Nevertheless, there would still be no good reason for accepting B+C as the primary grouping. Farris then makes the further point that, provided the B+C character *is* homoplastic, if *all* the B+A characters were homoplastic, the (AB)C grouping would not be false but merely unresolved. "Thus the phylogenetic parsimony criterion consists of nothing other than avoiding unnecessary ad hoc hypotheses of homoplasy". The method even allows for the possibility of the B+C character being synapomorphous: one is dealing in probabilities, and it is taken to be more probable that at least one homoplasy has occurred than that at least nine homoplasies have occurred. Thus as Sober (1983) says, following Farris's argument:

> Another way to see this point is to consider what would have to be true of the ten characteristics for either cladistic hypothesis to be refuted. (AB)C would be refuted only if the synapomorphy in characteristic 10 were a homology; A(BC) on the other hand, would be refuted if any of the characteristics 1–9 were. So even if homoplasies were plentiful, it would take an extremely special distribution of them to refute (AB)C; A(BC), on the other hand, is much more vulnerable.

Thus Farris's justification for the use of parsimony does not make the assumption that homoplasy is rare, or even that it is rarer than symplesiomorphy due to common ancestry. There must, however, be something in phylogenetic cladistics that makes "ad hoc hypotheses of homoplasy" objectionable whereas ad hoc hypotheses of synapomorphy are not. Farris (1983) explains the difference in that a genealogy explains resemblance between organisms in terms of common ancestry, but cannot explain homoplasy in those terms. Some other evolutionary hypothesis, or possibly a whole series of them, is necessary to explain homoplasy, and such explanations lie outside the scope of systematic methodology. Thus hypotheses of homoplasy are to be rejected on grounds of method rather than on grounds of probability. In phylogenetics synapomorphy requires one speciation event as an explanation. Homoplasy requires two speciation events to produce the homoplastic characters, *plus* some explanation, adaptive or otherwise, for the convergence or parallelism that results in different species acquiring a similar character.

Farris's defence of parsimony is a methodological one, but other approaches to the justification of parsimony are possible, either to criticise it or to support it. These include consistency, likelihood, and congruence. Before discussing them, however, I want to introduce phenetic phylogeny reconstruction, numerical cladistics, and molecular taxonomy, all of which bear on the problems involved. I therefore postpone further discussion of phylogenetic parsimony.

Transformed cladistics cuts the Gordian knot of phylogenetic parsimony with a vengeance. If the only axiom of cladistics is that the natural order of organisms is a divergent inclusive hierarchy and that hierarchy is recognised by taxic homology (i.e., synapomorphy), then synapomorphies are the only evidence by which the *a priori* hierarchy is made manifest, and homoplasies are simple mistakes – pseudo-synapomorphies misidentified as homologies. Provided that the hierarchy is axiomatic and symplesiomorphy has been recognised and rejected by out-group comparison and/or ontogeny, then parsimony is the only way of distinguishing homologies from "mistakes". If it were proposed that "mistakes" outnumbered homologies, the inference would be that the axiom of hierarchy was under attack. If that attack were successful, then the whole of transformed cladistics would collapse.

9

Classification and the reconstruction of phylogeny

As we saw in Chapter 7, phenetic clustering is based on the use of distance methods, however the original data are coded. Thus, as I pointed out at the beginning of Chapter 8, pheneticists have to use separate techniques for the construction of classifications and the reconstruction of phylogeny. The methods of "numerical cladistics" arose from within the school of phenetics as a result of attempts to reconstruct phylogeny, but its practitioners diverged from their pheneticist colleagues and converged on mainstream cladistics as a result of their conviction that *classification* should be phylogenetic. In this chapter, I describe the development of numerical cladistics and its eventual repudiation of the use of undifferentiated distance data in favour of reliance on apomorphy and parsimony.

While this was happening, from the 1960s onwards, various techniques of phylogenetic reconstruction involving the use of biochemical and molecular data were developing. These interacted with other schools of phylogeny reconstruction, and posed problems, notably those concerned with the use of distance data, in an acute form. In this chapter, I review some of these techniques, particularly as they concern the relationship between phylogeny and classification. Then in Chapter 10, I shall attempt some general summary of that relationship. But a complete statement of my position will be possible only after we have considered theories of evolutionary mechanism and something of philosophy.

I. The reconstruction of phylogeny

In looking at the use of parsimony in the previous chapter, we saw that, assuming phylogeny, its use in out-group comparison was to reduce the number of evolutionary steps required to explain the pattern of the stem group. In cladistic analysis of the "in-group", again assuming phylogeny, we also saw that parsimony was used, in effect, to minimise the number of hypothetical evolutionary events. A shared synapomorphy has, by definition, arisen only once; a shared homoplasy, at least twice. Thus both these uses of parsimony are "minimum evolutionary methods". The preferred phylogeny is the one involving the minimum hypothetical evolutionary change.

An early explicit usage of minimum evolutionary methods was proposed by Edwards and Cavalli-Sforza (1963, 1964). They were attempting to reconstruct the pattern of relationships among populations of *Homo sapiens* by using the differences in frequency as distance measures. To do this they had to assume evolution of blood-group polymorphisms by genetic drift, thus assuming that rates were not biased in any direction by selection (see Chapter 12, Sections I and II).

Camin and Sokal (1965) introduced the use of parsimony for discrete characters, using as their criterion the minimum number of evolutionary steps in the reconstructed phylogeny, each step being a character-state change. They made the assumption that evolution is irreversible. "Thus a descendant character state cannot revert to an ancestral character state".

This first study by Camin and Sokal included computation of a cladogram (*sensu* phenetics) of seven "Caminalcules", a group of hypothetical animals invented by Camin, using seven characters. They were coded with 0 representing the ancestral state, with numbers diverging from zero representing successive evolutionary steps. The authors made their cladogram to yield the minimum number of evolutionary steps, which are indicated on the internodes of the cladogram. Multiple origins of states are permitted, but the minimum step method will attempt to minimise them. The method is further explained by Sneath and Sokal (1973, pp. 332 – 41). Later, however, Sokal (1983) published the "true" phylogeny of the Caminalcules with "fossil" forms at the nodes and along the internodes, thus producing a cladogram that was a spanning tree rather than the original Steiner tree (Chapter 2, Section III).

The Camin–Sokal method allows no reversals towards the ancestral state. Another type of parsimony for character-state data is Dollo parsimony introduced by Le Quesne (1974, 1977) and said to be based on Dollo's "law" that evolution is irreversible. In this case the suggestion is that loss of a complex derived character state (1 → 0) may have occurred several times, but that each derived state (0 → 1) has arisen only once. The preferred phylogeny is then that in which the number of state reversions is minimised. Another way of dealing with apparent incompatibilities is polymorphism parsimony (Farris 1978; Felsenstein 1979). Apparent reversions are avoided by postulating a "01" state for any given character with the implicit assumption that an ancestral population was polymorphic for 0 and 1; descendant populations can then be either 0 or 1 without reversal. The most parsimonious tree is that which minimises the polymorphic nodes.

Another usage of phylogenetic parsimony was that introduced by Fitch and Margoliash (1967); here the number of reconstructed mutational steps is minimised. We shall consider that below with other techniques utilising biochemistry and molecular biology, but before doing so we must look at the development of numerical cladistics.

This has been due to the work of Dr. J. S. Farris and his colleagues. Its original basis was what is now known as Wagner groundplan-divergence analysis, and was developed by Wagner (1961). It is explained by Wiley (1981, pp. 176–8) with a more recent critique in Churchill, Wiley, and Hauser (1984) and has been used in its original form principally by botanists. Character states are coded as either plesiomorph (0) or apomorph (1). Wagner (in Wiley 1981) recommends that a transformation series be coded with "intermediate apomorphies" as fractions. Thus four states might be 0, 0.25, 0.5, 1. The results are presented as what is in effect a cladogram plotted on a series of numbered concentric semi-circles (Fig. 9.1). The centre of the circle represents the common ancestor of all the terminal taxa. Branch lengths diverging from the ancestor are plotted from the centre and represent the number of synapomorphies shared by the node or terminal taxon at the end of the branch.

The taxa being classified (the pheneticists' OTUs) are all terminal. The result is thus a cladogram in which the branch lengths are a numerical measure of the number of synapomorphies. The parsimony criterion is used to minimise aggregate branch length in what is in effect a rooted Steiner tree.

The origin of numerical cladistics was the development of algorithms to implement Wagner analysis. It is represented by the work of Kluge and Farris (1969), using character data from the families of Anura (frogs and toads). Methods for computing of Wagner trees were presented by Farris (1970). The computation of Wagner trees was then associated specifically with phylogenetic or Hennigian cladistics by Farris, Kluge, and Eckhardt (1970).

In discussing the development of phenetics in Chapter 7 (Section I), I introduced a simple distance coefficient for measuring the difference between two OTUs when at least some of the characters are numerical measures. This was the *Mean Character Difference* originally proposed by Cain and Harrison (1958):

Fig. 9.1. The Wagner groundplan-divergence method of phylogenetic analysis. Concentric semi-circles represent numbers of apomorph characters. The single line between internal nodes (open circles) represents shared apomorph characters. If shared branches are maximised, aggregate branch length is minimised. (Modified after Wiley 1981.)

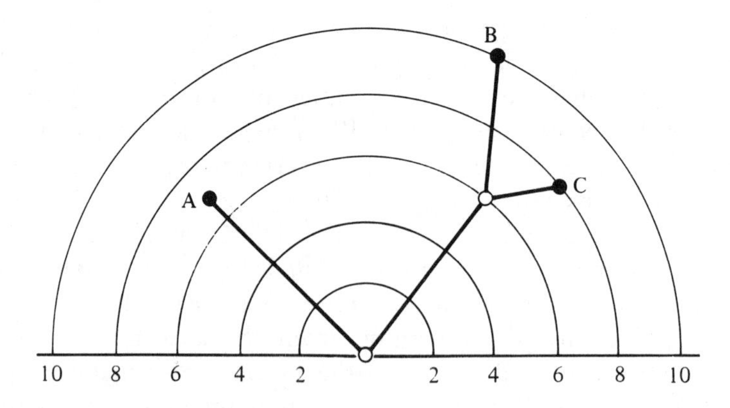

$$\text{M.C.D.} = \frac{1}{n}\sum_{i=1}^{n} |X_i\mathbf{j} - X_i\mathbf{k}|$$

Division by n is necessary if the distance between OTUs in character hyperspace is not to depend on the number of characters sampled. Furthermore, if the shortest distance in hyperspace is to be measured, the Euclidean distance is required, where R_i is the range of the ith character:

$$\frac{\sqrt{\sum_{i=1}^{n}\left(\frac{X_i\mathbf{j} - X_i\mathbf{k}}{R_i}\right)^2}}{n}$$

One then has a measure of the mean summed difference between all characters sampled, adjusted for the range of those characters, in character hyperspace.

Suppose, however, that one wanted to introduce parsimony into numerical taxonomy, going straight from polarised character difference to a dendrogram, as in the Wagner method. Then derived character states shared by two OTUs (synapomorphies) would be represented by a single branch between two nodes, whereas characters unique to each (autapomorphies) would be represented by separate branches; the more shared character states, the more parsimonious the dendrogram in terms of branch length. Thus any distance measure in an application of the Wagner method would take account of the number of characters, whether their derived states were shared (synapomorphy) or unshared (autapomorphy). Furthermore, Euclidean distances are no longer appropriate. One is not computing distances in hyperspace, but the sum of individual character differences along a branch. A very simple distance measure is then required, thus:

$$d(\mathbf{jk}) = \sum |X_i\mathbf{j} - X_i\mathbf{k}|$$

This is the *Manhattan Distance,* proposed as a measure by Kluge and Farris (1969). It is so called because each individual character difference is in effect parallel to an axis of the multidimensional graph of OTUs in hyperspace; it is a "city-block metric". The distance is a metric because it has *inter alia* a property known as the "triangle inequality". This takes the form $d(A,B) \leq d(A,C) + d(B,C)$. In the case of cladistic numerical taxonomy, A, B, and C are any three nodes (including terminal taxa) on a Wagner tree. It will be noticed that Euclidean distance in taxonomy is also a metric. This can be seen for a two-dimensional graph where each axis represents a character, with states of that character measured along the axis. The Euclidean distance between two OTUs then has a triangular relationship to the Manhattan distance: $d_E\ (\mathbf{j,k}) \leq d_M$, where d_M for two characters is $|X_1\mathbf{j} - X_1\mathbf{k}| + |X_2\mathbf{j} - X_2\mathbf{k}|$. Distances for characters X_1 and X_2 are at right angles to each other so that the Euclidean distance forms the hypotenuse of a right-angle triangle and the Manhattan distance is the sum of the other two sides. Originally the Manhattan distance was supposed to be usable for continuous characters, but subsequently all character coding was binary, so adjustments for range were unnecessary.

Clustering by the Wagner method results in either *rooted* or *unrooted* trees (the latter known as "Wagner networks"; Farris 1970). In the first case, characters are polarised by out-group comparison or otherwise, and computation begins from a *Hypothetical Taxonomic Unit* (HTU) having the ancestral condition to all the OTUs to be classified, coded as 0 for all characters. For binary coding, a morphocline of character states is coded as a series of binary characters. Multistate coding can be non-additive or additive, analogous to the two modes of phenetic coding illustrated in Chapter 7, Section I. For non-additive coding, the series of states is replaced by a series of variables, each of which is treated as though it were a separate character; the matrix then shows a series of variables each with a number of states corresponding to the number of variables. This is in essence similar to the coding shown in phenetics for qualitative multistate characters (red, yellow, blue). It can therefore be used for either polarised or non-polarised characters.

Additive coding presumes a morphocline, or more correctly a character-state tree, as the "morphocline" may not be a simple "*scala naturae*". The example given by Farris *et al.* (1970) has an ancestral and six successive derived states (Fig. 9.2) coded as:

state	ν_0	ν_1	ν_2	ν_3	ν_4	ν_5	ν_6
x_0	1	0	0	0	0	0	0
x_1	1	1	0	0	0	0	0
x_2	1	0	1	0	0	0	0
x_3	1	0	1	1	0	0	0
x_4	1	0	1	0	1	0	0
x_5	1	0	1	0	1	1	0
x_6	1	0	1	0	1	0	1

Thus an OTU with state x_6 scores 1 for the variables ν_0, ν_2, ν_4, and ν_6, because in addition to x_6 its phylogeny is presumed to have passed through the states x_0, x_2, x_4 (but not x_1, x_3, x_5). Additive coding is analogous to the coding for "continuous quantitative multistate characters" in phenetics (slightly pubescent . . . highly pubescent), but in the case of cladistics the morphocline or state tree is taken to represent phylogeny (or for pattern cladists, a hierarchy!).

In the Wagner method of Farris *et al.*, character state change is taken to be unrestricted. Multiple origins of characters and reversal to a more primitive state are not excluded *a priori*.

Fig. 9.2. A hypothetical character state tree, representing phylogenetic transformations of the primitive state X_0. (After Farris, Kluge, and Eckardt 1970.)

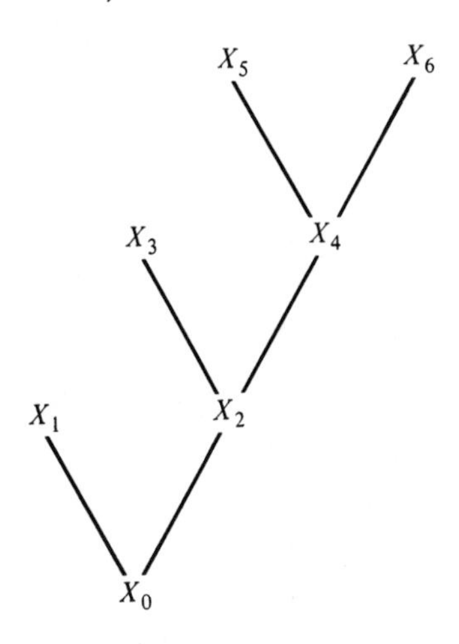

They are, however, minimised, so that the Wagner algorithms use parsimony in the same way as non-numerical cladistics. The basic algorithm for rooted trees is a quite simple one. In order to produce a cladogram that is a Steiner tree, Kluge and Farris (1969) introduced the concept of an "*interval*" representing the difference between any OTU and its immediate ancestor – that is, the nearest node on the cladogram to which that OTU is connected. The interval is then represented by the branch length between the OTU (e.g., B) and its ancestor (anc B). The algorithm is then as follows:

(1) Choose [or reconstruct by out-group comparison] an ancestor OTU.
(2) Find the OTU (B) that has the smallest difference [i.e., Manhattan distance $d(\text{B}, \text{anc B}) = \Sigma\,|X(\text{B}, i) - X(\text{anc B}, i)|$]. Connect it to the ancestor to form an interval.
(3) Find the unplaced OTU (A) that differs least from the ancestor.
(4) Find the interval from which the OTU identified in (3) differs least. The difference, d(A, int B), between OTU A and interval B is computed as follows:

$$d(\text{A}, \text{int B}) = \frac{d(\text{A},\text{B}) + d(\text{A}, \text{anc B}) - d(\text{B}, \text{anc B})}{2}$$

(5) Attach OTU A to the interval (int B) found in (4). To do this construct an intermediate, Y, and insert it into the tree. For each character, i, $X(\text{Y},i)$ is computed as the median of $X(\text{A}, i)$, $X(\text{B}, i)$, and $X(\text{anc B}, i)$.
(6) If any OTUs remain unplaced, go to (3). Otherwise stop.

Thus for the first cycle, one goes from:

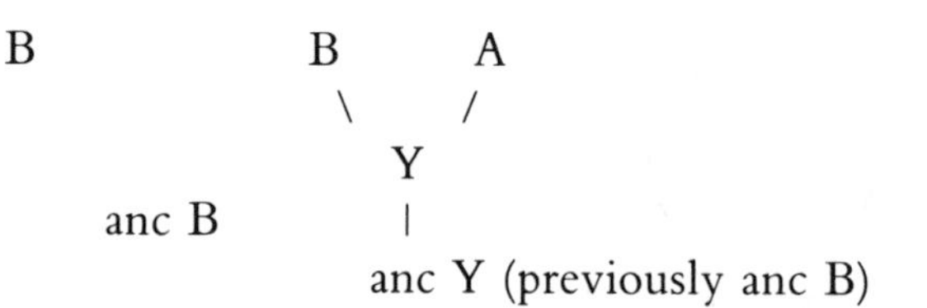

Wiley (1981, pp. 1182ff.) gives simple worked examples of the use of this basic algorithm, which forms the starting point of numerical cladistic computer programs.

The algorithm for a rootless tree ("Wagner network") allows one to proceed without finding or reconstructing an ancestral taxon, although characters must still be polarised, at least provisionally. The algorithm is a modification of that for a rooted tree, in that initially two OTUs are identified with the largest distance between them. As in the procedure for rooted trees, after the establishment of an initial interval, each OTU is added to the network so that its position is terminal, thus giving a Steiner tree. This involves the construction of an HTU at each stage. Thus one goes from

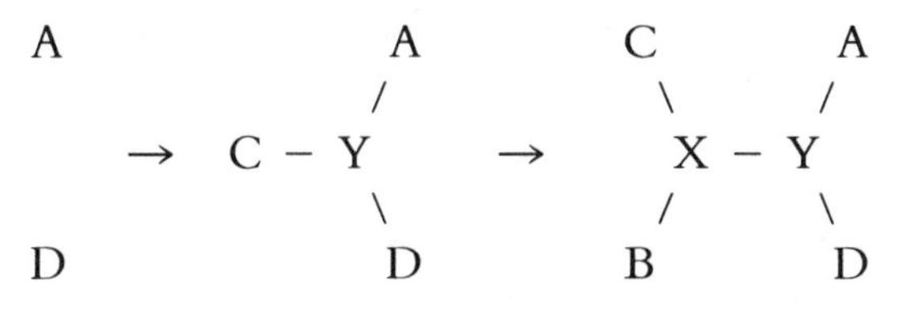

Wagner networks may be rooted by introducing a hypothetical ancestor or actual out-group as an OTU in the computation either at the time, or by doing so subsequently.

As I noted above, Farris, Kluge, and Eckardt (1970) made the explicit link between Wagner algorithms and Hennig's "phylogenetic systematics". This amounted to a statement that phylogeny reconstruction was the first stage in the construction of a classification and that the pattern of phylogeny, as reconstructed, should be isomorphic with the pattern of classification. Since then successful computer programs for numerical cladistics are usually based on the Wagner algorithm. Some of these are described and reviewed by Platnick (1987, 1989). Particular attention has been paid to cases where a number of trees are equally parsimonious.

Cladistic methods operate by minimising homoplasy. *Compatibility* methods deal with homoplasy by discarding the data represented by contradictory characters. The approach started with a simple test proposed by Le Quesne (1969), although Wilson (1965) is usually credited with having grasped the principle involved. If, in the terminal taxa being classified, two two-state characters are present in all possible combinations (i.e., 00, 01, 10, and 11), then convergent evolution of one of the characters must have occurred. If the polarity of the states is known (0 = primitive, 1 = derived), then the occurrence of 01, 10, and 11 is all that is required for the same conclusion. By use of these criteria, a pairwise comparison of all the characters across the terminal taxa is undertaken. The aim is to produce what is known as a "clique" of wholly compatible characters. In an ideal situation, therefore, characters showing the greatest state of incompatibility are eliminated successively from the data matrix until no homoplasy, represented by incompatibility, remains. Characters in the clique are then used for clustering. It is commonly the case, however, that the remnant clique contains too few characters to give a well-resolved tree, and potentially useful information is lost (Farris 1969; Hill 1975). One possible solution to this problem, as is often the case with numerical methods, is to construct a "proto-tree", in this case using the largest clique that can be found. If the taxa forming that tree are then divided into two or more groups according to the shape of the tree, and each group is treated separately, more characters compatible within each group may be found. "Sub-trees" based on a larger suite of characters may then be united to form a tree of all the taxa (Estabrook, Strauch, and Fiala 1977).

The development of compatibility analysis is reviewed by Felsenstein (1982) in a general review of numerical methods for reconstructing phylogeny, and by Meacham and Estabrook (1985). At this stage I want to make two further points about it. The first is that with which Meacham and Estabrook begin their review. They emphasise the idea that for a single character that appears in a number of states, a character state is a phylogenetic hypothesis. Each character state corresponds to one or more OTUs or terminal taxa (*Evolutionary Unit (EUs)* – to them!). Compatible state trees for other characters are corroborations of the original tree. Compatibility analysis aims to find the maximum number of compatible state trees for any group of taxa. The second point is that advocates of compatibility have concerned themselves with null hypotheses. Le Quesne (1972) calculated the probability of incompatibility of two two-state characters, both of which have the states distributed at random across a set of taxa. One such randomly distributed character will give no useful taxonomic information when tested for compatibility against any other character. If a multistate character, it will yield no coherent state tree. If all the characters of a group of taxa have their states distributed at random, there will be no detectable hierarchy for that group of taxa.

II. Molecular distance

We saw in Chapter 6 (Section IV) that Mayr (1969) claimed that the ideal criterion of naturalness in a classification was that of "genes-in-common". Mayr further claimed that traditional "evolutionary" classification, properly applied, would rate highly if judged by that criterion. At about the same time that Mayr was writing, a series of techniques was being developed that were claimed to have the capacity to investigate differences between the genomes of representatives of the taxa being classified, either directly or by way of the protein products coded by the genome. The claim, therefore, was that clustering was by "genes-in-common". I shall concentrate on four important methods among many. They form a two-by-two contingency table characterised by (1) whether the data are derived from protein or nucleic acid differences; and (2) whether the data are interpreted as distance measures between taxa, or as character state differences.

The best known techniques for measuring distances between proteins are based on immunology. If a protein, such as human serum albumin, is purified from blood and injected into another species, such as a laboratory rabbit, the "foreign" protein (the albumin) promotes the production of antibodies in the rabbit. Antiserum is then prepared from the rabbit blood. If this is mixed with albumin from another species related to man, the strength of the reaction is a measure of the closeness of resemblance between the human albumin and that of the other test species. A number of techniques can be used to measure the strength of the reaction, which is usually compared with that between the antiserum and albumin from the species (human) that yielded the original antigen. Techniques include quantitative precipitation (direct comparison of the degree of precipitation), immunodiffusion, turbidimetry, and immunoelectrophoresis. A particularly sensitive method is microcomplement fixation. "Complement", used in the calibration, is a complex series of blood serum proteins that act sequentially. It is prepared from mammalian blood and has two useful properties: (1) it locks on to the three-dimensional lattice of the antibody–antigen complex; and (2) free complement lyses specially prepared red blood cells, releasing haemoglobin. Thus in the complement-fixation test, the closer taxonomically the individuals from whom the antigens are derived, the more complement is taken up. Residual complement is then measured by determining the amount of haemoglobin released on the addition of prepared red cells. Thus the greater the amount of released haemoglobin (measured by spectrophotometry), the smaller the antigen–antibody reaction. An index of dissimilarity – the *Immunological Distance* (ID) – is given by the relative concentration of the test reaction (e.g., man vs. gorilla albumins) giving the same result as that for the homologous albumin (e.g., man vs. man). Thus the technique, like most other immunological techniques, gives a distance measure in pairwise comparisons between taxa.

An early apparent success in the use of microcomplement fixation was a study of the interrelationships of apes and man (Sarich and Wilson 1967). Not only did the authors produce a dendrogram representing the relationships of Old World monkeys (OWMs); represented by the mean ID of six species, the five extant ape genera, and man; they also calibrated the whole based on an estimate of 30 million years before the present (Myr B.P.). for the separation of OWMs and apes. They made the empirical claim that log ID is approximately proportional to the time of divergence (t) of any two species, so that where k is a constant, log ID = kt. If their claim is true (see below), then the palaeontological dating of a single node on the dendrogram would calibrate the whole.

Sarich and Wilson claimed that *Homo, Pan* (chimpanzee), and *Gorilla* had diverged as recently as 5 Myr ago. This was in direct contradiction to the then prevailing view that the fossil hominoid *Ramapithecus* (ca. 15–8 Myr B.P.) was a hominid – that is, more closely related to *Homo sapiens* than is any living ape species. Subsequently, it was shown from more complete fossil skull material that *Ramapithecus* was closely related if not co-generic with *Sivapithecus,* which in its turn was a member of the orang-utan clade (Andrews and Cronin 1982; Pilbeam 1982; but see Pilbeam *et al.* 1990). Thus there was no long "dating gap" between the time of divergence of hominids from African apes and the first appearance of fossil hominids (ca. 3.5–4.0 Myr), and the molecular evidence, corroborated by other molecular techniques, was represented as a triumph of molecular biology over palaeontology.

One of the corroborating techniques was that of DNA–DNA hybridization. This is represented as a closer approach to the "genes-in-common" criterion. The genetic code in the vast majority of organisms is embodied in DNA, present in most body cells as the famous "double helix", which forms the active part of each chromosome. The two strands in each case are united as complementary pairs by hydrogen bonds uniting the four possible bases of the DNA molecule. Thus links between strands are formed by the base adenine bonding to thymine, and cytosine to guanine. But the links are weak so that heating prepared fragments of DNA will separate the strands of the duplex without other damage. At a lower temperature, the strands will re-associate. If "heteroduplexes" – duplexes formed from single strands of DNA each from a different species – are formed in the laboratory, they will separate at a lower temperature than the natural homoduplexes because of non-bonding of bases along the DNA at many loci due to base mismatches. Thus the greater the number of mismatches, the lower the "melting" temperature. Again this gives a distance measure for the two species represented by the heteroduplex.

The use of DNA hybridization for the reconstruction of phylogeny was developed in the early 1960s, but came to prominence with the work of Sibley and Ahlquist in the 1980s in their attempt to reconstruct the phylogeny of birds. Summary accounts of their technique are contained in Sibley and Ahlquist (1986, 1987). Issue number 3 of volume 30 of the *Journal of Molecular Evolution* (1990) is a symposium on the subject. An important detail is that their distance measures are based only on "single copy" DNA, not on the repeated sequences present in all genomes. The DNA is extracted from cell nuclei (the nucleate red blood cells in the case of birds) and sheared into ca. 500 base lengths by ultrasound. Boiling followed by cooling causes multicopy sequences to reassociate first. These are filtered out by passing through a hydroxyapatite column which traps the reassociated duplexes. A small amount of the single-strand DNA is labelled with radioactive iodine. This "tracer" is then tested against a thousand-fold quantity of unlabelled single strand (the "driver") of (a) the same species, and (b) another species. The disparity between tracer and driver enhances the probability that any tracer strand will unite with a strand of driver. The duplexes (a) and (b) are then subjected to step heating in increments of 2.5°C from 55° to 95°C. Dissociated tracer at each temperature step is separated by holding unmelted duplexes in a hydroxyapatite column. The extracted single-strand tracer can then be assayed by its radioactivity. As a result, melting curves for homoduplexes and heteroduplexes against temperature can be constructed, and a "delta" value, analogous to depression of the freezing point, measured in each case.

Sibley and Ahlquist's ambitious project to reconstruct the phylogeny of all birds has attracted a large literature of criticism. There are concerns about the experimental error inherent in the technique overriding the very short distances between nodes in the authors' dendrograms, particularly when a whole dendrogram is based on only one tracer species (Templeton 1985; Cracraft 1987a). There are also concerns about the statistic used as a measure of melting point (Sarich, Schmid, and Marks 1989). Springer and Krajewski (1989) give an even-handed discussion of all the problems involved, but overriding all these critiques of the Sibley and Ahlquist techniques are the problems of using distance measures in phylogeny reconstruction. One major problem is the use of the concept of the "molecular clock", the assumption inherent in phenetic uses of distance measures in phylogeny reconstruction (as distinct from classification) that mutations in the DNA bases occur at a constant rate. This assumption becomes particularly important when distance data are used to construct a dendrogram and then the whole is put on a time base calibrated by a single geological or palaeontological datum represented by a particular node. This, of course, applies as much to protein distance data as to those of DNA.

For DNA hybridization Springer and Krajewski present the issues in the following way:

> (1) the so-called "uniform average rate" of molecular evolution once espoused by Sibley and Ahlquist for avian lineages; (2) the validity of phenogram-construction algorithms (i.e. those which construct groups of maximally *similar* taxa) for converting DNA distances into phylogenetic trees; and (3) the relative merits of "distance methods" (i.e., those methods which employ inter-taxon dissimilarity measures rather than character distributions, as raw data).

Springer and Krajewski then go on to say:

> The first is no longer an issue insofar as its strict validity in birds is concerned, since results from Sibley and Ahlquist's own laboratory have negated the claim (Catzeflis, Sheldon, Ahlquist and Sibley 1987; Sheldon, F.H. 1987). The second goes the

> way of the first: if rate variation is real, phenogram-building methods [i.e., phenetic clustering; see Chapter 7, Section II] are of manifestly limited utility in phylogenetic inference. This too has been recognised by Sibley and Ahlquist (Ahlquist *et al.*, 1987).

In response to Cracraft (1987a), Sibley and Ahlquist retreated somewhat to the claim that the *Uniform Average Rate (UAR)* was uniform but that the related "average genomic rate" (of nucleotide substitution) required correction for those birds who bred for the first time more than a year after hatching. For mammals the correction required brought the rate nearer to a measure per generation than a measure per year. The concept of the molecular clock is closely related to the neutral theory of molecular evolution (Chapter 12, Sections I and II; and Kimura 1983). Selection results in the stabilisation of pre-existing advantageous mutations, and the elimination or fixation of new ones according to the adaptive significance of their results. If there is no significant adaptive change as a result of nucleotide substitution, then DNA mutations should occur at a constant rate, hence the "clock", but there is still no agreement about its validity. The first two numbers of volume 26 of the *Journal of Molecular Evolution* (1987) are devoted to it, and a variety of opinions on the subject are represented in Patterson (1987).

The distance measures derived from immunology and DNA hybridization have the advantage that the number of comparisons made simultaneously, of corresponding amino acid sites in proteins from two species, or corresponding gene loci, is enormous. *In that respect,* with two provisos, it can be claimed that distance measures are well founded. The first proviso concerns the difference between *orthology* and *paralogy* (Fitch 1970). Paralogy is the result of gene duplication in phylogeny (Ohno 1970) followed by evolutionary divergence, in the case of active genes, to give different proteins. A well-known case is that of the globins, the haemoglobins and myoglobin, which have different molecular structures, but are coded by paralogous genes often in the same individual. A paralogous pair of genes (i.e., DNA sequences) in one species, corresponding to a single gene in another, will disrupt the formation of heteroduplexes between the two species. The other proviso is that no *xenology* (Gray and Fitch 1983) is involved. This is the transfer of fragments of DNA from distantly related species, which is known to occur (Syvanen 1987) but is unlikely to have a significant effect on DNA hybridization (Springer and Krajewski 1989).

There are, however, yet more cogent reasons for doubting the claim that even for closely related species, DNA hybridization data can be used to give accurate and reliable reconstructions of phylogeny. If the species are very close, experimental error may override measurement of distance (Cracraft 1987a), but for greater distances the inherent method must give results of diminishing accuracy. The theoretical upper limit is reached when there has been at least one mutation at every gene locus in one strand or the other; any remaining matches must then be due to homoplasy. In fact, other factors distort estimates of total mutation long before this limit is reached. An unknown number of loci are invariable. Also with only four possible bases at any site, given an equal probability, a 25 per cent match overall would represent a null hypothesis. Furthermore, when two species are being compared, less than 100 per cent of tracer will form heteroduplexes if there is more than ca. 20 per cent dissimilarity between strands (Springer and Krajewski 1989). Schmid and Marks (1990) discuss other limits to the technique.

Similar methodological objections apply to the use of immunological distance (Friday 1980). At the test level, there may not be a linear relationship between the number of amino-acid site matches in a pair of homologous proteins and the degree of antibody–antigen reactions. Also the linear relationship between log ID and time seems no longer to be valid (Read 1975; Corruccini *et al.* 1980; Andrews 1985), nor is there a straight-line relationship between log ID and DNA–DNA distances, so both cannot have a linear relationship with time (Corruccini *et al.* 1980; Ruvolo and Pilbeam 1985).

Because of all the variables involved, there has been considerable controversy regarding not only the clustering methods to be used for distance data but also the criteria by which the results should be judged. If the distances measured by DNA hybridisation or immunology reflected total evolutionary change (T) accurately, then distances in the pairwise matrix between species would be additive and the "triangle inequality" (see Section I, above) would apply. However some of the differences recorded (but unknown) in the data matrix would, at any rank, be plesiomorphous, and others apomorphous. Thus alone they would lead only to an unrooted tree. Inclusion of an unambiguous out-group would solve the problem. But when the recorded distances are not known accurately to reflect *T*, because of homoplasy, resulting from back mutation and parallel mutation, and "multiple hits" (more than one mutation at any one site), then actual nonidentity (*D*) will not be the same as *T*. Furthermore *d*, the measured distance, may not be an accurate measure of *D*.

Given all these uncertainties, a number of approaches to tree evaluation are possible. The three most popular are that of parsimony, with additive branch lengths, a statistical best-fit approach, and that of maximum likelihood. There has been vigorous debate about their merits, particularly the first and the last, notably by Farris (1972, 1981, 1985, 1986) and Felsenstein (1978, 1984, 1986, 1988a), extending into the question of the validity of any distance measures for phylogeny reconstruction.

Tree-building methods are reviewed by Felsenstein (1982) and for distance data by Springer and Krajewski (1989). If there is homoplasy, then the expectation is that $D < T$; and thus, if the tree is to represent phylogeny, its branch lengths between any two species (d') should be more than d. Also if branch lengths are to be additive, the triangle inequality must apply and negative branch lengths, arising out of calculations, cannot be allowed. Within these constraints a parsimony algorithm can be devised and tested by comparing the distances on the tree (d') with those on the data matrix (d) and by manipulating the former for the best match to the latter. In order to do this, a least squares criterion, summing all values of $(d - d')^2/d$, is used. For a perfect fit the sum is zero.

A second approach treats the best-fit tree as a statistical problem, so that additivity and the rejection of negative branch lengths are not mandatory. The branch lengths and tree topology are simply the best estimates of a real but unknown phylogenetic tree. A third approach is to propose some sort of hypothesis about the processes that generated the "true" tree and to test and improve the reconstructed tree against this. We shall consider this in the following section, in looking at protein sequence data. We have seen that all sorts of difficulty bedevil the use of distance measures in phylogeny construction even without the assumption of a molecular clock. It must be emphasised, however, that all the controversy refers to the reconstruction of phylogeny. If evolution is not assumed, distance measures, unlike character data, do not imply a Natural Hierarchy in the way that is taken *a priori* in "transformed cladistics" (Chapter 10).

III. Sequence data

Returning to molecular biology, the expectation would be that sequence data, whether for proteins or DNA, could be treated cladistically in the same manner as morphological characters. There is some irony, therefore, that an influential pioneering treatment of protein sequence data turned them into distance measures for phenetic tree construction. This was the work of Fitch and Margoliash (1967) on sequence data from the protein cytochrome *c*. Cytochrome *c* is active in cell respiration and is found in the mitochondria (see below) of both animals and plants and is found also in fungi. In all proteins the molecule is formed of one or more chains (polypeptides) of twenty possible amino acids. In the case of cytochrome *c*, the maximum length of the single chain is 112 amino acid residues (in wheat). Sequencing is accomplished by breaking the polypeptide with enzymes that split the molecule between known amino acids. This is done to produce fragments from one species in a number of operations so that the fragments overlap. They are then sequenced, an operation now done in a "protein sequencer". When this has been done for all the species concerned, all the reconstructed polypeptides can be aligned by a number of positions where the amino acids are invariant.

Fitch and Margoliash used cytochrome *c* data from a wide variety of sources to produce a dendrogram for 20 organisms of which 15 were vertebrates, but which also included 2 insects and

Fig. 9.3. Phylogenetic relationships of 3 fungi, 2 insects, and 15 vertebrates based on sequencing of cytochrome *c*. Branch lengths (numbered) represent "best-fit" estimated mutation distances. Position of each node ("apex") on the vertical axis represents the average distance of all the taxa descending from that node. (After Fitch and Margoliash, "Construction of Phylogenetic Trees," *Science*, 20 January 1967, vol. 155, pages 279–84, figure 2. Copyright 1967 by the AAAS.)

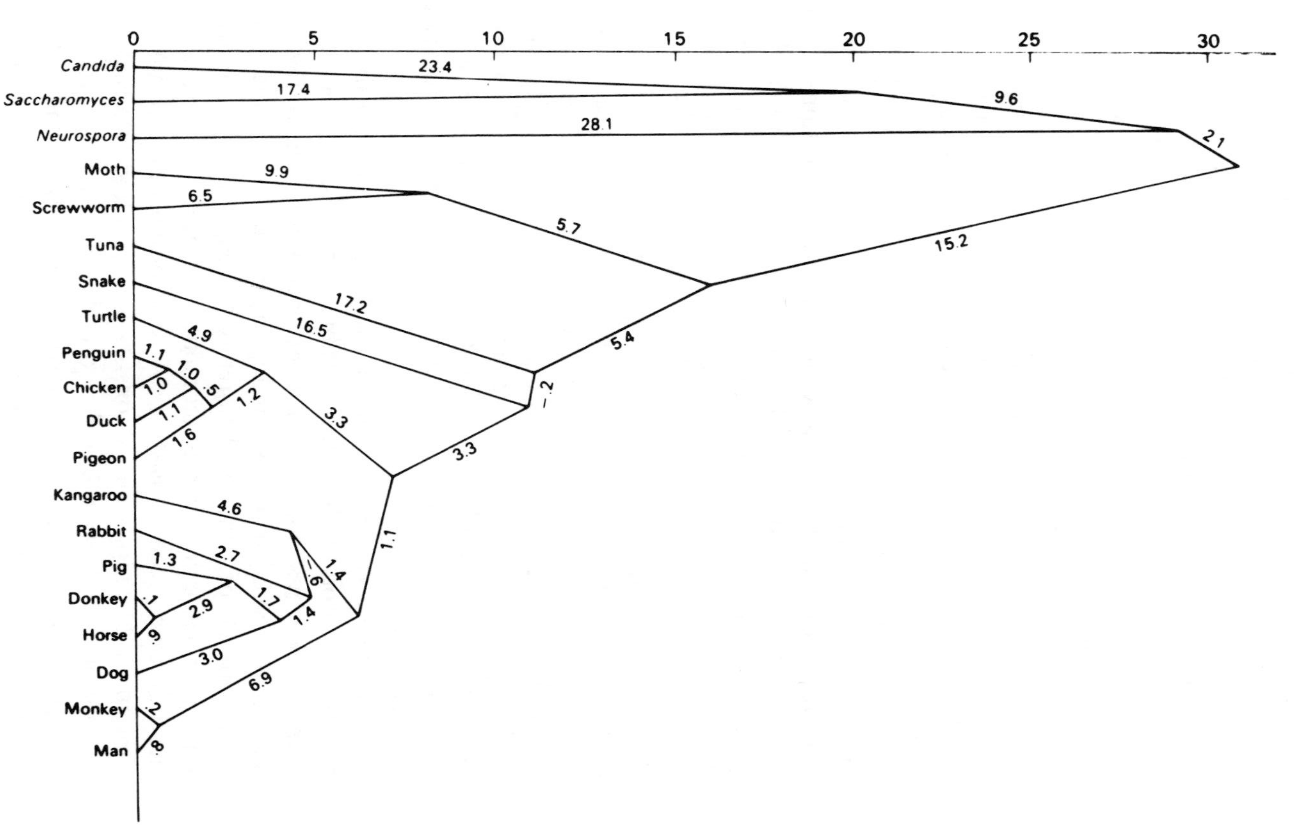
Average minimal mutation distance
0
5
10
15
20
25
30
Candida
Saccharomyces
Neurospora
Moth
Screwworm
Tuna
Snake
Turtle
Penguin
Chicken
Duck
Pigeon
Kangaroo
Rabbit
Pig
Donkey
Horse
Dog
Monkey
Man
23.4
17.4
9.6
28.1
2.1
9.9
6.5
5.7
15.2
17.2
16.5
5.4
4.9
1.1
1.0
1.0
.5
1.2
1.1
1.6
3.3
3.3
-.2
4.6
2.7
1.3
1.4
-.6
1.7
1.4
.1
2.9
.9
3.0
6.9
1.1
.2
.8

3 fungi (Fig. 9.3). Their technique was, first, to produce a pairwise matrix giving the "mutation values" for every one of the twenty amino acids against every other. These values were derived from the minimum number of mutations required in DNA nucleotides, coding for the amino acids, to change the coding from that for one amino acid to that for another. That matrix could then be used to produce a second one, with the minimum number of mutations separating the cytochrome *c* molecules of any pair of OTU organisms. This latter matrix was then used for tree construction. The constructed tree was produced by phenetic clustering and judged by comparing branch lengths between taxa on the tree with mutation distances on the pairwise matrix. To do this, Fitch and Margoliash generated forty trees and chose the one giving the best fit on the least squares criterion according to the percentage standard deviations formula. Their approach was therefore the statistical one.

At about the same time, however, cladistic algorithms for sequence data were introduced by Eck and Dayhoff (1966), Fitch (1971), and Moore, Barnabas, and Goodman (1973). The last authors provide a rigorous mathematical treatment, while Dayhoff (1969) gives a readable account of her algorithm. As in Fitch and Margoliash's original work, the intention was to measure branch lengths in terms of gene mutations rather than counting each amino acid difference as a unit of change. Dayhoff shows how the whole tree could then be calibrated by a geological event (an estimate – 400 Myr B.P. – for the divergence of the "bony fish" and mammal lines) on the assumption of a constant mutation rate. But very soon after that, the constancy of mutation rate began seriously to be questioned.

In their pioneering work, Fitch and Margoliash (1967) also used their tree method to show the relationships among globin molecules. The various globin chains concerned with oxygen transport in animals are coded for by paralogous genes. In humans, myoglobin consists of a single, elaborately folded chain, with an iron-containing heme group at its geometrical centre. There are several haemoglobins, each consisting of four polypeptide chains and heme units. Haemoglobin A consists of two α and two β chains; haemoglobin A_2 of two α and two δ chains, whereas foetal haemoglobin has two α and two γ chains. The α, β, γ, and δ chains are coded for by different genes that are paralogues of one another and of that coding for the myoglobin chain. Plants and invertebrates have a single globin. Fitch and Margoliash constructed a globin "phylogeny", using human α, β, γ, and δ haemoglobin and whale myoglobin.

Later, however, Langley and Fitch (1974) tested constancy of rates of nucleotide (DNA) substitution in α and β haemoglobin, cytochrome *c*, and fibrinopeptide protein sequences. The molecular clock was the null hypothesis, expressed in the form of an assumption that sequence differences are the result of stochastic processes of a Poisson nature (see below). The null hypothesis was rejected for all four proteins.

Goodman, Moore, and Matsuda (1975) also did a study of globin molecules with the intention of testing the molecular clock, using palaeontological data. The whole large cladogram represented both the phylogeny of the included species and the pattern descent of the three paralogous globin molecules from their "ancestral" molecule (Fig. 9.4). Examples of their datings include the branch length in the β haemoglobin lineage extending from the origin of tetrapods to the extant frog *Rana esculenta* (340 Myr to 0 yr) and the tetrapods' ancestor to the amniote ancestor (340 to 300 Myr). These geological estimates were then used with the estimated number of nucleotide substitutions along the same branches to give a substitution rate. The rate varied radically over different parts of the tree, but there was a clear pattern to this variation, with acceleration in rate for vertebrate globins during the period (500 to 400 Myr) while the different globin chains separated from one another. Once differentiated, the rate in both α and β chains slowed dramatically.

They saw the accelerated rate during separation as optimisation of the newly evolved chains by selection; the acceleration was then followed by a slowing of the rate as the selected configurations were conserved. Since then Dr. Morris Goodman and his colleagues have extended their investigation of protein sequences and have investigated rate changes in more detail (Goodman, Miyamoto, and Czelusniak 1987, and references therein). These rate changes were applied to other proteins such as cytochrome *c* and lens α crystallin A.

There were, however, other problems, analogous to those of distinguishing between homology and homoplasy in comparative anatomy. Apparent paralogy can be due, not only to gene dupli-

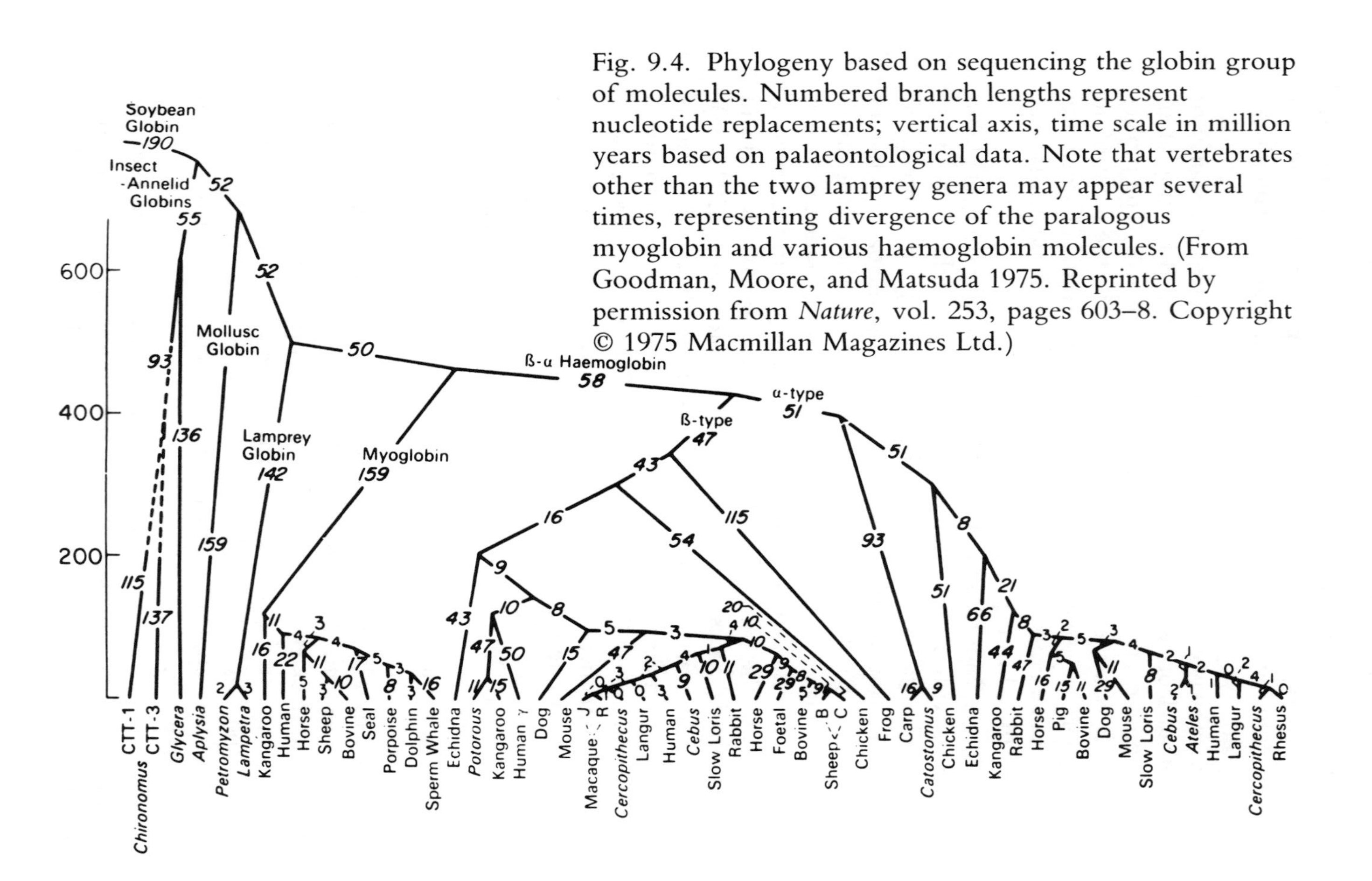

Fig. 9.4. Phylogeny based on sequencing the globin group of molecules. Numbered branch lengths represent nucleotide replacements; vertical axis, time scale in million years based on palaeontological data. Note that vertebrates other than the two lamprey genera may appear several times, representing divergence of the paralogous myoglobin and various haemoglobin molecules. (From Goodman, Moore, and Matsuda 1975. Reprinted by permission from *Nature*, vol. 253, pages 603–8. Copyright © 1975 Macmillan Magazines Ltd.)

cation and divergence of the proteins coded by the "daughter" genes, but also to gene conversion. Conversely, it appears that paralogous "sister genes" can converge to code for the same protein. Human α haemoglobin is coded for at two different gene loci. To minimise the misleading effects of paralogy, Goodman *et al.* have had recourse to what is in effect Patterson's criterion of congruence (Chapter 4, Section II). To do this they have constructed an enormous notional protein molecule by arranging a whole series of protein sequences in tandem for each OTU and treating them as a single molecule. The complete sequence (not, however, sampled in every OTU) was myoglobin, α and β haemoglobin, lens α crystallin A, cytochrome *c,* fibrinopeptides A and B, and ribonuclease.

Goodman *et al.* (1987) present a revised cladogram showing the phylogeny of globin molecules and also a cladogram for their amalgamated sequence. A striking detail in both cases concerns the position of the lungfish (Dipnoi). On both the α and β haemoglobin chain in the former, teleost fish cluster with tetrapods, with the South American lungfish *Lepidosiren* and the shark *Heterodontus* on the stem below the teleost–tetrapod cluster. The situation is similar for the amalgamated sequence. *Lepidosiren* and sharks appear as stem-group gnathostomes below a dichotomy into teleosts and tetrapods. Goodman *et al.* also point out that for the sequencing of parvalbumins the coelacanth *Latimeria chalumnae* occupies a similar position, as sister-group to teleosts plus tetrapods. These results are of particular interest in the light of our discussion (Chapter 7, Section IV) on the relationship of the Dipnoi. We shall return to them below.

Goodman *et al.* use parsimony as their criterion for the "best" cladogram, but globin molecules have also featured largely in the likelihood approach to phylogeny reconstruction. Felsenstein (1981) developed a method of estimating phylogenetic trees using maximum likelihood and applied it to nucleic acid sequence data. Bishop and Friday (1985) built on his approach to develop a likelihood method for nucleic acid or protein sequences. In order to apply their method to protein sequences, it is necessary, as before, to express each amino acid of the sequence as the triplet of nucleic acid bases (codon) that is taken to code for it. Because the genetic code is redundant, there is not a one-to-one relationship between amino acid and nucleic acid, in this case messenger RNA (mRNA). Notably the third base of each triplet is not specified by the amino acid, so further assumptions about probability have to be made, which are set out as a table by Bishop and Friday.

Given these complications, it is also assumed that mutation of any given nucleic acid base is a rare event and that its chance of mutation is dependent only on its present state, not its history (the Markov property). If, furthermore, mutation is "random", differences between sequences over time arrive according to a Poisson process.

Zuckerhandl and Pauling (1965) gave the probability that the *amino acid* at any given site would be unchanged over time t, as

$$P_{xx}(t) = e^{-ut}$$

where $P_{xx}(t)$ is the probability that state X will be present at the beginning and end of time t; e is the natural logarithmic constant ($e = 2.71828\ldots$), and u the rate of change at the site. The probability of *change* at the site is then

$$P_{xy}(t) = 1 - e^{-ut}$$

With four possible *nucleic acid bases* at any given site the assumption can be made that any possible change of a base is equally probable at the site (but see below). The probability of change from a *particular* base X to a particular base Y is then

$$P_{xy}(t) = \frac{1 - e^{-ut}}{4}$$

for four possible bases. The probability of finding the same base after time t is

$$P_{xx}(t) = \frac{1 + 3e^{-ut}}{4}$$

(with a different value of the rate constant u from Zuckerhandl and Pauling's original formulation).

Given this model of nucleotide change, the criterion by which competing trees are judged is that of likelihood. It is assumed that change at each site is independent of that of all others. If two orthologous DNA or RNA sequences are then compared, given the above probabilities, the associated likelihood is

$$L = [\ (1 + 3e^{-ut})/4]^{n-d}) \cdot [1 - e^{-ut})/4]^{d}$$

where n is the total number of sites and d the number of sites at which the bases differ (Bishop and Friday 1985). From this it is possible to calculate an estimate of relative divergence time between the two sequences, u being a constant:

$$\hat{ut} = -\frac{1}{2} \log_e \left(1 - \frac{4d}{3n}\right)$$

Thus the essence of the likelihood approach is to propose a hypothesis of evolutionary change (a Poisson process). That hypothesis is a probabilistic one in which change is a function of time at any given site. The differences between any two sequences of protein or nucleic acid, representing two terminal taxa, is also a function of time, so that distances between taxa on the tree are time-dependent, given the model of change. The maximum likelihood tree is then the one that maximises the probabilities of all the changes that can be inferred from it. Ideally one should find the tree that has an overall maximum likelihood, but, for more than a very small number of taxa, the computing time is unrealistic if not impracticable. Bishop and Friday (1987) describe three ways of getting around the problem.

The first "fairly coarse" method is to do a pairwise comparison over all taxa to give distance data using the *ut* estimator. Clustering is then by the phenetic unweighted pair-group method (see Chapter 7, Section II). The other two methods involve the more rigorous joint estimation of the tree, where a preliminary tree is refined by an iterative computer program until maximum likelihood is reached. The "big bang" pattern (Thompson 1975), with all the taxa connected by a single polytomy, then provides a baseline for judging the significance of a tree. Bishop and Friday follow the usage of the natural logarithm ($\log_e$) of the likelihood for a tree as the "support value". If the support value for an estimated tree is not appreciably greater than that for the corresponding big bang pattern, then there is no significant hierarchical structure in the data.

Bishop and Friday (1987) reported the results for two sets of data, each for six amniotes. The first set were myoglobin sequences for two mammals (man and opossum), one bird (chicken), and three reptiles (alligator, turtle, lizard). The best-supported trees clustered chicken and the mammals, and, of those, the two having the highest support level had an initial dichotomy between mammals plus chicken on one hand and the three reptiles on the other, suggesting support for Gardiner's (1982) views on amniote relationships (see Chapter 7, Section IV). The second data set consisted of α haemoglobin sequences for two mammals (human and kangaroo), two birds (chicken and goose), and two crocodilian reptiles (crocodile and caiman). This time three trees were virtually indistinguishable in their support level. The first clustered the mammals and birds as sister-groups (*pro* Gardiner), the second clustered crocodilians and birds (*anti* Gardiner), and the third had the three pairs as a trichotomy. Bishop and Friday (1988) reported the results of clustering more vertebrate protein sequences using likelihood. It was still the case, however, that the birds (chicken and goose) clustered with mammals (human and kangaroo) for both α and β haemoglobin. Bishop and Friday suggest that there may be a functional reason for this in warm-blooded vertebrates.

In using protein sequences to reconstruct phylogeny, an attempt is made to approximate the genes-in-common criterion. It might be thought, therefore, that direct sequencing of DNA would yield more valid results. This should particularly be the case for resolving relationships at low rank, such as investigation of phylogeny within the hominoids, because there is so little difference between orthologous protein molecules. Because of this, comparison of those protein molecules then involves looking at amino acid differences at individual sites rather than any statistical estimation of trees (Andrews 1987).

But nucleic acid sequencing, or rather the estimation of phylogenetic trees from the sequencing data, has its own problems. Most work in this respect has been concentrated on either mitochondrial or nuclear DNA. It could be said that change in the former is too fast and in the latter too slow for reliable tree construction. Mitochondrial DNA is more easily purified for sequencing than is nuclear DNA; it also lacks some of the complicating factors present in the latter, such as long repetitive sequences and the presence of introns – non-coding DNA sequences apparently inserted into the coding parts (exons) of the DNA chain.

But mitochondrial (mt) DNA has odd features of its own (Wilson *et al.* 1985). It contains the same four active bases as nuclear DNA, that is, the two purines adenine (A) and guanine (G), and the two pyrimidines thymine (T) and cytosine (C). But it appears to lack the repair mechanism present in nuclear DNA, which corrects errors in replication at cell division and also repairs DNA damage. Thus in mtDNA, mutation from one base to another (i.e., point mutation) is less constrained by function. As a result, *transitions,* from one purine to the other or from one pyrimidine to the other, are much commoner than *transversions,* from purine to pyrimidine or vice versa, for purely biochemical reasons. Next, in all organisms tested, mtDNA seems to descend only in the female line. Thus ancestor-descendant patterns of mtDNA inheritance represent clones: there is no recombination and, in a sense, if the mitochondria are considered as the separate organisms they probably once were, no speciation. Related to all this, mtDNA appears to be haploid, occurring as a single (usually circular) strand rather than a duplex.

Apart from sequencing, another DNA technique can be employed for the comparison of mtDNA from different organisms: restriction enzymes (references in Wilson *et al.* 1985) recognise a particular sequence of bases, usually four or six, along the DNA molecule and cut the molecule through that sequence (for a popular account, see Weinberg 1985).

For the actual sequencing of mtDNA, account has to be taken of the different substitution rates for bases that have different significance (Hasegawa, Kishino, and Yano 1985). In the triplet code for a specific amino acid, the third base of a codon is not usually specified by the amino acid. If, however, transfer (t) RNA is being coded, all three codon positions are specified in every case. Furthermore, the rate for transitions is greater than that for transversions at all these sites and at silent (non-coding) sites. With completely sequenced mtDNA fragments, the different rates can be taken into account in tree construction, using high-rate mutations for resolving low-rank taxa, or even looking at intraspecific variation, whereas slow-rate mutations can be used in dealing with high ranks or longer periods of reconstructed time. Another feature of mtDNA is important in the study of the evolution of populations at or below the species level. This has been used in the study of human evolution. If a population goes through a "bottleneck" in its history (i.e., a period of very low numbers), the variability of mtDNA is reduced to a much greater degree (one genome per breeding pair) than that of nuclear DNA (four genomes) (Wilson *et al.* 1985). With subsequent expansion and division into daughter populations, variation within populations will be high compared to that between populations. In humans this is the case between non-African populations, which, however, differ considerably from sub-Saharan African ones, where there is more variation between populations. These facts are taken, together with fossil data, to support an African origin for *Homo sapiens sapiens* with rapid dispersal world-wide rather than long periods of *in situ* evolution over the world (Stringer and Andrews 1988).

Both mtDNA and nuclear DNA sequences have been used extensively to unravel human phylogeny. It is generally, but not universally, agreed that the African apes and humans together form a natural group with the orang-utan as out-group (Andrews 1987). This grouping was supported by the immunology of Sarich and Wilson (1967) and by DNA hybridization (Sibley and Ahlquist 1984; Caccone and Powell 1989) as well as protein electrophoresis (Bruce and Ayala 1979). Resolving the relationships among gorilla, chimpanzee (two species), and humans is more difficult. Anatomical characters, notably knuckle-walking, which is unique to the African apes, suggests that the latter form a natural group with *Homo sapiens* as the out-group. Results from immunology and electrophoresis are equivocal, but DNA hybridisation has favoured a human–chimp clade. This conclusion is supported also by amino acid sequencing data (Goodman *et al.* 1983) and by mtDNA sequences using transversions only (analysed by Hasegawa *et al.* 1985).

The problem has been attacked with nuclear DNA sequencing; again it is the globin molecules, but this time via their coding DNAs, that have been used. The target has been the $\psi\eta$ gene – part of a sequence of β-related globin genes, but apparently inactive. The gene itself was sequenced by Koop *et al.* (1986), and its flanking regions were sequenced by Miyamoto, Slightom, and Goodman (1987). The sequences for the three genera differed so little that the site differences could be treated as unit characters for cladistic parsimony. The latter authors claimed that *Pan* (chimps) and *Homo* were sister-groups, with *Gorilla* as the out-group. They also inferred a much greater rate of substitution in the chimp line than in the *Homo* one, casting yet more doubt on even small-scale molecular clocks. Subsequently, Holmquist *et al.* (1988a, 1988b) supported this conclusion using three large data sets, but Kishino and Hasegawa (1989) put the ψη globin gene data (augmented by that from Maeda, Wu, Bliska, and Reneke 1988) together with mtDNA data. A tree produced by maximum-likelihood estimation does not resolve the *Gorilla–Pan–Homo* trichotomy.

In addition to mtDNA and nuclear DNA, a third type of nucleic acid – ribosomal (r) RNA – has come into prominence in sequencing in recent years. Ribosomes are small bodies associated with the endoplasmic reticulum in eukaryotes, which is a convoluted membrane present in the cytoplasm of the cell. They are, however, also present in prokaryotes, such as bacteria, and are the sites of protein synthesis. Ribosomes are made up of proteins and rRNA and function in groups (polysomes) in protein synthesis. Also, and importantly, rRNA base sequences do not appear to act as a template for the synthesis of either other nucleic acids or polypeptides, although their three-dimensional structure may be important. As a body the ribosome is divided into two subunits, the large and the small. The large subunit contains one very large rRNA molecule and one or two very small ones. The small subunit has a single rRNA molecule (the 16*S* molecule in prokaryotes; the 18*S* molecule, with about 1,800 bases, in eukaryotes).

The sequencing of 18*S* ribosomal RNA attracted considerable attention when Field *et al.* (1988) claimed that the Metazoa (animals other than sponges) were diphyletic. In a paper grandly entitled "Molecular Phylogeny of the Animal Kingdom", they clustered the coelenterates, hydra, and a sea anemone (Phylum Cnidaria), with a ciliate protist, yeast, and maize, whereas all other animals sequenced were on another ramus (the other coelenterate phylum, Ctenophora, was not sequenced). Field *et al.* were criticized by Nielsen (1989), Walker (1989), and Bode and Steele (1989) for this conclusion. In response Field *et al.* (1989) examined the statistical significance of their Cnidaria–ciliate–fungus–plant link by "bootstrap" resampling (Felsenstein 1985, 1988a) with the result that 54 per cent of resultant trees supported the grouping.

In their original clustering, Field *et al.* used a modified distance matrix. Lake (1990) reworked the same data by a technique of his own, which he calls "evolutionary parsimony" (Lake 1987), but which can be understood as a model amenable to likelihood estimation (Felsenstein 1988a, 1988b). It works by comparing unrooted trees for four taxa only. If such trees are resolved, they will have one internal branch connecting two reconstructed nodes.

A series of rival trees is generated for the four taxa, and any two taxa appearing as sister-groups in all significant trees could then be treated as one, and a fifth taxon added for the same procedure to be repeated until all taxa are grouped. Lake (1990) restored the integrity of the Metazoa but came to the startling conclusion that the arthropods, with their characteristic haemocoel, might be a paraphyletic group, which should also contain annelids, molluscs, and minor protostome phyla (see also Patterson 1989, 1990).

10

Is systematics independent?

The title of this chapter is taken from the title of Brady's (1985) paper "On the Independence of Systematics" (see Chapter 8, Section I) in which he seeks to show that as *explanandum* the pattern of classification has logical priority over the pattern of phylogeny as *explanans*. I have urged the logic of this stance several times so far in this book, but the stance, justifying transformed cladistics, is a difficult one for an evolutionary biologist to maintain. This seems even to be the case for transformed cladists, who at the time of writing are much less vehement in their rejection of phylogeny in taxonomic methodology than they were in the early 1980s. Pure transformed cladistics is appealing in its simplicity and austerity and in requiring no justification of its cladograms by appeals to the pattern or processes of phylogeny, but it is undermined by two principal factors. The first is the lack of any explicit justification of the axiom of hierarchy which must underlie an independent taxonomy claiming to elucidate a natural pattern. The second is the claim, if correct, that the pattern of a cladistic classification can be refuted, or at least questioned dangerously, by appeals to fossil evidence and/or the adaptive explanation of characters.

In discussing these problems, this chapter will act as a critical summary of all the data and ideas I have presented so far which bear on the nature of the classification of living organisms. In Chapter 9, I described methods of phylogeny reconstruction and the types of molecular data on which they could be based, but left in the air any discussion of the merits of the methods and the validity of their assumptions. We take these up again in the first section of this chapter. After that I shall utilise material from all the previous chapters to comment on the nature of the relationship between classification and evolution. This discussion will lead to consideration of the proposed mechanisms of evolution in Chapters 11 and 12.

I. Distances, parsimony, compatibility, or likelihood?

All the methods of reconstructing trees cited in the previous chapter are explicitly concerned with the reconstruction of phylogeny. Thus any valid criticism of them depends on a demonstration that they may result in a tree that does not represent the true phylogeny. This is very likely to be the case for reconstruction from a limited data set of characters, but, given more data, such as a longer amino acid or DNA base sequence, it might be argued that the estimated tree should converge on the correct result. A method that guarantees that this will happen is *consistent* in the sense of that term used by statisticians (Felsenstein 1988a). Because inevitably, the true tree is unknown, three strategies exist for criticizing phylogenetic methods. The first, positive, one is to adopt some criterion of fit of the tree to the data, and then to show that a particular method does, or does not, yield acceptable results in the light of that criterion. This often involves comparison of distances read from the data matrix with total branch lengths between taxa on the estimated tree, using statistics such as the least squares method (Chapter 9, Section II). The second, negative, strategy is to propose a hypothetical case for the "true" phylogeny and then to demonstrate that a particular method either fails to retrieve it from the hypothetical data or is inconsistent with it if those data are added to it. The third strategy, also negative, is to set up some epistemological criterion or criteria, and then to show either that the method is at odds with the criteria, or alternatively, that given the data and the method, the criteria simply cannot be applied.

We have seen some of the difficulties inherent in the use of distance measures, whether these are derived from immunological data, DNA–DNA hybridization, or the conversion of protein sequences into nucleic acid distances. I noted in Chapter 9, Section II, that for unknown (or unused in the previous case) base differences between two DNA or RNA strands, the actual total evolutionary change (*T*) represented was underestimated by the actual nonidentity (*D*), however the latter was measured, because of (1) homoplasy, (2) more than one change at a given base site, and/or (3) reversion to the original nucleotide (base), with the three *not* mutually exclusive. Felsenstein (1988a) shows how these would affect additivity even under a simple model of change like that of Jukes and Cantor (1969):

> [W]hen we expect 10% nucleotide sequence difference between nodes A and B on a tree, and a further 10% between B and C, then . . . we expect that 1% of the sites have been changed twice between A and C. One third of these double changes will cause reversion to the original nucleotide [for only four nucleotides], so that the net difference between the sequences A and [C] is expected to be not 20% . . . but 19.67%. Thus the branch lengths will not be additive. . . . This . . . becomes severe with larger differences between sequences. As two DNA sequences become very far apart in the tree, the branch length between them should rise towards infinity, but their sequence difference cannot rise above 100%, and in fact will approach 75% under the Jukes–Cantor assumption. With more realistic models of nucleotide substitution, involving unequal frequencies of the four bases [and different rates of transition and transversion!] the problem becomes even worse.

Felsenstein and Farris agree on the inherent problems but differ on not just the solution but also the possibility of a satisfactory solution. Felsenstein (1984, 1986) urges the statistical approach noted above, but Farris's views have hardened against the possibility of the accurate and thus consistent estimation of correct trees from distance data. Farris (1972) had proposed the "distance Wagner" method for estimating trees from distance data in which unrooted trees were produced by manipulating the distance data to give branch lengths directly, instead of calculating Manhattan distances (Chapter 9, Section I). But the method retained the triangle inequality and, to allow for homoplasy, the condition (Chapter 9, Section II) that

$$d \leq d'$$

as well as the prohibition of negative branch lengths. It was assumed that branch lengths were metric.

If the molecular clock was assumed, then branch lengths would also be proportional to time, and the *ultrametric inequality*

$$d(\mathrm{i,j}) \leq \max\,[d\ (\mathrm{i,k}),\ d(\mathrm{j,k})]$$

(for any three taxa i,j,k) would also apply. We have seen, however, that the assumption of a strict molecular clock for either immunological or DNA–DNA distance data is invalid. Distance data did not directly give trees whose lengths were ultrametric (see below). Nor were they metric: many such trees had branch lengths (including negative lengths) that violated the triangle inequality if adjusted to give the closest match to the data matrix. Thus Farris (1981) urged that distance data should be regarded with suspicion, and particularly that protein sequence data should not be recast as distance data, as Fitch and Margoliash (1967) had originally done. Farris had proposed the epistemological criterion of parsimony for trees based on shared derived characters along branches of the tree, and thus "real" branch lengths, and showed that there was no reason to believe that trees from distance data could conform to it – a combination of the first and third strategies referred to above. We have already noted Felsenstein's (1984) response that branch lengths should be regarded as statistical estimates rather than metric distances.

Meanwhile Felsenstein (1978) had attacked parsimony and compatibility by use of the second strategy – inventing hypothetical cases where these two methods could be shown to be inconsistent. Felsenstein accepted the criteria of statistical inference in validating trees as in his later defence of distance methods. He then proposed specific hypothetical trees, with probabilities of change along each branch, which could be varied to demonstrate that a particular method could in certain circumstances fail to converge on the "true" tree. The same three-taxon tree served for Camin–Sokal parsimony and compatibility (Fig. 10.1), while an unrooted four-taxon tree was used to test Wagner methods (Fig. 10.1). For Camin–Sokal parsimony (Chapter 9, Section I), he employed their assumption that for all two-state characters, change could occur only from the ancestral to the derived state ($0 \rightarrow 1$) with no reversals. The probabilities of change in Figure 10.1 are indicated by P, Q, and R and apply over all characters. Felsenstein then

Fig. 10.1. Testing parsimony and compatibility methods for consistency by means of hypothetical trees. (a) A phylogeny with three terminal taxa, ABC: PQR are the branch probabilities of change from 0 to 1. (b) An unrooted tree (Wagner network). (c) Related values of P and Q for which the Camin–Sokal parsimony method is consistent (C) and inconsistent (NC) for the phylogeny in (a). For further information see text. (After Felsenstein 1978. Published by permission of the Society of Systematic Zoology, National Museum of Natural History, Washington, D.C.)

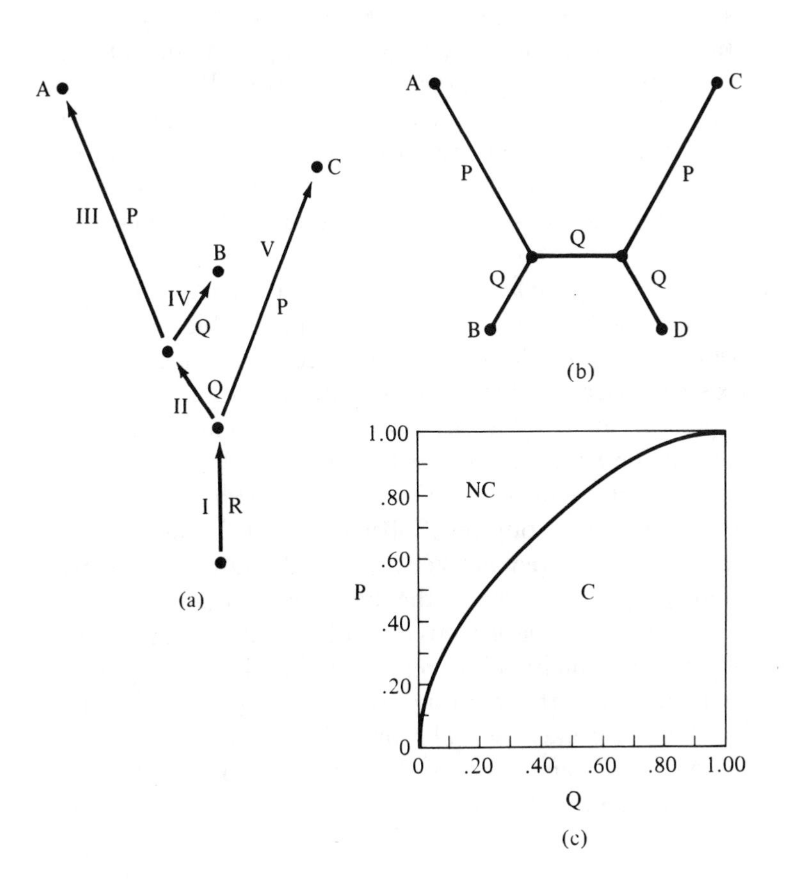

showed that these were a set of related values of P and Q for which the Camin–Sokal method fails to be consistent (Fig. 10.1). With the same tree he showed that compatibility methods gave the correct clustering (AB)C if and only if $n_{110} \geq n_{101}, n_{011}$, where the three digits (110, 101, or 011 in this case) represent the character states in A,B,C, respectively, and *n* represents the number of characters having that distribution of states over the three taxa. Felsenstein also noted that the unequal sums of branch-length probabilities between each of the three taxa A,B,C and the root of the tree could be due either to the three taxa being non-contemporary or to different rates of change along the branches. For the Wagner network the precondition for consistency was

$$P^2 \leq Q(1 - Q)$$

with the acceptance that reversals $(1 \rightarrow 0)$ were allowed and assigned the same probabilities as $(0 \rightarrow 1)$. In all three of these invented cases, the method "failed" when the probability of parallel change along two long branches of the tree (P and P) was greater than that of change along a single short branch Q. Felsenstein concluded by accepting that his examples were simplified and extreme and that likelihood methods, which would have passed the test, require postulating untestable hypotheses of change.

Farris (1983), in his defence of phylogenetic parsimony (see Chapter 9, Section I), attacked the use of improbable models to invalidate taxonomic methods; he would certainly have agreed with Felsenstein's last conclusion. Contributing to the debate, an important point is made by Sober (1985). Felsenstein's hypothetical three-taxon tree was a representation of phylogeny (i.e., a "tree" in the cladists' sense), not a cladogram (*sensu* cladistics) or diagram of taxonomic grouping (Sober does not discuss Felsenstein's hypothetical Wagner network). A cladogram or tree diagram would simply have been the best representation of the character data, but a *phylogenetic* tree would involve branch transition probabilities, which are not represented in the tree topology (i.e., its branching pattern), although that topology would be the aim of the clustering exercise. The probabilities are *nuisance parameters,* whose values are unknown and, while affecting the final result, are not represented in that result. If the aim is the reconstruction of phylogeny, then a realistic estimation of nuisance parameters is the principal methodological problem. In the absence of a comprehensive model of evolutionary change, one solution is to make the assumption that evolutionary change along the branches of a phylogeny has followed a stochastic pattern. The *a priori* assumption of a Poisson process of change in the likelihood estimation of trees (Chapter 9, Section III) is one such strategy.

Sober (1988) is a book-length, rigorous discussion of the logic of tree reconstruction in which, inevitably, nuisance parameters play an important part. Much of Sober's discussion is concerned with the competing claims of likelihood and parsimony to validity in phylogeny reconstruction. Originally, Farris (1973) had argued that parsimony, as used in his Wagner methods, gave a maximum likelihood result, but Felsenstein (1973) would accept only that parsimony (in this case Camin–Sokal parsimony) gave this result when the probability of change along the branches of the tree was very small. The rarity of change was a sufficient condition for parsimony and likelihood to give the same result. The assumption here was that the likelihood method was consistent throughout, but this need not be the case. Sober (1988, pp. 172ff.) makes the point that likelihood, correctly applied, describes which hypotheses are best supported by the evidence; if the evidence is misleading, the best-supported hypothesis will be false:

> Likelihood does not provide a "rule of acceptance". It does not say that the best-supported hypothesis ought to be accepted as true. Likelihood has much more modest pretensions; it is a "rule of evaluation", simply indicating which hypotheses are best supported by the data. . . . Likelihood is no more misleading when it fails to be statistically consistent than is an interpreter who correctly reports what an utterance means when *the utterance* is misleading.

Sober goes on to suggest that the reason that convergence on the correct result – that is, consistency – has been assumed to be a necessary feature of all statistical inference is that statisticians also concern themselves with the design of optimal experiments. In the case of taxonomy, it is difficult to see how the option of designing an optimal experiment could present itself. One can choose the method and source of taxonomic data (as in the choice of protein or nucleic acid supplemented by the choice between distance or sequence data noted above, or as in the choice of morphological characters), but the analogy with an experiment in the physical sciences is a false one (Chapter 14). Sober (1988, pp. 182–3) sums up his discussion of consistency thus:

> Likelihood does not require statistical consistency. It follows that parsimony's failure of consistency in a hypothetical model does not answer the question of whether parsimony and likelihood coincide. . . . Parsimony aims at saying which hypothesis is best supported by the evidence. Its function is the same as that possessed by the likelihood concept itself. This does not mean that parsimony has a likelihood rationale, but only that parsimony should not be judged by standards that are alien to the likelihood concept. The demand for consistency embodies just such a standard.
>
> With considerations of likelihood and considerations of consistency now clearly separated, we should see that the question of the connection of likelihood and parsimony is still open.

The problem is further discussed by Goldman (1990).

We may now return to the question posed at the beginning of this section: "Distances, parsimony, compatibility, or likelihood?" Firstly, it must be admitted that the choice represented by the question does not appear to be between equivalent entities. In talking about distances in Chapter 9, I was talking mainly about distance *data,* in which the actual unit sequence differences among protein molecules or DNA strands are simply not available for inspection or manipulation, and thus cladistic autapomorph, symplesiomorph, and homoplastic characters cannot be distinguished. The second point to make, however, is that phenetic clustering employs distances that may be derived from data where those distinctions could have been made. In the sphere of molecular taxonomy, this approach is represented by Fitch and Margoliash (1967) (but not by most of their subsequent work). But it is also the approach of phenetic clustering of morphological characters. Here I agree with cladists of all persuasions in the following senses. (1) *If* an irregular, divergent, inclusive hierarchy (Chapter 8, Section I) is accepted *a priori,* either because phylogeny is assumed or for any other reason, *then* turning character states into phenetic distances is an under-use of the data (except where polythetic taxa are concerned; see below). (2) *If,* furthermore, it is agreed that the pattern of classification and the pattern of reconstructed phylogeny should be intimately related, whatever the nature of that relationship, *then* there seems little to be said for a school of thought that employs different techniques for classification and phylogeny reconstruction. Like most evolutionists my inclination is to accept the premise embodied in (2), but at a deeper level that premise depends on one's attitude to the premise embodied in (1) – that is, the *explanandum–explanans* problem which has kept recurring throughout this book. I attempt a partial resolution of that problem in this chapter, but there is still the separate but related problem of the use of distance *data*.

There is one set of circumstances (given the fact of phylogeny *a priori*) under which phenetic clustering would retrieve a phylogeny. All phenograms, because of the nature of the clustering techniques, are characterised by the ultrametric inequality

$$d(\mathbf{i},\mathbf{j}) \leq \max\,(d(\mathbf{i},\mathbf{k}),\ d(\mathbf{j},\mathbf{k}))$$

(see above). The visual representation of this is that (1) the axis of rank (conventionally the vertical axis) of a phenogram is a linear measure of dissimilarity (Chapter 2); and (2) all OTUs occupy the same level and thus the same rank in the phenogram (Fig. 2.11). But this is not necessarily the case for the original pairwise matrix of similarity of OTUs. The matching of branch lengths on the phenogram and the original distances on the similarity (or dissimilarity) matrix is represented by the co-phenetic correlation coefficient (Chapter 7, Section II). If for distance data, such as DNA–DNA hybridisation or immunological distances, that coefficient approached 100 per cent in every case in which the data were full enough to show consistency, then distance measures would also be ultrametric. In that case, the "molecular clock" could be inferred to be a chronometer, and reversals and multiple hits would either be absent or occur with such regularity as not to affect the result. Under those circumstances, the phenetic clustering of distance data might yield the true pattern of phylogeny. But we know that this is not the case. The chronometric molecular clock has been rejected. Also with different rates of transition and transversion in nucleotide substitution, multiple hits, reversions to the same base, and the limit on the possible measurable difference between two DNA strands and only the same four possible "states" to each character (i.e., site), even the concept of homoplasy in DNA sequences is difficult. If one adds to this the experimental errors inherent in immunology and DNA hybridisation, which in the latter case may over-ride the small calculated distances for closely related organisms, then the robust trust in the latter method, expressed by Sibley and Ahlquist (e.g., 1987), seems misplaced. Perhaps an analogy with the radioactive dating of rocks is apt. Depending on the half-life of the radioisotope used, if the period to be measured is too long, then inaccuracy will result because of the small quantity of the original element left, with an absolute limit beyond which it is not measurable. If the period is too short, there will be too little of the measured breakdown product. If all other factors are favourable, there will be a band of geological time in between, over which an acceptable dating can be expected. For DNA hybridisation there will be a limited spectrum of genetic distances over which a modest expectation of acceptable results is reasonable.

We have seen that while distance *data* have been most commonly derived from biochemical investigations (with the measurement of anatomical features as an exception), distance *methods* – that is, phenetic clustering – are equally applicable to binary unit characters, whether these are two-state characters or the results of binary coding of qualitative multistate characters. *Compatibility methods,* however, deal only in unit characters (or states of a character). For multistate characters, additive binary coding, based on a character state tree, is used for preference. Similarly, where possible, the binary coding is derived from the polarisation of characters (i.e., 0 = primitive, 1 = derived). In all these respects, compatibility is similar to numerical cladistics. Furthermore, the core of the technique, the elimination of characters from clustering whose states can be shown to have arisen in parallel with those of other characters, is closely related to cladistic parsimony, and clustering of the taxa using the compatible "clique" (Chapter 9, Section I) of characters can be achieved via the Wagner methods (Gauld and Underwood 1986). Nevertheless, compatibility analysis is usually rejected by cladists (e.g., Farris 1983).

In talking about compatibility methods in Chapter 9, we saw that a null hypothesis of no hierarchical structure in the data had been proposed by Le Quesne (1972; also Meacham 1981). I shall refer to this below, but Farris points out that it is not valid to extrapolate from the "high probability" of the null model to a criterion in which the "least likely" clique is the one to be selected. A pattern of character distribution that rejects the null model requires some new hypothesis that then makes the observations more probable. In the brief discussion of cliques, we also saw that when a wholly compatible clique had been produced, the number of remaining characters was often too small to give a well-resolved tree. The suggested cure was to produce "sub-trees" from the constituent terminal taxa and then unite them in a single aggregate tree. But the "cure" exposes the faults that cladists find in compatibility. Characters rejected in clique analysis (whether for tree or sub-tree) are rejected as homoplastic. In cladistics, homoplasy, like synapomorphy, refers to a particular rank and pair of taxa. The very similar skulls of the wolf and the "marsupial wolf" (*Thylacinus*) both represent a complex series of interrelated characters that are homoplastic with respect to those of the other species. This homoplasy, tested by congruence (Chapter 4, Section II) and interpreted as convergent evolution, is so characterised because the characters do not unite *Thylacinus cynocephalus,* a dasyurid marsupial, and *Canis lupus,* a canid carnivore and a placental mammal, in a taxon that excludes mammals without them. But the characters homoplastic between the two species also serve to unite *C. lupus* with the jackal *C. aureus* in the same genus; with *Vulpes vulpes,* the red fox, in the same family (Canidae); and with *Panthera leo,* the lion, in the same order (Carnivora). As Farris puts it: "The defect of cliques is just that they treat every conclusion of homoplasy as if it implied universal homoplasy". A "one tree" compatibility analysis aiming to classify a group of mammals including, *inter alia,* all the species mentioned above would forbid altogether the use of the characters homoplastic between marsupial and placental wolf. Thus while clique analysis is a parsimony technique, it is not a *ranked* parsimony technique.

It also has another defect important with respect to the theme of this book. In the next section, I discuss the assumptions behind the *a priori* acceptance of the pattern of classification as a divergent inclusive hierarchy. But we have already seen that classical and Linnean taxonomy accepted the hierarchy as representing the correct way of systematising knowledge (Chapter 6); "evolutionary" taxonomy as a representation of the conflation of anagenesis and cladogenesis (Chapter 6, Section IV); phenetics as the best way of summarising taxonomic data, justified by a criterion of "naturalness" (Chapter 7, Section II); and Hennigian cladistics as a representation of the pattern of cladogenesis (Chapter 7, Section III). Exponents of compatibility accept the Hennigian justification. Each state tree is a hypothesis of phylogeny; a series of compatible state trees (a clique) is a well-corroborated hypothesis. But supposing one wanted to investigate whether the "natural" pattern is indeed an inclusive divergent hierarchy. All techniques, as well as accepting the pattern *a priori,* are designed to produce divergent dendrograms whatever the true pattern, but character states incompatible with that pattern are retained within the system so that an (as yet unformulated) technique for investigating the existence and extent of contrary patterns would have the data to work with. Compatibility rejects all data that might yield hypotheses of reticulation beyond recall.

I shall comment briefly on likelihood. As we saw in Chapter 9 (Section III), there are practical difficulties in dealing with any more than a few taxa using likelihood methods. As far as theory is concerned, however, the method has obvious advantages. Starting with a simple model of change, as in Zuckerhandl and Pauling's (1965) simple Poisson model, the model of change can be modified, as in assigning a different rate to the sites representing the third nucleotide of each triplet of the genetic code. But the basis of the technique is still that the model of change is a stochastic one. This can, however, be seen in a different light – not so much as an admission of ignorance about the mode of change, but as a basis for investigating the nature of that change. As Bishop and Friday (1988) put it:

> The assumptions made are fairly robust, but they are subject to revision as estimates of the appropriate parameters become available. This is all that one could (scientifically) expect. In contrast, the assumptions about change underlying maximum parsimony estimation are deterministic, being based on the predominance of divergent change.... Both approaches use phylogenetic information contained in molecular sequences, but we feel that the probabilistic models are more in keeping with what is known about evolutionary processes and more importantly are well placed to be modified in the light of experimental evidence.

Thus, apart from likelihood providing an experimental baseline, we are, with parsimony, once again confronted with the *a priori* hierarchy. I shall turn to that now.

II. The hierarchy

In a perfect world, or one made perfect for transformed cladists, the natural arrangement of organisms would have the following properties:

(1) The arrangement would form an inclusive, divergent hierarchy.
(2) That arrangement could therefore be represented by an "ordinally stratified hierarchical clustering" (Jardine and Sibson 1971).
(3) All taxa at every rank (including the whole Biota) would be monophyletic in the transformed cladistic sense.
(4) Each taxon at every rank could be characterised by at least one apomorph and thus a unique character, which, unless that taxon was terminal, would unite the two sister-taxa of immediately lower rank which that taxon included.
(5) Apomorph characters of every taxon could be distinguished unambiguously by (probably ontogenetic) means that did not depend on any pre-existing classification.

If all these provisos were true, cladistics would be the best, in fact the only valid, method of producing a natural classification, and there would be no necessity to validate any given classification by reference to a hypothesis of phylogeny or to theories of the mechanism of evolution. But the world has not been made perfect for transformed cladistics, and we now need to ask whether the fact that some of the above propositions are false, or their truth unknown or unknowable, is sufficient to invalidate, first, the assumptions of transformed cladistics, and second, cladistic methodology. I shall take the numbered propositions in order.

(1) We have seen (Chapter 3, Section IV, and elsewhere) that some plant species are known to be of hybrid origin, so that a true representation of their phylogeny would include reticulation. There is also evidence of hybrid origin of animal species. White (1978) describes possible cases in reptiles, amphibia, insects, and planarian flatworms. Notable is the case of the European edible frog *Rana esculenta* which exists in dynamic equilibrium with its two parent species (Uzzell, Gunther, and Berger 1977). The existence of hybrid species yields a double dilemma: first, as I have said, it is evidence that the pattern of *phylogeny* is not universally divergent, but second, it immediately reopens the question of the

independence of the pattern of classification from the pattern of phylogeny. Does reticulation in phylogeny force one to infer reticulation in the pattern of classification? If so, not only is the divergent hierarchy refuted, but also the independence of systematics as an *a priori* principle cannot be maintained.

(2) The second proposition depends on the truth of the first. The reason for stating it separately is that even if the pattern of classification as a divergent hierarchy is rejected as an empirical phenomenon, it is still reasonable to accept it as a methodological proposition. If this were done, and the independence of systematics were maintained, there would be nothing to choose *in that respect* between transformed cladistics and phenetics. After all, neither would claim validation by phylogeny or any other inferred natural phenomenon.

(3) *Monophyly* in phylogenetic cladistics has the literal meaning of descent from a common ancestor. In transformed cladistics, despite the use of the same methodological criteria, it refers only to the structure of the hierarchy and the apomorph features that characterise it. Once again we have the dilemma of phylogeny. In the case of animal phyla, it has been regarded, probably throughout this century, as difficult to produce a hierarchical clustering of high ranks, although various patterns of sub-kingdoms and super-phyla have been suggested (e.g., Clark and Panchen 1971). Willmer (1990) recently discussed invertebrate relationships, but again presuming phylogeny: her book is entitled *Patterns in Animal Evolution*. Willmer cites evidence of relationship in terms of the fossil record, embryology, and larval forms, as well as characters drawn from comparative and functional anatomy, cell ultrastructure, chemistry, and genetics. She then concludes that no hierarchical scheme with largely congruent characters is possible among animal phyla. She does not, however, attempt any cladistic analysis, or explain any alternative taxonomic methodology, so her conclusion that no hierarchical clustering of invertebrate phyla is possible is little more than a cry of despair, and the whole book has been severely criticized (Fitzhugh 1990; Pease 1990).

We have seen (Chapter 9, Section III) that tracing the phylogeny of metazoan animals by ribosomal RNA sequencing led to conflicting results. The monophyletic origin of metazoan phyla (phylogenetics) or the monophyletic clustering of metazoa (transformed cladistics) seems little more than a dogma produced by extrapolation from the clustered membership of lower ranks.

(4) The pattern of unique apomorph characters is violated in two respects. First, homoplasy undoubtedly occurs, and the most important development in cladistics since Hennig is the use of parsimony (and the related concept of congruence), whatever one thinks of its validity. The second violation, if correct, is the recognition of polythetic groups (Chapter 6, Section III). In phylogenetics homoplasy is taken to be evidence of parallel or convergent evolution, but in transformed cladistics some different explanation, if any, is required. One early attitude was to regard a character homoplastic between two species (as tested by parsimony) as a mistake:

> [W]hereas phenetic methods are willing to accept incongruence between characters as a feature of the real world, cladistic methods regard the discovery of apparent incongruence as an indication that the taxonomist has made a mistake.
>
> (*Platnick 1980*)

Thus the character in the two species was not really "the same" despite the fact that it has perhaps been branded as homoplastic as a result only of parsimony analysis. Thus was the hierarchy taken as *a priori*. Brady (1985), in defending the independence of systematics and setting up the *explanandum/explanans* distinction, was more cautious. In some cases the hierarchy was clear, but (repeating part of the quotation in Chapter 8, Section I):

> About the groups that we cannot resolve we may say nothing, for our knowledge is insufficient to make a judgement as to whether these groups are somehow different or simply have not been resolved *as yet*. Our claims to knowledge must rest upon the well resolved groups, and these are most interesting to the investigator.

Even apart from its echoes of Wittgenstein (1922) – "Whereof one cannot speak, thereon one must remain silent" – I find this odd. After setting out the *explanandum–explanans* argument very clearly, Brady declines to postulate that the natural order is a comprehensive hierarchy. And that is the only reason for claiming, according to transformed cladists, that the cladistic method yields a natural classification.

Another factor to be considered under (4) is the existence and frequency of polythetic groups. We saw in Chapter 6 (Section III) that Mayr (1982) regarded the recognition of polythetic taxa as an important advance in pre-Darwinian taxonomy. Polythetic taxa are defined by Sneath and Sokal (1973) thus: "in a *polythetic group,* organisms are placed together that have the greatest number of shared character states, and no single state is either essential to group membership or is sufficient to make an organism a member of that group". Thus polythetic taxa, if they exist as natural entities, have no autapomorphies. The theoretical properties of such taxa were explored by Beckner (1959). His more analytical definition is paraphrased by Simpson (1961a, p. 94) as follows:

1. Each individual has a large but unspecified number of a set of properties occurring in the aggregate as a whole.
2. Each of those properties is possessed by large numbers of those individuals.
3. No one of those properties is possessed by every individual in the aggregate.

It should be obvious that the members of such a taxon will be grouped together by phenetic clustering techniques, but not by cladistics. The question that arises is whether "natural" (meaning what?) polythetic groups are sufficiently common to invalidate clustering by synapomorphy. Considerable importance has been attached to the concept of polythetic groups by Beckner, Simpson, Mayr, Sneath, and Sokal amongst others, but examples of such groups, at least among sexually reproducing organisms, seem more difficult to come by. Mayr (1969, p. 83) cites the supra-generic taxa of fleas (Siphonaptera) "that can be reliably defined only by combinations of characters, each one of which may also occur outside the given taxon or may occasionally be absent in a member of the taxon (Holland 1964)". An important point here is the nature of the aggregate set of characters of which each constituent sub-taxon has some but not all. If the characterisation of the polythetic taxon is by aggregate similarity – that is, phenetic – most of the character set will probably be the equivalent of symplesiomorphies; they will occur in taxa of higher rank, rather than being a set unique to the polythetic taxon.

Nevertheless, as we saw in Chapter 8 (Section I), Platnick (1985), as a "transformed cladist", accepted that a species could be polythetic (and in the phenetic sense!) in that "as terminal taxa species must have unique *sets* of apomorphic characters, but need not have any autapomorphies." But the characters comprising such a set, if not unique, must characterise a supra-specific taxon. As we saw in Chapter 8, Platnick seems to regard the category species as unique in that respect. Nevertheless, it seems a reasonable working assumption that what one might describe as *cladistic* polythetic groups – that is, having a unique set of unique characters but no autapomorphies – do not exist above the species level.

(5) The last of our provisos for the cladists' perfect world was the practical one: that apomorph characters could be distinguished unequivocally by means (probably ontogenetic) not dependent on a pre-existing classification. This again was discussed in Chapter 8 (Section II). I concluded there that "polarisation of characters by ontogeny represents an ideal beset with difficulties". Here I believe that there are two points to be made: First, it is claimed that divergence of different species in ontogeny, von Baer's third and fourth laws, validates the *a priori* hierarchy independently of phylogeny. This involves an inductive generalisation (Chapter 13) from known ontogenies to those of all organisms. Even setting aside the fact that the ontogeny of fossil characters is not available, and the point made by Scott-Ram (1990, pp. 164–5) that characters such as chromosome number do not have ontogenies, I do not believe this generalisation is sufficiently sound to validate the hierarchy. Nevertheless, my second point is that, as method, polarisation by ontogeny may well work in many cases and could be further developed alongside out-group comparison.

So what of the *a priori* hierarchy? Scott-Ram (1990) provides a lengthy critique of transformed cladistics. I shall not attempt a summary, but shall pick on some points important to our theme. His principal aim was to show that transformed cladistics "does not achieve what it claims and that it either implicitly assumes a Platonic World View, or is unintelligible without taking into account evolutionary processes – the very process it claims to reject".

To understand Scott-Ram's argument, we need to make a threefold distinction in each of two different philosophical domains. The first domain includes the difference between what, to be consistent with discussions in our Chapter 6 and elsewhere, I shall term essentialism, idealism, and "Platonism". The second domain includes the distinction among ontology, epistemology, and methodology. In Chapter 6 (Section I), I characterised essentialism in taxonomy as the "act of faith" in asserting that all taxa have essences – that is, in cladist terms, autapomorphies that confer reality on them. For reasons that are unclear to me, Scott-Ram (1990, pp. 155–6) denies that transformed cladistics is essentialist. I now wish to distinguish between idealism and "Platonism" for the purposes of the present argument, although Scott-Ram (quite rightly) would regard them as two aspects of the same theory. "Idealism" I shall characterise here as a belief in the reality of higher taxa based on the concept of the reality of archetypes or *Baupläne* (Chapter 4, Section I; Chapter 6, Section I). In this sense Richard Owen and the transcendental morphologists were idealists. But the other aspect of the beliefs of Plato, here distinguished as "Platonism" (*s.s.*), is that the *ideas* (or "Forms") were timeless abstractions. Observable entities were then contingent (in time), but also imperfect and *less real* images of the *ideas.* Together idealism and "Platonism" (*s.s.*) represent true Platonism as in the *Timaeus* (see Chapter 6, Section I) and in Plato's famous metaphor of the Cave. Scott-Ram's contention is that by rejecting the validation of the hierarchy by phylogeny, transformed cladists must be idealists and "Platonists" (*s.s.*). Thus the actual organisms being classified are apprehended only as observables that are contingent (and evolving) in contrast to the timeless reality of taxa. This is a repeat of Beatty's (1982) point (see Chapter 8, Section I). It also, according to Scott-Ram, poses difficulties for the use of ontogeny, a time-based process, in validating the hierarchy.

But we now must turn to our second philosophical domain. If the transformed cladists accept the hierarchy *a priori,* are they making an ontological claim – that the hierarchy has a real existence "out there"; an epistemological claim – that our understanding of the natural world includes a picture of the natural order of organisms as an inclusive hierarchy; or merely a methodological statement – that we shall accept the hierarchy *a priori* so that the cladograms we produce have the same pattern as the classifications we construct from them? If the last is their stance, one can suggest three possible justifications for it. The first is that of Linnaeus and of his forerunners back to Porphyry and even to Plato, that logical division, the "correct" method of classification, yields a hierarchy. The second is that cladistics is a method of cluster analysis superior to any other but yields only hierarchical results in contrast to the several methods of expressing phenetic clustering. The third is in the same spirit as Bishop and Friday's justification of stochastic models plus likelihood estimation quoted above (Chapter 9, Section III): the hierarchy provides a baseline, a point of departure from which the true nature of the pattern may be investigated.

But transformed cladists must be making more than a methodological claim; otherwise they could not assert that instances of homoplasy, if not rejected, are "mistakes". Nor, perhaps more importantly, could the *explanandum–explanans* justification (not made explicit by Scott-Ram, who does not cite Brady 1985) be maintained.

I do not believe there is a solution to the problem of the hierarchy that satisfies the canons of philosophical rigour, but I do believe that a working solution is essential. In talking about phenetics and compatibility analysis, I emphasised the idea of taxonomic null hypotheses. Clusters will be produced by either technique from random character distributions. Their practitioners have therefore addressed themselves to distinguishing significant from random distributions: the latter do not imply any natural hierarchy. The claim of cladistics, either phylogenetic or transformed, to produce a natural classification (however defined) depends totally on the independent existence of a natural hierarchy (however explained). If the hierarchy, based on a hierarchy of homologies, exists, then cladistic methodology is not just the best, but the only valid technique for revealing it. The hierarchy must be established as an independent entity; independent, that is, of any theory of the existence of phylogeny and of any theory of evolutionary mechanism. The reason is the *explanandum/explanans* problem. Thus the independence of the hierarchy is a necessary validating claim on behalf of transformed cladistics. But the existence of a hierarchy embracing all living things is in doubt (e.g., Brady 1985; Willmer 1990), and in any case, the only valid method of revealing it is that of cladistics, which depends absolutely on its *a priori* existence. This dilemma is insoluble.

To say this does not validate a return to *phylogenetic* cladistics, but if it can be established that cladistics can yield a natural classification only *in particular cases* by the use of evolutionary hypotheses based on data extrinsic to the synapomorphy scheme, then phylogenetics would have to be reconsidered.

III. Cladograms, trees, and scenarios

The distinction between the terms "cladogram" and "tree" (as used by cladists) was explicit in Hennig (1966) but as "(phylogenetic) diagram" and "phylogenetic tree", respectively. We have seen, however, that Schaeffer, Hecht, and Eldredge (1972) made the distinction clear by using the terms "cladogram" and "tree". I cited Eldredge's (1979b) subsequent definitions in Chapter 3, Section III. The term "phylogenetic scenario" was introduced in a manuscript by Dr G. Nelson and subsequently discussed together with "cladogram" and "tree" by Tattersall and Eldredge (1977). It is defined by Eldredge (1979b) as "*a phylogenetic tree with an overlay of adaptational narrative*". All cladists probably see the order of logical priority as:

Cladogram → phylogenetic tree → phylogenetic scenario

I discussed the problem of whether there was more than one tree to every cladogram in Chapter 8. One extreme position is taken by Wiley (1979a, 1981) who believes that, with the exception of trees derived from a three-taxon cladogram with an unresolved trichotomy, the branching pattern of a tree must be isomorphic with that of the cladogram from which it is derived; thus one cladogram – one tree. The other extreme is represented by Platnick (1980, 1985) and Nelson and Platnick (1981), who invented the new game of "component analysis" in which they show, following simple rules, how many trees they think can be derived from a given single cladogram. In fact, there is no particular reason why this controversy should be of any practical importance. Transformed cladists rarely convert their cladograms into trees, and the status of Wagner trees (Chapter 9) in this argument is ambiguous.

What is important, however, is that a cladogram is not a hypothesis of phylogeny in its own right. It is, in Nelson's phrase, "a synapomorphy scheme". To propose it as a phylogeny involves the proposition of an additional hypothesis. This phylogenetic hypothesis, the tree, can be represented graphically in a number of ways, identically to the cladogram, with either the terminal taxa or all the nodes of the cladogram extended on a time basis to give the geological range of the real and hypothetical taxa involved, but with the connecting branches indicating only relationship, or with the branches themselves representing the persistence of clades in geological time (Fig. 8.1). Scenarios then represent narratives based on trees, not necessarily presented as diagrams. Eldredge (1979b) says of them:

> Scenarios are inductive narratives . . . concocted to explain how some particular configuration of events . . . took place. The hallmark of such narratives is the analysis of the adaptive significance of evolutionary changes in size, form, and structure. . . . [T]hey are mostly fairy tales constructed of a maze of untestable propositions concerning selection function, niche utilization, and community integration, and alas, do not generally represent good science.

We have seen the general paradox of the relationship between the *a priori* hierarchy and the general hypothesis of phylogeny. The parallel particular paradox is that of the relationship between cladogram and scenario. I see no reason to dispute the cladist dogma that cladograms are logically prior to trees, but can they be tested by scenarios? This question is particularly acute when the role of fossils in taxonomy is discussed.

We saw in Chapter 7 that reconstructed *Stufenreihen* derived from fossil specimens can be used to test cladograms derived from extant taxa only. Thus Kemp (1988a) and Gauthier *et al.* (1988) refute Gardiner's clade Haemothermia by demonstrating that stem-group mammals and stem-group birds lack the supposed synapomorphies uniting crown-group mammals and crown-group birds (Chapter 7, Section IV). Apart from claiming that Kemp and Gauthier *et al.* were correct, of which I am convinced, there are other claims that can be made about this methodology. First, *contra* Patterson (1981b), fossils can overturn groupings based only on extant taxa. Second, it could be said that in the case of the mammal-like reptiles, a transformational homology, that of the middle ear complex, was used to establish the *Stufenreihe* rather than the Patterson–Rosen technique of cumulative (mammalian) autapomorphies. Third, it could be claimed that Kemp's analysis was a case of a tree refuting a cladogram. I believe, however, that both these latter claims are at best equivocal. The whole argument can be cast in terms of transformed cladistics, with the *Stufenreihe* appearing as a Hennigian-comb cladogram.

The other case I cited of fossil evidence apparently refuting a (transformed!) cladogram was that concerning the relationship of Dipnoi and Tetrapoda (Panchen and Smithson 1987). Here a scenario, the reconstructed mode of life and respiratory physiology of early (Devonian) dipnoans, was used to suggest that the apparent synapomorphies of Dipnoi and Tetrapoda were in fact homoplastic. I now want to investigate the theoretical implications of this conclusion a little further.

Paralleling the epistemological sequence of cladogram → tree → scenario, one can construct a methodological sequence in systematics:

> **Description of organisms ⇄ identification of characters (taxic homologies) → classifications → reconstruction of phylogeny**

But we have seen (Chapter 5, Section I) that the interpretation of fossils must depend on reference to their supposed living relatives, and further, that reconstruction of their functional morphology and mode of life depends in part on reconstruction of their phylogeny (trees → scenarios). The sequence then becomes:

> **Description of fossil(s) ⇄ comparison with supposed living relatives ⇄ characters → classification → phylogeny → functional morphology (and other scenarios)**

But if Dr Smithson and I are correct, we have established a "feedback loop" from scenarios to characters. The cladistic inference of a sister-group relationship between Dipnoi and Tetrapoda was refuted by an attack on the only apparent synapomorphous characters uniting them, and that attack was based on a fossil scenario.

The particular scenario in this case is also one of reconstructed functional morphology and is thus related to Ridley's (1986) claim that "natural selection" must be used in the construction of cladograms (Chapter 8, Section I). I noted that Ridley's claim was really that adaptation had to be taken into account in classification and that the reconstructed functional morphology of fossils is a special case of adaptation. Earlier in his book (his chap. 8), Ridley suggests adaptation as a criterion for polarising morphoclines, particularly for distinguishing rudimentary → fully developed and functional, from fully developed and functional → vestigial. All the stages in the first should be adaptive, but those in the second need not be so. The claims he makes for this technique are modest ones:

> The question, in a field like this, is not whether a method gives us certain conclusions. It is whether it gives us any information at all. If it does it should be used, and thought should be given to its improvement. The functional criterion should therefore be admitted to the repertory of cladistic techniques.

So far in this book I have been concerned with the relationship between phylogeny and classification, so that in discussing homology and methods of classification I have been discussing the first step in the reconstruction of phylogeny. In Chapter 9, I then reinforced this by a critique of molecular methods which, it is claimed, reconstruct phylogeny directly. I have also tried to disentangle the *explanandum* of evolution from the evidence for evolution in Chapters 4 and 5. But the phylogeny of the biota on this Earth represents a unique, though inconceivably complex, series of events, and thus all the characters of extant organisms, in so far as they differ from those of other organisms, are contingent. Any generalisations made about those characters are then perhaps inductive and not like the supposed universal-law-like statements of physical science. We may then ask, Are there then any laws in Biology? In the remaining chapters, I hope to say something about this problem. One universal feature of organisms is their adaptation to their environment. Perhaps then, accepting that evolution has occurred, theories explaining the origin of adaptations have the status of natural laws. I devote the next two chapters to such theories *en route* to a discussion of the nature of Biology as a science.

11

Mechanisms of evolution: Darwinism and its rivals

It is not my intention, in these two chapters on theories of evolutionary mechanism, to give a detailed history. For a more complete account readers should consult Mayr's *Growth of Biological Thought* (1982), Bowler's (1983) account of rival theories to Darwinism from the publication of the *Origin* in 1859 until the development of the "Neo-Darwinian" synthetic theory and finally the multi-author work on the development of the synthesis itself (Mayr and Provine 1980). The pre-Darwinian history of evolutionary theory in nineteenth-century Britain is treated in its sociological context by Desmond (1989).

These two chapters have four principal themes: in Chapter 11, (1) a discussion of Lamarck's theory of mechanism, representing the first complete, if flawed, evolutionary theory; (2) a critique of Darwin's and Wallace's theory of Natural Selection; (3) the contrasting theories supported "post Darwin" and referred to above; and, in Chapter 12, (4) a critique of the Synthetic Theory and reactions to it, thus bringing the account up to date.

In passing, however, we may refer to the theories of Etienne Geoffroy Saint-Hilaire and Robert Chambers, both of whom have been discussed in previous chapters. Geoffroy (1772–1844) speculated about evolution late in life (Mayr 1982). He accepted anagenetic change in separately created lineages by the direct action of the environment during ontogeny, proposing a type of radical environmentalism which, as we shall see, was rejected by Lamarck. Chambers, in *Vestiges of the Natural History of Creation* (1844), saw evolution (and ontogeny) as the unfolding of the divine plan. One was an environmentalist theory, the other internalist. Another antithesis we shall see in theories of mechanism is whether, to use Mayr's terms, inheritance is regarded as "hard", with no "inheritance of acquired characters", or "soft".

Both Lamarck and Darwin (and Wallace) were concerned to explain the phenomenon of adaptation in scientific terms, but theories of mechanism should also provide an explanation of evolutionary change and thus of phylogeny. We have already noted that there are two accepted components ("modes"; Simpson 1944, 1953) of evolution: phyletic evolution or anagenesis and speciation or cladogenesis – to which Simpson added a third, "quantum evolution", of doubtful status but related to the concept of punctuated equilibrium. A complete body of theory should explain both modes, as well as provide a causal theory of adaptation. Both Darwinism and the modern theory are incomplete in that respect. This may be one of the reasons that publication of the *Origin* was followed by a number of rival more or less scientific theories claiming to explain anagenesis and cladogenesis. The result was "the eclipse of Darwinism" – which is the title of Bowler's book – so that the Darwinian theory re-emerged triumphant only with the development of the "synthetic theory" in the middle third of this century.

In looking at theories of mechanism, therefore, we shall be considering their adequacy in explaining all three phenomena: adaptation, anagenesis, and cladogenesis. There is, however, another way of considering the plausibility of theories of mechanism – that is, by looking at the nature of the factors that constitute that mechanism. Near the beginning of Chapter 3, we noted that virtually all such theories are derived from two classes of factors: intrinsic or internal factors and extrinsic or environmental ones. So we may ask, (1) Are the intrinsic factors "random" or directional? (A third possibility is constrained by historical or developmental features.) (2) Is directionality imposed, if at all, by extrinsic factors? (3) Is inheritance "hard" or "soft"? (4) If the resulting evolution is adaptive, how does the enhancement of adaptation occur?

In talking about environmentalist versus internalist theories of evolution above, we were referring to one of the antitheses which Gould (1977a) saw as one of the three great questions about the history of life that have pre-occupied palaeontologists from pre-evolutionary times, Another concerns the antithesis between "phyletic gradualism" and "punctuated equilibria" (Eldredge and Gould 1972; Gould and Eldredge 1977). The third was directionalist versus steady-statist. In Gould's own words:

1. Does the history of life have definite directions; does time have an arrow specified by some vectorial property of the organic world (increasing complexity of structure, or numbers of species, . . .). [Is it] *directionalist* [or] *steady-statist*[?]
2. . . . Does the external environment and its alterations set the course of change, or does change arise from some independent and internal dynamic within organisms themselves? . . . [T]he belief in external control [is] *environmentalist* . . . [C]laims for an inherent cause of change [are] *internalist.*
3. What is the tempo of organic change? Does it proceed gradually in a continuous and stately fashion, or is it episodic? . . . I will label stately, Lyellian change as *gradualist* and episodic views as *punctuational.*

Gould asserts rightly that "the formulation of these questions preceded evolutionary thought and found no resolution within the Darwinian paradigm"; and that "the major contemporary issues in palaeobiology represent the latest reclothing of these ancient questions".

I. Lamarck and "Lamarckism"

In Gould's categories Lamarck's theory of mechanism is characterised as *steady-statist, internalist,* and *gradualist.* In answer to the questions I posed above, (1) the intrinsic causal factors of evolutionary change are directional; (2) direction is imposed by extrinsic, environmental factors only to the extent that adaptation is improved with respect to contingent ecological conditions; but this causes the deviation of organisms from the innate overall direction of evolution; (3) inheritance is "soft", but with important qualifications; (4) the enhancement of adaptation is the result of the response of the organism to environmental conditions.

More specifically, there were three major factors in Lamarck's theory. The first factor was spontaneous generation; the new organisms so produced were his monads (Chapter 3), which were the only organisms for which Lamarck allowed the possibility of extinction, apart from extinction due to human agency. The second factor was the innate tendency for successive descendants of each *Monas* to evolve, in a gradualistic manner, up the *scala naturae,* eventually reaching the status of mankind. The most extraordinary feature of Lamarck's scheme, as Simpson (1961b) pointed out, is that Lamarck's is a steady-state theory of evolution, with constant spontaneous generation of monads, their evolution up the *scala naturae,* and the degradation, in each generation, of organised beings into their basic constituents at death. So while "*la marche de la nature*" has continued from time immemorial, there is no net change in the state of the biota with time, except that due to extrinsic factors. Without the latter, Gould's characterisation of Lamarck's theory as *steady-statist, internalist,* and *gradualist* is justified. "Lower" organisms exist only because their monad forebears were generated more recently than those of mankind. The direction of evolution is set by "some independent and internal dynamic within organisms", . . . and the *scala naturae,* combined with the concept of plenitude, can represent only a gradualist pattern of phylogeny.

But Lamarck's name is always associated with the third major factor in his theory of mechanism, the "inheritance of acquired characters". He used this factor not only to explain the phenomenon of adaptation to environmental conditions, but also to rationalise deviations in phylogeny from the straight and narrow path of the *scala naturae*. In *Philosophie Zoologique* (1809), Lamarck expressed his theory of the inheritance of acquired characters together with that of the use and disuse of organs in the form of two laws:

> *First Law:* In every animal that has not reached the end of its development, the more frequent and sustained use of any organ will strengthen this organ little by little, develop it, enlarge it, and give it a power proportionate to the duration of this use; while the constant disuse of such an organ will insensibly weaken it, deteriorate it, progressively diminish its faculties, and finally cause it to disappear.
>
> *Second Law:* All that nature has caused individuals to gain or lose by the influence of the circumstances to which their race has been exposed for a long time and consequently, by the influence of a predominant use or constant disuse of an organ or part, is conserved through generations in the new individuals descending

> from them, provided that these acquired changes are common to the two sexes or to those which have produced these new individuals.

In *Histoire Naturelle des Animaux sans Vertèbres*... (1815–22), the two laws have become four, although the principles which they represent are implicit in *Philosophie Zoologique.* Laws numbers three and four are more succinct statements of the original two. The new laws are as follows:

> *First Law:* Life by its own forces tends continuously to increase the volume of all bodies which possess it, and to extend the dimensions of their parts up to a limit which it itself imposes.
>
> *Second Law:* The production of a new organ in an animal's body results from a newly arisen need [*besoin*] which continues to make itself felt, and from a new movement that this need brings about and maintains.
>
> (*quoted from the 1837 edition: my translation*)

Thus in the *Histoire Naturelle,* the first law embodies the innate tendency to evolve, while the second concerns the origin of new organs, required as a separate factor from the use and disuse of parts and the inheritance of acquired characteristics embodied in the original two laws. The four laws represent a complete theory of evolutionary mechanism.

One important qualification to the inheritance of acquired characters is emphasised by Lamarck in the *Histoire Naturelle* and also in the *Philosophie Zoologique.* Thus in the latter (p. 107 in Elliot's translation):

> I must now explain what I mean by this statement: *the environment affects the shape and organisation of animals*.... It is true if this statement were to be taken literally, I should be convicted of an error; for, whatever the environment may do, it does not work any direct modification whatever in the shape and organisation of animals.
>
> But great alterations in the environment of animals lead to great alterations in their needs, and these alterations in their needs necessarily leads to others in their activities. Now if the new needs become permanent the animals then adopt new habits which last as long as the needs that evoke them.

If one accepts "the inheritance of acquired characters", it seems to me that there are three categories of acquired characters that could be held to be inherited. The first comprise mutilations or the results of accidents; these are absolutely ruled out by Lamarck. Thus Weismann's famous experiment of cutting off the tails of mice in successive generations to see whether there was an inherited loss or reduction was no test of Lamarckism. This type of inheritance was, however, accepted by some "Neo-Lamarckists" (and also was allowed as a possibility by Darwin). The second category made possible the environmental modification of plants and *animaux apathiques.* We noted in Chapter 3 (Section I) Lamarck's division of the Animal Kingdom into three grades – *animaux apathiques, sensibles,* and *intelligents.* Plants and the lowest grade of animals were said to lack a nervous system. Thus in the case of the insensitive animals, the causes of all their activities were external to them, resulting in purely mechanistic responses in the "subtle fluids" which they contained; similarly in plants.

The third category of acquired characters are those inherited by the higher animals. It is only here that the full Lamarckian mechanism appears to come into play, although Lamarck's own views changed with time and are in some cases inconsistent. Here "use and disuse" produces change as a result of the individual response of generations of animals to their environment, so that the response reifies an adaptive need. Even in this case Lamarck saw the process as a largely mechanistic one in which behaviour resulted in almost every case from instinctive patterns produced by a *sentiment intérieur,* so that only the higher vertebrates were capable of voluntary action, and then only rarely. Thus Cuvier's famous parody of Lamarck's views in a funeral eulogy was particularly egregious. It was based on the ambiguous word *besoin* in the second of Lamarck's four laws, which, if interpreted as desires or wishes rather than needs, has animals willing themselves new organs.

In Britain, but not always in France and elsewhere, Lamarck's views were decried by the scientific establishment (Desmond 1989). They were popular in Edinburgh and London among radicals in the medical profession in the late 1820s and in the 1830s, with Robert Grant, who taught Darwin in Edinburgh but later moved to London, as the leading figure. Lyell, however, represents the establishment view in his *Principles of Geology* (1830–3). He gives a very full and mostly fair account of Lamarck's views, which he rejects as a special creationist. But his ridicule is reserved for the idea, in Lamarck's theory, that function gives rise to form, rather than form dictating function. This, too, was Cuvier's main objection to Lamarck (Chapter 2): taxonomic differences may all have reflected differences in function, but the latter were anatomically preordained, they did not evolve. The difference in view is clearly set out by Lamarck himself (but his smug assertion, on this as well as other occasions, that he could prove his case were he of a mind to, probably didn't help):

> I could prove that it is not at all the form either of the body or its parts that gives rise to habits, to the way of life of animals, but that to the contrary it is the habits, the way of life and all the influential circumstances that have with time established the form of the bodies and the parts of animals. With new forms, new facilities have been acquired, and little by little nature has arrived at the state where we see her now.
>
> (*Lamarck 1801, p. 13; Burkhardt's 1977 transl.*)

II. Natural Selection: Darwin and Wallace

The currently accepted theory of an evolutionary mechanism is that of Natural Selection, proposed in its original form by Charles Darwin (1859) and independently by Alfred Russel Wallace (Darwin and Wallace 1858). It is not surprising, however, that the modern theory – "Neo-Darwinism" or the Synthetic Theory – differs radically from that proposed 130 years ago. What is surprising is that, like Darwin and Wallace's original theory, the synthetic theory does not explain or constrain the pattern of phylogeny reconstructed from the phenomenon of natural classification. Furthermore, the modern synthesis introduced an unsatisfactory feature that has led to the accusation, now a wearisome cliché, that the theory of natural selection is a "tautology". This is emphatically not the case for Darwin's and Wallace's original theory, as I shall show in this section. There may, however, be something in the charge with respect to the synthetic theory; we shall consider that later, in Chapter 12.

One of the most important tasks that both Darwin and Wallace set themselves in proposing the theory of natural selection was to explain the origin of adaptations, and it is perhaps the emphasis on this that has made their theory, and its modern counterpart, an incomplete one. But the explanation of adaptation was very necessary, both to counter Lamarck's unacceptable theory and Cuvier's critique of it, and to accommodate the long preoccupation of English naturalists with adaptation as evidence of the existence of God, as in John Ray's *The Wisdom of God Manifested in the Works of the Creation* (1691). More immediate, as far as Darwin was concerned, was Paley's *Natural Theology: Or Evidences of the Existence and Attributes of the Deity Collected from the Appearances of Nature* (1802). Despite the fact that in many ways Paley's work was derivative, particularly from Ray, and that the whole "argument from design" had been severely criticised by David Hume in *Dialogues concerning Natural Religion* (1779), Paley's *Natural Theology* (1802) was immensely influential. It revived a tradition that led to the *Bridgewater Treatises.*

Francis Henry Egerton, the eighth (and last) Earl of Bridgewater (died 1829), left a bequest of £8,000 to the Royal Society for a treatise or treatises to be written

> on the Power, Wisdom and Goodness of God as manifested in the Creation; illustrating such work by all reasonable argument, as for instance the variety and formation of God's creatures in the animal, vegetable and mineral kingdoms; the effect of digestion, and thereby of conversion; the construction of the hand of man, and an infinite variety of other arguments; and also by discoveries ancient and modern, in arts, sciences and the whole extent of literature.

Eight Bridgewater Treatises were published between 1833 and 1840, thus falling within the period (1831–44) between Darwin's setting out on the voyage of the *Beagle* and his writing of his first extended (but private) essay on the Theory of Natural Selection. The authors included William Buckland with his treatise on *Geology and Mineralogy Considered with Reference to Natural Theology* (1836); the physiologist Sir Charles Bell (with his work on the hand); William Whewell (see Chapter 13); Peter Mark Roget (of *Thesaurus* fame) on animal and vegetable physiology; and the entomologist William Kirby (on instinct).

Thus natural theology was a potent force in the early part of the nineteenth century, and any naturalistic explanation of adaptation, leading to a "positivist" theory of evolutionary mechanism, had to tackle it head-on. A Lamarckian type of theory, with unknown intrinsic forces responding to environmental pressures, would not do: the organising principle had to be exclusively extrinsic, with "hard" inheritance.

The theory of Natural Selection is usually presented as a series of logical propositions and inferences from them. If the initial propositions, all of which are matters of empirical observation, are true, then natural selection follows. Thus Mayr (1982) presents the theory as follows:

Fact 1: All species have such great potential fertility that their population size would increase exponentially . . . if all individuals that are born would again reproduce successfully.

Fact 2: Except for minor annual fluctuations and occasional major fluctuations, populations normally display stability.

Fact 3: Natural resources are limited. In a stable environment they remain relatively constant.

Inference 1: Since more individuals are produced than can be supported by the available resources but population size remains stable, it means that there must be a fierce struggle for existence among the individuals of a population, resulting in the survival of only a part, often a very small part, of the progeny of each generation.

These facts derived from population ecology lead to important conclusions when combined with certain genetic facts.

Fact 4: No two individuals are exactly the same; rather, every population displays enormous variability.

Fact 5: Much of this variation is heritable.

Inference 2: Survival in the struggle for existence is not random but depends in part on the hereditary constitution of the surviving individuals. This unequal survival constitutes a process of natural selection.

Inference 3: Over the generations this process of natural selection will lead to a continuing gradual change of populations, that is, to evolution and to the production of new species.

This formulation is chosen from a recent account of an acknowledged leader in evolutionary theory, but I suggest that to do justice to Darwin and Wallace even with their slight knowledge of heredity, Mayr's statement of the theory must be amended to give it logical rigour. Where I have retained them, his original propositions are slightly abridged (but note rewording of inference 1); – significant additions and amendments are in italics:

Fact 1: All species have the reproductive potential for exponential growth in population size.

Fact 2: Population numbers normally display stability with limited fluctuations.

Fact 3: Natural resources are limited and, in a constant environment, remain stable.

Inference 1: Since more individuals are produced than can be supported by the available resources, but population size (normally) remains stable, only an (often very small) part of the progeny of each generation survive to reproduce.

Fact 4: *There is (phenotypic) variation between individual members of some or all populations.*

Fact 5: *Some of the variation is heritable.*

Fact 6: *Some of the variation represents differences in degree of adaptedness to current environmental conditions.*

Hypothesis 1: *Heritable differences in adaptedness occur between members of the same population in some or all populations.*

Hypothesis 2: *Within a population better adapted individuals have a better chance of surviving to achieve their reproductive potential.*

Hypothesis 3: *Heritable improved adaptedness (potential or actual) appears spontaneously in individuals within some or all populations.*

Inference 2: "*In the struggle for existence*" (Inference 1) *better adapted individuals increase in frequency in a population with time:* **this differential increase is Natural Selection.**

The Theory: *Adaptive anagenesis (i.e., adaptive phyletic evolution) is the result of Natural Selection, acting on spontaneously occurring adaptive features.*

This seemingly pedantic expansion of Mayr's formulation has been proposed to achieve two ends: first, to make all the assumptions of the theory explicit; and second, to refute the suggestion that Darwin's and Wallace's theory was "tautologous". I have retained Mayr's usage of "fact" and introduced three "hypotheses", but all are empirical statements that most biologists would accept. Given their correctness, inferences 1 and 2 follow, and the phenomenon represented by inference 2 is labelled "Natural Selection", so that a definition of natural selection is implicit in the two inferences.

My hypothesis 1 does not follow deductively from facts 5 and 6. If only *some* variation is heritable and only *some* is of adaptive significance, it does not follow deductively that there are individuals with heritable differences of adaptive significance. Thus hypothesis 1 has empirical content and should be testable. Hypothesis 2 is necessary to exclude the possibility that anagenetic evolution proceeds by some internal mechanism, orthogenesis, or racial life cycles (see below), so that although adaptation patently occurs, it is not translated into differential survival. Similarly, for selection to occur, there must be some correlation between adaptation and reproductive success. Furthermore, if evolution is to continue without the supply of adapted variants becoming exhausted, new ones must arise as proposed in hypothesis 3.

Inference 2 and the Theory correspond to inferences 2 and 3, respectively, of Mayr, reworded to take account of the revised propositions that precede them, but there is an important difference: "the production of new species" cannot be inferred from anything that has gone before. It is a different hypothesis that cladogenesis – the splitting of one species into two or more – is the result of natural selection.

In my amended propositions I have used several (probably confusing) words related to the concept of adaptation. This I believe to be necessary absolutely to refute the charge that natural selection is "tautologous". The charge of tautology seems originally to have arisen, at least in part, with Darwin's adoption of Herbert Spencer's phrase "survival of the fittest" in the fifth edition of the *Origin* (Darwin 1869). Then the theory, or rather its operative principle, can be parodied as "natural selection is the survival of the fittest: how is the degree of fitness judged? – by differential survival!"

Even if the theory did contain such a circular argument at its centre, it would still not be tautologous for two reasons: The first is a matter of definition, discussed by Sober (1984). In logic, a tautology is a *logical truth* that is *truth-functionally true*. A logical truth is a proposition that is true by virtue of its logical structure; symbolically:

$$(x)\ (x = x)$$

or, "whatever the nature of x, it is true in every case that x equals x". A *tautology* is a special case of logical truth that is true by virtue of the truth-functional terms it contains. Truth-functional terms are words such as "and" (conjunction), "or" (disjunction), "not" (negation), the conditional ("*if p then q*"), and the biconditional ("*p if and only if q*"). An example would be "either it is raining or it is not" ($p \vee -p$) (e.g., Quine 1966).

It cannot be claimed that the theory of natural selection is of this nature. Presumably the claim is that, as parodied above, it is either analytic, as in "all bachelors are unmarried", where the statement is true by virtue of the meaning of the words alone, or that it is synthetic (i.e., saying something about the world) but self-evidently true, as in the statement (if true) that "nothing can be completely red and completely green at the same time". Propositions of this latter kind are necessarily true or comprise *a priori* knowledge.*

However, even if it were true that the theory contained the circular argument at its core, it would still not be *a priori*. The *definition* of natural selection would be unsatisfactory, but the *theory* that natural selection can result in anagenetic evolution would still both be synthetic and have empirical content. It would amount to a statement that "differential reproduction in populations can result in anagenetic evolution", which is certainly not a necessary truth.

All this much debated ground can be bypassed by a clear picture of the role of adaptation in Darwin's theory. The var-

*I am grateful to Dr Jane Heal and Mr Geoffrey Midgley for making these distinctions clear to me.

ious terms derived from the concept in my statement of the theory are used in the senses suggested by Burian (1983), who gives a clear and definitive account of the subject. Burian distinguishes *adaptiveness* (of a trait) or *adaptedness* (of an organism), as two ahistorical descriptive concepts, from *adaptation,* a more onerous historical one:

> Fleetness contributes to the adaptedness of a deer . . . if, and only if. . . , it contributes to the solution of a problem posed to the deer – for example, escaping predation. Fleetness is an adaptation of the deer if, and only if, the deer's fleetness has been molded by a historical process in which relative fleetness of earlier deer helped shape the fleetness of current deer.

Burian also equates "adaptedness" of an organism with "engineering fitness" to emphasise the analogy of "adaptation" and design. Thus according to Ospovat (1981) Darwin in his earliest speculations on evolution believed that organisms had "limited [by design constraints] perfect adaptedness" or "absolute engineering fitness". Darwin later moved to believing that organisms had only "relative [to that of other organisms or specified states] engineering fitness". The concept of "adaptedness" or "engineering fitness" is distinguished by Burian from two types of (traditionally but unfortunately named) "Darwinian fitness", which are defined in terms of reproductive success. We shall consider these in talking about "Neo-Darwinism" or the Synthetic Theory. Meanwhile let us note Burian's succinct conclusion:

> For Darwin, natural selection is systematic differential reproduction due to the superior engineering fitness of certain available variants – due, that is, to the relatively better design of the favoured organisms.

Darwin and Wallace agreed that natural selection resulted in gradualistic anagenesis [although Wallace was never happy about the term "natural selection" and in his original paper (Wallace, in Darwin and Wallace 1858) rejected comparison with selective breeding of domestic animals]. They also agreed that there was a graded series from varieties to full species in evolution, and Darwin (1859) makes the point in the *Origin* that

> [n]atural selection also leads to divergence of character; for more living beings can be supported on the same area the more they diverge in structure, habits and constitution, of which we see proof by looking at the inhabitants of any small spot or at naturalised productions.

From this principle of divergence both he and Wallace assumed that sympatric speciation could occur, with one species splitting without geographical isolation of the potential daughter species.

Darwin, however, always accepted that selection "has been the main but not exclusive means of modification", that "variations neither useful nor injurious would not be affected by natural selection", and that

> there are many unknown laws of correlation of growth, which when one part of the organisation is modified through variation, and the modifications are accumulated by natural selection for the good of the being, will cause other modifications, often of the most unexpected nature.

He was also at pains to emphasise that selection produces adaptations that are adequate to circumstances rather than perfect (Burian's relative engineering fitness):

> No country can be named in which all the native inhabitants are now so perfectly adapted to each other and to the physical conditions under which they live that none of them could anyhow be improved; for in all countries, the natives have been so far conquered by naturalised productions, that they have allowed foreigners to take possession of the land.

Wallace, on the other hand, had, by the 1870s, become what would now be described as a "pan-selectionist", rejecting use and disuse and the inheritance of acquired characters, rejecting Darwin's (1871) sexual selection as a separate mechanism, and asserting the selective origin of isolating mechanisms between species. The startling exception was his refusal to believe in the natural origin of the human mind (Wallace 1870).

Gould (1977a) characterised natural selection, or rather Darwin, as environmentalist and gradualist and ambiguous between directionalism and steady-statism: there was no inherent guarantee of direction in the theory, but it did not prohibit general improvement in design, and the hierarchical arrangement of phylogeny, supported by both Darwin and Wallace (Chapters 2 and 3), was irreversible.

The answer to the other questions I posed at the beginning of this chapter are that (1) the intrinsic factors are taken to be "random" with respect to the direction of selection; (2) any directionality is imposed by the environment; (3) inheritance was taken to be "hard" by Wallace, with Darwin agnostic on the issue; and (4) the enhancement of adaptation was by competition under environmental duress.

III. Post-*Origin* theories

As has been recorded many times, publication of *On the Origin of Species* convinced most naturalists, as well as many of the lay public, that evolution had occurred, but natural selection fared less well. Huxley's comment to Darwin that "you have loaded yourself with unnecessary difficulty in adopting *Natura non facit saltum* so unreservedly", is well known, but despite this saltational theories were relatively unimportant until the beginning of the twentieth century. Mayr (1982) lists Nägeli, His, Kölliker, W. H. Harvey, Mivart, and Galton, none of them field naturalists, as nineteenth-century saltationists in addition to Huxley.

The strongest movement, however, was back to something very close to Lamarck's theory. Haeckel claimed to be reconciling Darwin and Lamarck, but a series of mostly North American palaeontologists developed Lamarck's theories almost as though the *Origin* had never been published. Both the internalist and the environmentalist components of Lamarck's theory were echoed, perhaps because of the lack of any developed phylogenetic theory in the *Origin*. A number of variants of each component can be recognised. There are three categories of purely directionalist-internalist theories. These are orthogenesis, finalism, and racial life cycles. A belief in the "inheritance of acquired characters" was often added to each.

The term "orthogenesis" has been used in two ways: first, to represent an innate tendency to evolve towards a future (usually adaptive) "target", here referred to as "finalism"; and second, to represent innate evolutionary trends that could carry the evolving organism towards a non-adaptive state and even result in extinction. It is the latter sense in which I use the term "orthogenesis" here.

Suggested examples of orthogenesis originally came almost entirely from palaeontologists. Lang (1923) cited overcoiling of one valve in the Triassic oyster *Gryphaea* to a point (according to him) when old individuals late in the series were unable to open their valves. In this case, not only is there no reason to suggest an orthogenetic mechanism, but also there is no evidence in the fossil record of such a trend at all (Hallam 1968; Gould 1972; Hallam and Gould 1975). Schindewolf (1950), like others before him, claimed to show orthogenetic enlargement of the upper canine tooth in sabre-tooth cats, but his series was not in the correct time sequence, and once again the trend did not actually exist (Simpson 1953). In the case of the so-called Irish elk *Megaloceros,* there is in fact a trend for increasing body size and with it a geometrical increase (positive allometry) of the span of the antlers of adult males. Simpson (1944, 1953) suggested that increase in body size was adaptive but was linked genetically to antler size. Selection was therefore supposed to drive the evolving lineage towards a net optimum in which sub-optimal body size was balanced by super-optimal antler span. Gould (1974), on the other hand, interpreted the huge antlers as adaptive in their own right in intraspecific competition between males, in which antler span determined status.

As Gould (1977b) pointed out, the evolutionary ideas of Edward Drinker Cope proposed a directional component and one causing adaptive deviations from the general direction of evolution. His theory of directional evolution involved changes in the rate of development in ontogeny. Acceleration in development allowed terminal addition of new ontogenetic stages; the result was recapitulation of the type that Haeckel championed (Chapter 3). Cope (1896) also postulated rarer retardation in development, resulting in paedomorphosis, retrogressive evolution, or degeneracy. The results of this he described as "catagenesis", in contrast to "anagenesis" produced by acceleration and terminal addition. New characteristics were at first held to arise by an internal teleological or finalistic mechanism (Cope 1870); later, however, Cope reversed his position. He then held (Cope 1896) that new characters arose by a "Lamarckian" mechanism but that the acceleration and retardation were controlled by some internal "peculiar species of energy" and were thus orthogenetic.

Finalism in evolutionary theory is of two types: Cope's early theory of the origin of new characters before they were adaptively useful was extended by Henry Fairfield Osborn with his invention of "Aristogenesis". Study of the palaeontology of horses, proboscideans, and titanotheres convinced him that new characters (incipient tooth cusps, titanothere nasal horns) arose mediated by "aristogenes" to become adaptively useful only generations later (Osborn 1934), an extreme case of "pre-adaptation" which reverses causality. This is the first type of finalism. The second is where evolution is supposed to approach a terminal target, usually the origin of humankind. It is not clear whether Lamarck's internal perfecting principle was seen as finalistic in this sense; that of Teilhard de Chardin (1959), with his "omega point", certainly was.

The theory of racial life cycles is most notably associated with the name of the American palaeontologist Alpheus Hyatt (Gould 1977b), who championed the idea of acceleration and thus recapitulation in evolution. According to Hyatt, however, degeneracy was not caused by retardation, but by the terminal addition of senile features as each lineage went through the successive phases of phyletic youth, maturity, old age, senescence, and thus to extinction. Towards the end of the phyletic life cycle the mature phases of ontogeny began to drop out of the individual development.

"Lamarckism" in the popular sense is also represented by Cope in his book *The Primary Factors of Organic Evolution* (1896). He used the terms "kinetogenesis" and "physiogenesis" for the two categories of "acquired characters" which he took to be inherited. Kinetogenesis was the acquisition of characters by the active response of an animal to environmental change, resulting in the origin of new organs or changes in proportions and relative growth. The inheritance of kinetogenetic effects was thus strictly comparable to Lamarck's mechanism for higher animals. Physiogenesis was in part comparable to the mechanism that Lamarck proposed for lower animals or plants. There were effects resulting from the imposition of an environmental change on animals that produced a response, but not one due to a change in behaviour.

It is not clear that Cope excluded "pseudo-Lamarckism" – that is, the inheritance of accidental characters, mutilations, and so forth. In the case of kinetogenesis, Cope cited the control of bone growth by the stresses produced by muscular action. An example of physiogenesis in plants was ecophenotypic variation; analogous effects in animals included variation in the shells of molluscs, environmentally controlled variation in the colour of the pupae of Lepidoptera, the colours of flatfish, and the loss or reduction in size of eyes in cave-living animals. All are real phenomena; the error was in assuming the inheritance of what were previously purely phenotypic features.

Darwin's "provisional hypothesis of pangenesis" in the second volume of *The Variation of Plants and Animals Under Domestication* (1868) also provided for the inheritance of physiogenetic and kinetogenetic changes by the migration of gemmules or "pangenes" from the affected parts of the body to the individual's gonads. Weismann's experiments involving amputation of the tails of mice, referred to above, refuted the migration part of Darwin's theory (for mice!), but more influential was his theory of the germ-plasm (Weismann 1892) in which he saw the germ-cells as segregated and inviolate, but giving rise to individual organisms ("*soma*") in each generation, and thus as a refutation of "Lamarckism".

Lamarckism (real and "pseudo") persisted into the twentieth century, but suffered a sharp decline after the First World War with the development of genetics (Bowler 1983). A number of authors in the twentieth century have claimed Lamarckian inheritance, particularly of behavioural traits (see Fothergill 1952), some of them reaching the level of the ridiculous (Cannon 1958, 1959). More recent proposals of a wholly or partially Lamarckian mode of inheritance in Metazoa (Steele 1979) have not been corroborated, but there is now a suggestion of environmentally directed mutation in bacteria (Cairns, Overbaugh, and Miller 1988). Classic experiments by Luria and Delbruck (1943) and by Lederberg and Lederberg (1952) on phage-resistant mutants of the bacterium *Escherichia coli* showed that mutant resistant clones that survived on phage virus-impregnated culture medium had arisen at different times *before* contact with the virus. They were not produced by the stimulus of the phage. Cairns *et al.*, however, claim that nonlethal stress will produce appropriate mutants. Thus they claim that strains of *E. coli* unable to utilise particular sugars will produce mutants able to do so in a medium where they have to utilise those sugars.

Before turning to the evolutionary theories of the early geneticists, we should note the unique contribution of D'Arcy Thompson (1917). His ideas of "laws of growth" resulting from mechanical and geometrical constraints are in part reminiscent of Cuvier's ideas that all taxonomic features reflect adaptive necessity, and in part, in their insistence on the derivation of body plans from one another by geometrical laws, reminiscent of the *Baupläne* of idealistic morphology.

The development of saltational theories of evolution ushered in, and was reinforced by, the beginning of modern genetics during the earliest years of this century. William Bateson (1861–1926) is usually recognised as the first important figure in this movement, but Bowler (1983) suggests that Bateson was influenced by W. K. Brooks, whose *Law of Heredity* (1883) included a section on saltation, citing the opinions of Galton and Huxley. Bateson collected cases of discontinuous variation on the assumption that internally driven saltation was the mechanism of evolution. His data were presented in *Materials for the Study of Variation* (1894). He was particularly interested in variation in meristic characters within species, the number of petals in a flower, the number of body segments in an animal, where they could not easily be intermediate forms. Eventually Bateson's ideas on evolution degenerated to a form of preordained orthogenesis, with characters becoming manifest in a set order by mutations that unblocked them from previous inhibitors. Ironically, Bateson, who coined the word "genetics", never accepted the chromosome theory of heredity. Contemporary with Bateson was the origin of the school of biometry of Karl Pearson and W. F. R. Weldon, who attempted the study of continuous variation to support gradualistic evolution (Provine 1971).

The discovery of Mendel's work by de Vries, Correns, and Tschermak in 1900 gave a tremendous boost to the theories of saltation. De Vries (1889) had already done "Mendelian" studies of his own and, shortly after produced his *Die Mutationstheorie* (1901–3), based heavily on the abnormal behaviour of the flowering plant *Oenothera lamarckiana* (see Darlington in Mayr and Provine 1980). According to de Vries, *Oenothera* was the only organism he could find in an active evolutionary phase, producing spectacular "mutations", which led, according to him, to a new sub-species, or even to speciation, in one jump. De Vries's theory received support from a famous series of experiments by Johannsen (1903, 1909) in which he bred pure self-fertilised lines of beans. In each generation there was a considerable variation in the size of the seed in spite of the fact that all had the same "genotype" (Johannsen's term). He saw this variation as that on which Darwin's theory depended and thought he had demonstrated that small differences among individuals, on which selection was supposed to act, were simply not heritable. Johannsen also coined the term "phenotype", but used it to represent the average of a phenotypic series rather than the individual expression of the genotype. As Mayr (1982) noted, Johannsen's concepts were entirely typological, as were those of most early laboratory geneticists: population thinking was entirely foreign to him. This was true also of Thomas Hunt Morgan, whose team at Columbia University established the identity of the genes of mutation and recombination with the genes ordered on chromosomes, by the technique of chromosome mapping. At first, Morgan (1903) accepted a theory of evolution not dissimilar to Bateson's, but later (Morgan 1916) accepted evolution by selection of gene mutations, which were scaled down from those of de Vries but still did not correspond to Darwin's gradualism.

Thus because of the nature of the development of early genetics, there occurred that unfortunate split between experimental genetics and natural history which delayed the development, or rather the general acceptance, of the synthetic theory until the 1930s and 1940s. Naturalists, notably palaeontologists, retained their belief in gradualism [with the notable exception of Schindewolf (1950) and some of his German colleagues] and took refuge in internalist theories of mechanism. The reconciliation between the two types of biology occurred in the work of the authors of the synthesis, but unhappily the split remains to this day between those who pursue "normal science" in both disciplines (Chapter 13).

12

Mechanisms of evolution: The Synthetic Theory

The story of the development of the synthetic theory of evolution, named from Julian Huxley's contribution, *Evolution: The Modern Synthesis* (1942), has been told many times, notably in Mayr (1982) and in the collective work, Mayr and Provine (1980). As Mayr has frequently emphasised, population thinking – seeing anagenesis as the establishment of new forms in interbreeding populations – plus the biological species concept, traceable to John Ray in the seventeenth century (Ray 1686; see Chapter 14), were necessary before the theory could be developed. Theoretical population genetics – the mathematical modelling of genes in populations – was developed with different emphases by Fisher, Haldane, and Sewall Wright. Contemporary with them, Chetverikov and his colleagues in Russia were not only developing parallel ideas but testing them on wild populations of the fruit-fly *Drosophila,* Morgan's choice of experimental animal. Dobzhansky, not a member of Chetverikov's group but strongly influenced by them, continued his work in the United States in a series of papers on *D. pseudoobscura* from 1938 to 1976 and also produced what is usually considered the first great book of the synthesis, *Genetics and the Origin of Species,* in 1937. The others usually recognised are Huxley's; Mayr's *Systematics and the Origin of Species* (1942); Rensch (1947), translated in its second edition as *Evolution Above the Species Level* (1959); and Simpson's *Tempo and Mode in Evolution* (1944), considerably changed (Gould 1980c) and reissued as *The Major Features of Evolution* (Simpson 1953).

All the elements of the synthesis were, however, established long before this. The origin of species by allopatric speciation, developed by Mayr (1942, 1963), was suggested by David Starr Jordon in 1905, but has a pedigree going back to von Buch (1825; Mayr 1982). The Hardy–Weinberg equilibrium as the basis of theoretical population genetics (see below) was established in 1908. Yule (1902) suggested that continuous variation in a population could be caused by a series of genes affecting one character. Selection experiments on such a character, the pigmented back stripe on "hooded" rats, were reported by Castle and Phillips (1914). In the same year, Muller (1914) suggested that gene modifiers were being selected, and this was accepted by Castle in 1919. The concept of modifiers played an important part in Fisher's theory of evolutionary mechanism as summarised in his *Genetical Theory of Natural Selection* (1930). But Fisher (1918) had already modelled genotype–environment interactions and the types of interaction among genes. Lewontin (1980) points out that two groups of geneticists were influenced by Fisher's work: the theoretical population geneticists, to whom the fundamental theorem of natural selection – "The rate of increase in fitness of any organism at any time is equal to its genetic variance at that time" – is his most stimulating contribution; and the naturalists, to whom gene modifiers, the evolution of dominance, and the maintenance of polymorphism by selection are the inspiration of their work. The most vociferous supporters of the synthetic theory are drawn from the two traditions so produced. The latter tradition is exemplified by Ford's *Ecological Genetics* (first edition, 1964).

Sewall Wright (1931) was concerned not just with selection but also with all the evolutionary factors affecting populations, large and small; hence the "shifting balance" theory, allowing a small population to move from one peak of adaptation to a higher one, a process involving an element of non-selective evolution – genetic drift.

The two terms "Neo-Darwinism" and "Synthetic Theory" are often treated as if they were synonymous, but in fact "Neo-Darwinism" was originally applied by Romanes (1896) to the theory of Natural Selection as interpreted by Weismann, whose theory of the germ-plasm dictated that heredity was always "hard" (Chapter 11, Section III). There could be no transmission of information from the body ("soma") of an organism to the hereditary material (in the "germ-plasm") and thus no "inheritance of acquired characters". This strong version of natural selection was supported by a number of naturalists, notably Wallace and Poulton, through the beginning of the century.

The main logical structure of the Synthetic Theory is the same as that set out for Darwin's and Wallace's original theory of Natural Selection (in Chapter 11, Section II), but, of course, it incorporates the knowledge of heredity and the models of population genetics acquired and developed since. As Mayr has repeatedly emphasised, theoretical population genetics had to be paralleled by the naturalists' knowledge of the heterogeneity of natural populations. This gave a special importance to the study of polymorphism in populations, in *Drosophila* by Dobzhansky, in a number of organisms by Ford, following Fisher's work, and later in the banded snails of the genus *Cepaea* by Cain, Sheppard, Currey, and others, and also notably in mimetic butterflies, a preoccupation of Alfred Russel Wallace and other nineteenth-century naturalists. The interpretation of mimicry had also been a sparring ground for supporters and opponents of natural selection ever since Bates (1862) first explained the phenomenon of mimicry in terms of natural selection (see Ford 1975).

The models of population geneticists are concerned primarily with the frequency of alternative alleles at the same gene locus, so that the early models of selection were concerned with the rate of "fixation" of an allele [so that all its rivals at a locus were eliminated from the (hypothetical) population]. Naturalists, on the other hand, were concerned to show that character differences that they found in heterogeneous populations were of adaptive significance. This later degenerated into an attempt to show that all characters in all organisms were adaptive, notably in the case of observation and experiment on animal behaviour, a preoccupation stigmatised by Gould and Lewontin (1979) as the "Panglossian paradigm".

I. "The Modern Synthesis"

We can characterise the natural selection of the Synthetic Theory as follows:

> *Natural Selection occurs in any genetically heterogeneous population which has excess reproductive potential in a given environment. It is*

the enhanced probability of reproductive success of those members of the population whose heritable phenotypic differences from others adapt them better to that environment.

The theory of natural selection is then that (1) natural selection is a *necessary* condition for adaptive anagenetic change, and (more controversially) (2) natural selection is a necessary condition for cladogenesis (speciation).

More specifically we may list the assumptions made within the Synthetic Theory, mostly by population geneticists:

(1) That all evolutionary phenomena are, in principle, explicable in terms of (a) the generation of genetic variation through mutation and recombination; and (b) changes within populations in genotype and/or karyotype frequency due to gene flow, genetic drift, and selection.

(2) That differences in "Darwinian fitness" among different genotypes representing different combinations of alleles at the same locus reflect differences in degree of adaptiveness (*sensu* Burian 1983; see Chapter 11, Section II, above) produced by the phenotypic effects of those genotypes.

(3) That, at a single gene locus, if one allele goes to fixation, *in most cases* that allele is of adaptive significance relative to its rivals within the gene pool of the population. That allele may thus be described as having been "selected". The elimination of an allele from a population, or its approximation to an equilibrium frequency, is also *normally* to be explained by selection.

(4) That the frequency of mutation and direction of mutation of alleles at adaptively significant loci are neither the cause, nor the result, of the prevailing direction of anagenetic change.

(5) That *purely* phenotypic differences in one generation do not become genotypic differences in the next (i.e., inheritance is "hard", without the "inheritance of acquired characters").

While the Synthetic Theory sees adaptive anagenetic change as a result of selection, with genetic drift assigned a (usually minor) role, there is no unanimity amongst its advocates about the importance of selection in cladogenesis. The critical contribution to theories of speciation within the paradigm of the Synthetic Theory was made by Mayr (1942, 1963). Since his work it has been generally agreed that, among sexually reproducing organisms, speciation is usually allopatric; that is, that it requires a period of geographical isolation to prevent gene flow before a single species can split into two or more. The debate was reopened by theoretical models of sympatric speciation proposed by Maynard Smith (1966), and various mechanisms, some involving chromosome rearrangements, were postulated by White (1978). These models as well as some proposed examples were reviewed and dismissed by Futuyma and Mayer (1980), but the debate continues (Diehl and Bush 1989; Grant and Grant 1989; Tauber and Tauber 1989). Nor is it agreed whether there is direct selection to enhance genetic isolation between incipient species if they become secondarily sympatric ("reinforcement" or the "Wallace effect"). If so, there would be selection against the production of hybrids to produce genetic isolation; otherwise genetic isolation would be an epiphenomenon resulting from the two incipient species having evolved apart under different regimes of selection while still in allopatry (Butlin 1989; Coyne and Orr 1989).

Another controversy now perhaps of the recent past concerned the degree to which protein polymorphism, particularly that of enzymes, consisted of differences that were "invisible" to natural selection. The technique of gel electrophoresis showed an unexpected degree of heterozygosity, represented by two or more forms of what was functionally the same enzyme, coded at many loci in populations of *Drosophila* (Hubby and Lewontin 1966; Lewontin and Hubby 1966) and man (Harris 1966). This conflicted with what Lewontin (1974) referred to as the "classical" view of the genotype and posed problems of "genetic load" – severely sub-optimal fitness as modelled by population geneticists on the assumption that each enzyme form coded at the same gene locus had a different "Darwinian fitness" (see below). Attempts were made to show experimentally that two forms of an enzyme ("allozymes"; Prakash, Lewontin, and Hubby 1969) were maintained at equilibrium by frequency-dependent selection, whichever was commoner being at a disadvantage (Kojima and Yarbrough 1967; Kojima and Tobari 1969). These results, however, are not generally accepted (refs. in Kimura 1983), and the emerging consensus is that at least some apparently non-functional differences between alleles are of no selective significance: the "neutral theory of molecular evolution" (Kimura). Thus Wright's concept of genetic drift is extended from short-term evolution in small populations to long-term evolution in all populations. "Neutralists", however, accept the selection of adaptive traits as in the Synthetic Theory.

II. Criticisms of the Synthetic Theory

The "neutral theory of evolution" is spelled out by Kimura (1983) as "the theory that at the molecular level evolutionary changes and polymorphisms are mainly due to mutations that are nearly enough neutral with respect to natural selection that their behavior and fate are mainly determined by mutation and random drift". The neutral theory was originally thought of as outside the paradigm of the Synthetic Theory, as the title of the important paper by King and Jukes (1969) – "Non-Darwinian Evolution" – suggests. More recently, however, the Neo-Darwinists have accepted the neutralist theory into the fold of the Synthetic paradigm as an extension of genetic drift. One reason for this is perhaps that the neutralist theory led to the concept of the "Molecular Clock" (Chapter 9). But the history of the neutral theory is also one case of the type of scenario in which the advocates of a new theory often assert that it supercedes Neo-Darwinism, whereas its opponents, usually dogmatic Neo-Darwinists, attempt to refute or belittle it. Later the argument shifts to an assertion that the new theory is in fact part of the Synthetic paradigm after all.

Criticisms of the Synthetic Theory can be grouped together under three headings:

(1) the accusation of "tautology" (again!) and that the theory is untestable.

(2) that factors other than "random" mutation and recombination, selection, and drift must be taken into account in explaining anagenesis; suggestions include constraints on genotype and structure as well as ontogenetic factors.

(3) that natural selection has nothing to say about the pattern of phylogeny and that higher level processes or emergent properties may override selection in macro-evolution.

The accusation of tautology can be directed against the Synthetic Theory because of the way in which "fitness" is characterised and because of the extraordinary but common equation of "evolution" with change in gene frequency (e.g., Wilson and Bossert 1971, p. 20). The simplest case of the modelling of natural selection in population genetics is that in which a single gene locus of a hypothetical organism is taken to have two possible alleles, often designated *A* and *a*. Within a population of that organism the three resulting genotypes (*AA*, *Aa*, *aa*) are all taken to be present and are assigned *frequencies* (as a fraction of the total) and *fitnesses*. Conventionally the fittest genotype has a fitness of one; that of the others is relative to that of the fittest and thus may lie anywhere between zero and one. From the data, changes in genotype frequency from one generation to the next can be calculated, and thus changes in gene frequency (the frequencies *A* and *a*) follow. In this simplest of models, the changes of frequency are equated with evolution, and the whole is regarded as a model of selection.

But the model has no empirical content: the assigned fitnesses represent differential reproduction; natural selection is treated as (nothing but) differential reproduction; the result of selection is differential reproduction of alleles – that is, changes in gene frequency. So the charge of tautology depends, as in Darwin and Wallace's theory, on the characterisation of fitness. It is a charge that seems to have been accepted in the past by some leading evolutionists:

> The phrase "survival of the fittest" is something of a tautology. So are most mathematical theorems. There is no harm in stating the same truth in two different ways.
>
> (*Haldane 1935*)

> Natural selection, which was at first considered as though it were a hypothesis that was in need of experimental or observational confirmation, turns out on closer inspection to be a tautology. . . . It states that the fittest individuals in a population (defined as those which leave the most offspring) will leave the most offspring. Once the statement is made, its truth is apparent. This fact in no way reduces the magnitude of Darwin's achievement . . .
>
> (*Waddington 1960*)

> I propose . . . to define selection . . . as *anything tending to produce systematic, heritable change in populations between one generation and the next.*
>
> (*Simpson 1953; italics in original*)

As Brady (1980) notes, Haldane appears to confuse the aim of mathematics, which is the expansion of analytic statements from axioms, with causal explanation in science, which is synthetic. Simpson's definition, on the other hand, embodies all the circularity considered above but for the key word "systematic", but then "systematic" presumably means not produced by stochastic processes such as drift – that is, in the absence of other suggested mechanisms – "produced by selection".

Natural selection, like all the phenomena studied by population genetics, is modelled as an analytic expansion of the axiom represented by the Hardy–Weinberg "Law" (Falconer 1960):

> *In a large panmictic population, with random mating and without mutation, selection or migration, both genotype frequencies and the gene frequencies will remain constant from generation to generation; and the genotype frequencies will be determined by the gene frequencies.*

For the simplest case, referred to above, the gene frequencies are signified by p (freq. A) and q (freq. a), and the *Hardy–Weinberg equilibrium* is:

$$\frac{AA}{p^2} \quad \frac{Aa}{2pq} \quad \frac{aa}{q^2}$$

where $p + q = 1$ and $p^2 + 2pq + q^2 = (p + q)^2$. But in population genetics fitnesses are merely assigned, so that in talking about selection in action there is a remaining problem with the concept of fitness. It has been tackled most successfully by Mills and Beatty (1979) and Burian (1983).

Mills and Beatty point out that standard texts on evolutionary genetics define fitness in terms of the "*actual* number of offspring left by an individual or type relative to the actual contribution of some reference individual or type". They then provide what they describe as a "*propensity*" interpretation of "fitness". For individuals (Fitness_1) is the *expected* number of offspring of an individual (analogous to the actuarial expectation of life from insurance companies). Their (Fitness_2), then, represents the average Fitness_1 of all members of a particular class within a population (e.g., all those with genotype AA). This is then easily converted to the relative fitness of population genetics.

Burian also distinguishes between fitness as *actual* survival and reproduction, and fitness as *expected* survival and reproduction, using the terms "realized fitness" ("tautological fitness") and "expected fitness", respectively:

> An organism (or class of organisms sharing some property) has higher realised fitness in environment E than alternative organisms (or classes or organisms) if, and only if, its actual rate of reproductive success is higher than those of the alternatives. A type of organism (or other replicating entity) has higher expected fitness than its competitors in environment E if, and only if, it has an objective propensity to out-produce them in E.

Both realized fitness and expected fitness are included within the term "Darwinian fitness". Burian suggests that expected fitness can be measured in the laboratory by replicating population cages in which the genotype concerned is allowed to compete with its rival(s). There would then be the potential to investigate the adaptive features ("engineering fitness") that distinguish the genotype whose fitness is being measured against those of its rivals. This was done for karyotypes of *Drosophila pseudoobscura* by Dobzhansky and his colleagues (Dobzhansky, in Lewontin *et al.* 1981) and by Kettlewell (1973; see this section, below) for colour morphs of the moth *Biston betularia* in alternative environments. In *D. pseudoobscura* a correlation was shown between the equilibrium frequencies of two competing karyotypes and temperature and nutritional factors. In the case of *B. betularia,* estimates of the frequency of three morphs at two sites showed that the *typica* form, apparently cryptic against tree-trunks with foliose lichens, was in the overwhelming majority at an unpolluted site where lichens occurred; the almost black *carbonaria,* absent at the unpolluted site, was in the majority at a site with soot-covered lichen-free trees; and a third morph, *insularia,* was present at a low frequency in both.

It is worth dwelling a little longer on the case of *Biston betularia.* In British secondary schools and institutions of higher education, and no doubt elsewhere, it is the cliché example of a demonstration of natural selection. Basing his work on reviews of the phenomenon of "industrial mechanism" by E. B. Ford (1937, 1940, 1945), Kettlewell (1955a, 1955b, 1956) conducted an elegant series of observations and experiments to demonstrate that the colour and pattern of both the *typica* and *carbonaria* forms were adaptive as camouflage against bird predators. The work, including the population estimates noted above, is summarised in Kettlewell (1973) and Ford (1975). More recent work is summarised by Berry (1990) and Mani (1990).

All the original results showed convincingly that the apparent camouflage of *carbonaria* in polluted woodland and of *typica* in unpolluted woodland was in fact adaptive, *assuming that the resting position of the moths during the day was on exposed tree trunks* as they were planted by Kettlewell and his colleagues (see Mikkola 1979, 1984a; Liebert and Brakefield 1987). But we must ask, Are the results a test of adaptation, natural selection, or both? With the proviso noted above, they are certainly a demonstration of adaptation. The colour of each morph does protect it against predation while on the appropriate background.

I believe that the correct formulation of Kettlewell's work, as summarised so far, is that it is a *demonstration* of natural selection but not a test of the theory. The experiments on *B. betularia* corroborate the idea that the morphs are cryptic, and that crypsis protects them from predation; furthermore, the work identified the predators. Differences in "Darwinian fitness" were demonstrated and correlated with a demonstration of the adaptiveness of the colouration. It is therefore reasonable to assert that natural selection was seen to occur. Can it be stated that selection was responsible for the genetic dominance and expression of the *carbonaria* gene?

Fortunately the history of the moth from the mid-nineteenth century is well recorded (Kettlewell 1973); and recent work (Clarke, Mani, and Wynne 1985; Cook, Mani, and Varley 1986; Clarke, Clarke, and Dawkins 1990) has shown the retreat of *carbonaria* with the reduction of air pollution. Furthermore, it is known from museum specimens that *carbonaria* specimens (but of the same genotype?) from the nineteenth century were much less uniformly dark, particularly on the hind wings, than typical extant heterozygotes; so that there appears to have been an evolution of the expression of the *carbonaria* gene. Kettlewell (1965, 1973), by crossing into genetic backgrounds from *carbonaria*-free populations in Cornwall and Canada, showed that the genetic dominance of *carbonaria* over *typica* must be due to modifying genes, which are difficient in Cornwall and absent in Canada (but see West 1977; Mikkola 1984b). Thus it can be postulated that the evolution of both the expression and dominance of the *carbonaria* gene, as well as its frequency, has taken place by selection in polluted urban areas by bird predators.

In characterising both Darwin's and Wallace's original theory and the Synthetic Theory, we concluded that, having defined natural selection, the Theory of Natural Selection was that natural selection acting on spontaneously occurring adaptive features (due to gene and chromosome mutations, and recombination) can result in adaptive anagenesis, or, more strongly, that *natural selection is a necessary condition of adaptive anagenesis.* It will be noted, however, that two hypotheses can be derived from this last statement (bearing in mind genetic drift):

(1) Natural selection is a necessary condition in the enhancement of adaptation

(2) Natural selection is a condition for anagenetic evolution.*

In the particular case of *Biston betularia,* it can be proposed that adaptation has been enhanced by selection, in accord with our hypothesis number (1). Unfortunately, this and other examples of the action of selection (resistance to pesticides in insects and warfarin in rats, heavy metal resistance in plants; see Dobzhansky *et al.* 1977, pp. 121ff.) are always intra-specific and frequently reversible. They provide no corroboration of hypothesis (2).

This is unfortunate because it allows a somewhat different accusation of "tautology". If natural selection is accepted as the mechanism by which complex adapted structures arise, then its function is, or was, to produce an evolving correlation of adapted parts, functional throughout, over enormous periods of time. Dawkins's *The Blind Watchmaker* (1986) is a very readable account of how such complex adaptations *could* have arisen by selection. But demonstrations of selection in action merely show the selection of pre-existing adaptations – that is, favourable mutations. We can then take our characterisation of selection (Section I): *Natural Selection . . . is the enhanced probability of reproductive success of those members of the population whose heritable phenotypic differences from others* [with pre-existing adaptations] *adapt them better to that environment.* Thus Natural Selection (the enhanced probability of reproductive success of favourable adaptations) is a necessary condition of the reproductive success of favourable adaptations; or,

*There is the temptation to state that "natural selection is a *cause* of evolution"; but strictly speaking, only events cause other events.

more succinctly, *ceteris paribus,* expected fitness becomes realised fitness! This conclusion is essentially the same as the *a priori* conclusion of population genetics that expected change in gene frequency (assigned fitness) results in the change in gene frequency expected ("evolution").

Darwin and Wallace's Theory of Natural Selection proposed that selection results in irreversible trans-specific evolution. It is almost certainly correct, but it has yet to be demonstrated.

III. Developmental constraints and selection

The second of the three categories of criticisms of the Synthetic Theory, suggested in Section II, was that factors other than "random" mutation and recombination, selection, and drift must be taken into account in explaining anagenesis (suggestions include constraints on genotype and structure, and ontogenetic factors). We also noted previously the concept of the "Panglossian paradigm" used by Gould and Lewontin (1979). As a counter to the telling of "just-so stories" about adaptation and selection, of which they give several particularly egregious examples, Gould and Lewontin present "an incomplete hierarchy of alternatives to immediate adaptation for the explanation of form, function, and behaviour". Their first major category is that of cases where there is neither adaptation nor selection. They cite, first, genetic drift, perhaps now more acceptable to those who labour within the Panglossian paradigm than it was then; and second, the fixation of non-optimal alleles in small populations with weak selection, an extension of drift; and third, the loss of new mutations before selection had had time to act (also by drift).

Their second major category includes cases where there is no adaptation and no selection on the "part at issue"; the form of that part is a result of selection directed at some other, but correlated, part of the body. These cases encompass Darwin's "mysterious" laws of correlation of growth – now classified as (a) pleiotropy, (b) allometry, (c) the "material compensation" of Rensch (1959), and (d) finally correlations that arise from mechanical necessity (D'Arcy Thompson 1961, p. 275), thus demonstrating, in the spirit of Cuvier, that organisms are integrated wholes, not aggregate of "characters" each produced by a separate part of the genome. Gould and Lewontin also cite cases of paedomorphosis (reproduction in morphologically immature individuals) and progenesis (rapid maturation resulting in dwarf individuals) which they suggest are more probably to be explained as a by-product of selection for rapid turn-over of generations rather than as selection for the anatomical symptoms.

Their third category is that of the decoupling of selection and adaptation. Lewontin (1979) gave a hypothetical example of selection without adaptation in a mutation that doubles fecundity. This would "sweep through a population rapidly" but, if other factors limited population size, would simply result, after fixation, in twice as many eggs being produced for a constant population size.

Adaptation without selection was exemplified by ecophenotypic differences in corals and sponges, in which the form of the colony is moulded by the water-flow regime in which it lives.

Their fourth category is where both selection and adaptation occur but related species occupy different "adaptive peaks". The differences between these different adaptive optima may be of no selective significance, being the end-products of different evolutionary histories. In the West Indian land snail *Cerion,* specimens living on rocky, windy coasts have white, thick, squat shells; but there are two known developmental pathways to the first feature, two to the second, and three styles of allometry leading to the third.

Gould and Lewontin's fifth and last category is where there is both adaptation and selection, but the adaptation is constrained by structure, development, or past history.

The theme of constraints is taken up by Maynard Smith *et al.* (1985). They define their subject as follows:

> A developmental constraint is a bias on the production of variant phenotypes or a limitation on phenotypic variability caused by the structure, character, composition, or dynamics of the developmental system.

They quote Thomas Henry Huxley in a letter to Romanes as evidence of the long preoccupation of evolutionists with the concept:

> It is quite conceivable that every species tends to produce varieties of a limited number and kind, and that the effect of natural

> selection is to favour the development of some of these, while it opposes the development of others, along their predetermined lines of modification.

Maynard Smith *et al.* suggest a division of constraints into universal and local. Universal constraints are those due to the laws of physics. Local constraints, although the authors do not quite say so, may be regarded as a class of taxonomic characters. They cite the inability of most monocotyledons to produce secondary thickening, so that palm trees have unbranched trunks, and the fact that the amino acids incorporated into all proteins are levorotatory isomers. Both these cases can be seen as contingent, the results of quirks of evolutionary history. Local constraints can also be seen to be the results of *developmental* history. The authors note particularly the differentiation of the many cell-types in multicellular organisms. These arise, not all at once, from a single progenitor, but in a branching pathway analogous to patterns of classification, as each cell type can give rise only to a small number of others. This is in fact a particular phase of Arthur's (1984, 1988) "morphogenetic tree" (see next section).

Maynard Smith *et al.* then discuss the origin of constraints, distinguishing those that are taxonomic or contingent – those where "*[p]henotypes [are] accessible or inaccessible, given a particular developmental mechanism*" – from those where "*phenotypes [are] accessible or inaccessible, given any developmental mechanism*". The latter include geometrically similar pigment patterns, which occur on feathers and on the coats of mammals, and other convergent patterns which they suggest could represent the solution of a wave equation, "even if it is hard to say what was waving". They also suggest three other categories:

1. Constraints due to the course of selection that has resulted in the present adaptation of organisms: "[I]t is unlikely that kangaroos will evolve adaptations to bipedal running, because the initial changes in that direction would be maladaptive". This category results in "orthoselection" (Simpson 1953). This first category includes the second as a special case.
2. Constraints resulting from canalising selection (Waddington 1957) in which the expression of new mutants is progressively reduced by stabilising selection.

3. Constraints that depend on the availability of new mutants that are dependent on current mutational pathways.

Maynard Smith *et al.* continue with a discussion of the identification of developmental constraints and their distinction from "selective constraints". The distinction is most clearly seen in the case of adaptive convergence (homoplasy to the taxonomist), where homoplastic resemblance can be attributed to similar modes of life: "swifts and swallows; jerboas, kangaroos, and cape jumping hares; tunafish, ichthyosaurs, and porpoises". Developmental constraints are to be identified by the distribution of realised forms in morphological space, but here the difficulty is distinguishing constraint from "chance wanderings of actual evolutionary lineages" (Raup and Gould 1974).

Genetic analysis yields even more convincing data. Maynard Smith and Sondhi (1960) showed that although absence of one ocellus or the other of a bilateral pair in *Drosophila* (i.e., the asymmetry) was heritable, one could not select specifically for absence of the left or right ocellus. Knowledge of developmental mechanisms also makes it possible to identify some types of constraint. Lande (1978) and Alberch and Gale (1983, 1985) have shown that if digits are reduced or lost in extant amphibia, the order of their loss, due, for example, to the experimental use of mitosis inhibitors, is remarkably congruent between remotely related species within the Anura, and, similarly within the Urodela. Furthermore, whereas a frog (anuran) will lose the thumb, a salamander (urodele) will lose a post-axial digit as a result of similar manipulation. This reflects differences in the processes of morphogenesis in the two amphibian orders.

Of course, no evolutionary biologist, however pan-selectionist or "Panglossian", would deny the existence of constraints on the ability of natural selection to fill morphospace with adapted organisms. The more rigorous the constraints, however, the more negative natural selection appears to be, *in extremis* merely culling the least adapted of a series of preordained forms. Accepting the latter situation would represent the other extreme from pan-selectionism of a spectrum of possible attitudes to the interaction of selection and internal factors.

It is possible to plead, however, that natural selection, and even evolution, are the wrong sorts of theories to deal with the data of embryology and comparative anatomy. This plea is made by

Goodwin (1984) in his suggestion for a *generative paradigm* as a better way of looking at the diversity of organisms rather than "the tracing of genealogies of temporal patterns of inheritance":

> This is so deeply entrenched in modern consciousness that it has become commonsense to 'explain' the presence of six toes on a cat by simply pointing out that the cat's mother has six toes. This is like saying that the reason why the earth now follows an elliptical trajectory around the sun is that last year it followed an elliptical trajectory. Although there is an element of truth in both of these statements, it it will be acknowledged that the latter is not very acceptable as a scientific explanation.

Rather Goodwin urges the use of "mathematical models embodying principles of spatial organisation which satisfy the long-recognised need in developmental biology for high level relational order [and which] are now available in the form of field theories of morphogenesis".

He illustrates his thesis by a study of the development of the tetrapod limb, firstly by pointing out the difficulty of using a phylogenetic criterion of homology in determining the equivalence of digits. If one tetrapod has only four digits on the hand, and another, taken as representing a more primitive condition, has five, by what criteria does one assert that a particular numbered digit has been lost (or, as in the case of the polydactylous cat above, gained)? The conventional explanation is that each (say) skeletal element is programmed in the genome, not in the naive sense of one gene per bone, but in a way that preserves continuity of specification from one generation to the next (Wolpert 1969, 1971). Goodwin draws attention to a different series of models derived from his own work and that of others. Studies of normal development, and of the results of experimental manipulation, mostly in amphibia and the chick, show that the skeletal *anlagen* are laid down from proximal to distal in the limb bud so that some sort of gradient or diffusion of a morphogen can be suggested. Thus transformations from one limb to another, if seen in evolutionary terms, are not to be seen as the gain or loss of individual elements but as a change in the morphogenetic determinants. In the same spirit, Oster *et al.* (1988) describe work mainly on the limb of *Ambystoma,* a urodele, and review problems of tetrapod limb development, in terms of mechanisms of morphogenesis and developmental constraints. In arguing for his "generative paradigm", Goodwin is not attempting to refute the theory of evolution by natural selection, but suggesting that the proper scientific approach to problems of morphogenesis is not that of genetic and phyletic scenarios expounded in terms of natural selection, but that of the study of experimental comparative anatomy to discover general principles of development.

IV. Macroevolution

Our third suggested criticism of the Synthetic Theory (Section II) was that natural selection has nothing to say about the pattern of phylogeny and that higher-level processes or emergent properties may override selection in macroevolution. In a review dealing with the second part of this criticism, Hecht and Hoffman (1986) formulate the basic questions as follows:

(1) What are the patterns of macroevolution?
(2) What are macroevolutionary explanations for these patterns?
(3) Are these macroevolutionary explanations irreducible to explanations based entirely on microevolutionary processes?
(4) Are such irreducible explanations supported by examples from nature?

To these we may add the first part of our criticism expanded thus: the Theory of Evolution explains the natural pattern of classification as a pattern of descent – that is, phylogeny – but the Synthetic Theory neither predicts nor prohibits any pattern (or class of patterns) and is thus no explanation of evolution.

Hecht and Hoffman (1986) say that "[m]acroevolutionary patterns include distribution of character states among taxa, rates of change of character states within clades, rates of change in and interrelationships among origination, extinction and diversity of taxa, concordance of distribution patterns in time and space among clades, etc." They then go on to say, "It is only these patterns of macroevolution that can be observed in nature".

It is patent, however, that not even this is true. "The distribution of character states among taxa", if one assumes that the constitution of the taxa is given, might be said to constitute a part of the data from which macroevolutionary patterns are inferred.

Their other patterns, however, concerning rates of change in characters, and statements about the origin, history, distribution, and extinction of clades or taxa (if the two are distinguishable), are all inferences from long chains of reasoning based on both geological and biological data. If this were not the case, the battle over the frequency in the fossil record of patterns of "punctuated equilibria", prominent in Hecht and Hoffman's review, would be merely an argument about data, not their interpretation.

"Punctuated equilibria", the morphological stasis of fossil species over long periods punctuated by apparent instantaneous change, is believed by Eldredge and Gould (1972; Gould and Eldredge 1977) to be the commonest pattern of phylogeny seen in the stratigraphic fossil record. A contrary view is that of Gingerich (1979, 1983, 1985) that a dense fossil record in a well-established stratigraphic section would show a pattern of "phyletic gradualism". Stasis over considerable lengths of geological time does seem established for many fossil organisms, but its explanation is still controversial.

Thus the "punctuations" – sudden changes in morphology as seen in the fossil record – need not coincide with speciation events, as Eldredge and Gould believed (Malmgren, Berggren, and Lohmann 1983). Furthermore, several explanations of apparent stasis are possible. Stasis could be illusory: the static "species" lineage could be one in which anagenetic change occurred – a change that was biochemical or physiological, or that concerned "soft anatomy" – or a succession of species differing only in such features could be involved. If the stasis is real, the following explanations might be suggested:

(1) The evolutionary potential of the species was exhausted. This is an impossible situation unless the species became extinct without issue – and improbable even then.
(2) The selective potential of the species was constrained by "orthoselection", resulting in adequate adaptation so that any change in the direction of adaptation was maladaptive (see previous section).
(3) Stabilising selection, resulting in canalisation of the phenotype, outweighed selection as a result of extrinsic factors that produced "tracking of the environment".

These three explanations lie within the paradigm of the Synthetic Theory. As far as I know, no theory outside the scope of the

Synthetic Theory has been offered as a causal mechanism to explain the data of stasis alone. All such theories make the assumption that the punctuations (absence of data) correspond to speciation events. When they first presented the theory of punctuated equilibria, Eldredge and Gould (1972) presented their view of the data as follows:

> The history of evolution is not one of stately unfolding, but a story of homeostatic equilibria disturbed only "rarely" (i.e. rather often in the fullness of time) by rapid and episodic events of speciation.

The "rapid and episodic" speciation events were explained by allopatric speciation caused by the separation of peripheral isolates from the main species stock according to Mayr's (1963) scenario. Later Eldredge and Gould's theory was more specifically identified with Mayr's (1954) peripatric speciation – that is, allopatric speciation associated with peripheral isolates and the "founder effect". The Founder Effect is the principle that peripheral isolates will represent an impoverished and atypical sample of the species gene pool, thus giving a "kick-start" to genetic differentiation of the new species.

However, in their original paper, Eldredge and Gould (1972) suggested that "adaptations to local conditions by peripheral isolates are stochastic with respect to long-term, net directional change (trends) within a higher taxon as a whole". They saw this as an extension of what they (Gould and Eldredge 1977) subsequently termed "Wright's Rule", after Sewall Wright's (1967) suggestion that as mutations are stochastic with respect to selection within a population, so speciation is stochastic with respect to the origin of higher taxa. This "decoupling" of micro- and macroevolution was seized upon by Stanley (1975) to propose the term "species selection" for the mechanism explaining long-term trends in clades during phylogeny. Thus, while evolutionary adaptation was to be explained by natural selection, the fact that "the world contains millions of species of beetles and only a handful of pogonophorans" was not to be explained "simply because beetles represent a successful adaptive design. Propensity for speciation and resistance to extinction must also be considered" (Eldredge and Gould 1988). Stanley's species selection thus sought to explain "trends" not as persistent anagenetic evolution (phyletic gradualism) but as a bias in the pattern of phylogeny of a clade because some species within that clade speciated at a higher rate or were longer-lived.

Discussion of the theory of punctuated equilibria has been continued vigorously and sometimes acrimoniously ever since and continues today (e.g., P. R. Sheldon 1987; Eldredge and Gould 1988). Full references are given in two recent reviews (both anti) – Hecht and Hoffman (1986), referred to above, and Kellogg (1988). In reviewing Eldredge and Gould's writings on the subject, Kellogg claims to identify a "brief but intense infatuation (Gould 1977[c], 1980[d])" of Gould for the systematic mutations of Goldschmidt which "even its staunchest proponents have abandoned (Gould, 1985:5; Gould and Eldredge 1986: 144–145)".

Goldschmidt has been the *bête noir* of advocates of the Synthetic Theory ever since the publication of his book *The Material Basis of Evolution* in 1940. His view was that there was total decoupling between evolutionary change within species ("microevolution") and the mode of speciation, and thus also all trans-specific evolution ("macroevolution"). He believed that speciation was due to macromutation in the form of major rearrangement of the karyotype – that is, chromosome mutation. Goldschmidt admitted that most such mutations, if they had major phenotypic effects, would produce "monsters", but that, in his most famous phrase, some of those could be "hopeful monsters". Hopeful monsters would be "preadapted" to some new environment so that the macromutation would constitute instant speciation. Radical macromutation could even produce new *Baupläne* and thus new major taxa.

But most proponents and opponents of punctuated equilibria reject any saltational theory of speciation and agree that if the punctuations of the fossil record represent speciation, while geologically instantaneous, they probably represent adequate animal generation time (one hears little about punctuation in plants!) for normal allopatric speciation to occur. There thus remain two principal areas of contention: first, whether stasis is consistent with the expectation of the Synthetic Theory, and second, whether some higher-level emergent property should be regarded as a supplementary evolutionary mechanism to natural selection. A reasoned negative answer from two orthodox evolutionists is given to both these questions by Stebbins and Ayala (1981), who nevertheless agree that in one important sense macroevolution is decoupled from microevolution.

> But evolutionary principles are compatible with both gradualism and punctualism; therefore, logically they entail neither. Thus, macroevolution and microevolution are decoupled in the important sense that macroevolutionary patterns cannot be deduced from microevolutionary principles.

Gould (1985), however, claims that Stebbins and Ayala have missed the point. The first claim of Gould and Eldredge's theory was the preponderance of stasis in the fossil record, but the second claim was that some mechanism of species sorting (Stanley's "species selection") was characteristic of evolution beyond the species level. Vrba (1984), in her "effect hypothesis", suggested that different probabilities of speciation resided in species gene pools as a side effect of the nature of species adaptations to their environments; hence species selection. Gould therefore suggested that there were three tiers of evolutionary events, corresponding to three scales of evolutionary time. Darwinian mechanisms prevail at the first tier – that of "ecological time"; species sorting at the second – that of "normal geological time", yielding the phenomenon of punctuated equilibria and thus the trends seen in the fossil record. The third tier is that of periodic mass extinctions.

If punctuated equilibrium was the principal palaeontological and evolutionary controversy of the 1970s, extinction was that of the eighties. Within the controversy there are two principal areas of contention, first, whether episodes of mass extinction had extra-terrestrial causes, and second, whether there is a significant periodicity in extinction events.

The theory of the extra-terrestrial origin of extinction events came to prominence with the discovery of an anomolously high level of the metallic element iridium in sediments at or near the Cretaceous/Tertiary (K/T) boundary, notorious as the time at which the dinosaurs (among many other taxa) are said to have become extinct (Alvarez *et al.* 1980). The iridium was said to be of extra-terrestrial origin and probably the result of meteorite impact. Other evidence of a catastrophic event at the K/T boundary includes soot-rich carbon, suggesting a global fire (Wolbach *et al.* 1985, 1988); and shocked grains of quartz and other minerals, suggesting large and sudden pressure changes (Bohor *et al.* 1984,

1985). Two controversies emerge: whether the extinction event was catastrophic or due to the accumulation of normal geological events (such as changes in sea level, tectonic plate movements, and climatic change) and whether, if catastrophic, extra-terrestrial causes may be invoked. The evidence is reviewed in a book on the subject of extinction by Stanley (1987) and by Officer *et al.* (1987). Both settle for terrestrial causes including changes in sea level and volcanic events, both related to plate tectonics.

Raup and Sepkoski (1982) showed that there had been four or five episodes of mass extinction recorded in the marine fossil record since the beginning of the Phanerozoic (Cambrian onwards) and later attempted to show that from the end-Permian event (ca. 250 Myr) there had been a regular series of events at ca. 26-Myr intervals (Raup and Sepkoski 1984). Physicists rushed in with a series of extra-terrestrial explanations (Davis, Hut, and Muller 1984; Rampino and Stothers 1984; Whitmire and Jackson 1984; Whitmire and Matese 1985). The proposition of periodicity has, however, been subject to damaging criticism. It is founded on a data base, compiled by Sepkoski (1982), of the ranges to geological stage level of marine fossil families from a number of sources. Hoffman (1985) suggested that the periodicity, even if real, might result from purely random processes related to the average length of geological stages. More damagingly, Patterson and Smith (1987) showed that for fish and echinoderms (about 20 per cent of Sepkoski's families), only about 25 per cent of the "families" were valid taxa. Plotting the extinctions for these 25 per cent gave no periodicity, whereas for the invalid taxa the periodicity was enhanced. One possible explanation for their remarkable results is that the apparent periodicity is an artefact of the way in which geological stages are defined, but Stigler and Wagner (1987) showed that the periodicity could be a statistical artefact.

Thus, although the major Phanerozoic extinction events – Late Ordovician and Permian, Late Triassic and K/T – undoubtedly occurred, together with a prolonged and complex event during the Devonian (Scrutton 1988), a regular periodicity is not recorded before the Permian and is very doubtful since. Nevertheless, in Raup and Sepkoski's (1982) phrase, "a number of mass extinctions have 'reset' major parts of the evolutionary system during the Phanerozoic" and thus constitute Gould's "third tier". Jablonski (1986a) has suggested geographical and ecological features that might have resulted in survival of taxa through an extinction event (natural selection under abnormal circumstances), but these are at odds with adaptations selected during "background" times, are at odds with "species selection" (if that is a real phenomenon), and refer to whole clades rather than to selection within species.

Another way of looking at emergent properties in macroevolution and thus in the pattern of phylogeny is represented by the term *Bauplan*. Valentine (1986) seems to equate *Baupläne* with the basic plans of Metazoan phyla and refers to the distinctive plans of classes and orders as "*Unterbaupläne*". He is thus able to say that no new *Baupläne* have arisen since the Paleozoic, and that most appear from the record to have originated near the time of the Precambrian/Cambrian boundary. A *Bauplan,* according to Valentine, not only constitutes the aggregate autapomorphies of a phylum but also "contains both ancestral and derived characters, the former commonly indicating alliances with the next higher taxon [i.e., synapomorphies uniting the phylum with its sister-taxon], the latter indicating membership in the clade". Thus a definition should perhaps include the phrase *all the taxonomically significant characters of a taxon of high rank,* but there is also the sense in which, taken as a whole, a *Bauplan* is a distinctive way in which the members of a clade have solved their adaptive problems. Thus two different *Baupläne* may either embody adaptive solutions to different environments, or different solutions to the same range of environments. A snail and a caterpillar feeding on the same leaf have profound anatomical differences which relate to the respective clades to which they belong, and thus (evolution accepted) to their phylogenetic history.

Valentine attempts to explain why the *Baupläne* represented by the Metazoan phyla should have appeared so soon after the first appearance of the Metazoa. On the extreme view that phyletic gradualism prevailed from the Precambrian onwards, the anatomical divergence between any two organisms should be proportional to the time separating them from their most recent common ancestor; but the Metazoan phyla differ so strikingly from one another when they first appear in the record that this appears not to be the case. Thus the pattern is comparable to punctuated equilibria, the punctuation representing the rapid origin of the phyla, and stasis representing the persistence of their respective *Baupläne* to the present day. Valentine points out, however, that a mechanism like Stanley's species selection is very improbable. He gives reasons for believing that for species the relative completeness of the known fossil record is roughly similar from the early Paleozoic to that from more recent strata, with low provinciality in the past compensating for the greater poverty of the sample. If he is correct (Bambach 1986), there was a much greater range of phyla, many of which are now extinct, in the early Paleozoic, but they were extremely species-poor clades. This does not fit the pattern of the selection of species-rich clades of Stanley's theory.

A different approach to the *Bauplan* problem is suggested by Arthur's (1984, 1988) development of the concept of the "morphogenetic tree". In its simplest and most abstract form, this is presented as a bifurcating dendrogram (spanning tree) (Fig. 12.1)representing the ontogeny of an individual organism. The tree can be viewed in three ways:

(1) As a pattern of causal links (the vector lines) connecting morphogenetic events such as embryological inductions, cytoplasmic localisations, gradients, and so on, which are represented by all but the terminal nodes of the tree; the whole then represents the transference of heterogeneities (such as cell differentiation) from one stage of ontogeny to another and also an increase in the number of simultaneous heterogeneities.

(2) With the subterminal nodes representing the action of alleles at genetic loci (Arthur's "D-loci") bearing genes responsible for morphogenesis.

(3) As a basis for considering the frequency, and magnitude of effect, of mutations at each rank level in the tree.

This last version is illustrated as a three-dimensional diagram (Fig. 12.1) in which mutations at early-acting D-loci, because they influence a greater subsequent hierarachy of events, have a greater average magnitude of effect. The third dimension is introduced to show the frequency distribution of magnitudes at each locus, shown as a normal distribution. All D-gene mutations Arthur refers to as "phase changes"; he cites as an example of an early-acting phase change the mutation from dextral to sinistral torsion of the shell and body anatomy of a gastropod (snail). Two other types of mutation are also envisaged: (1) structural change (of the tree) with the addition or deletion of branches (with their subbranches if any) to give increases or decreases, respectively, in the

Fig. 12.1. Two graphic versions of the "morphogenetic tree" of development in its simplest form. (a) can represent either embryological development with causal links (arrows), or the action of morphogenetic genes; (b) represents the effect of mutation at various stages, with the magnitude of action of each mutation as a frequency distribution (see text). (After W. Arthur 1988, *A Theory of the Evolution of Development*. Copyright © 1988. Reprinted by permission of John Wiley & Sons, Ltd.)

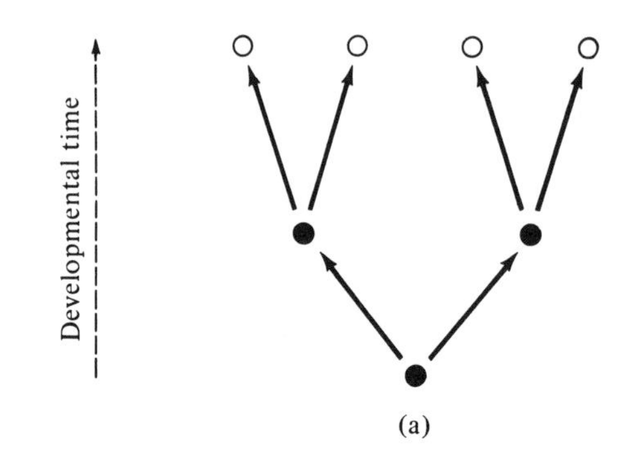

(a)

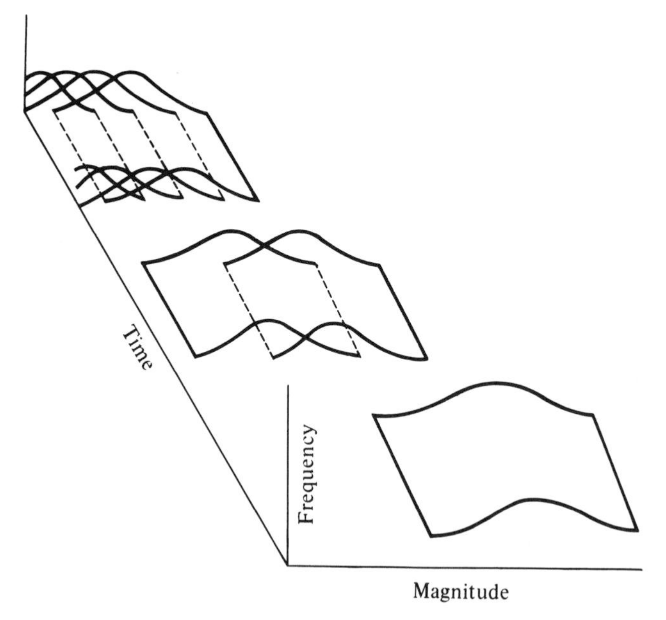

(b)

complexity of the resulting organisms; (2) distortional changes of the relative vertical positions of nodes on different parallel branches, resulting in heterochrony (Gould 1977b).

The importance of the whole morphogenetic tree idea to evolutionary theory is that mutation of early-acting D-genes produces greater phenotypic effects, but is by the same token less likely to result in viable individuals, whereas that of later-acting genes will have a lesser effect but a greater probability of survival; hence Arthur's concept of a spectrum of *Baupläne*. A new major body plan may have arisen as the result of the natural selection of a rare viable mutant resulting from mutation of an early-acting D-gene and all the subsequent necessary modifications of development. Less radical changes would occur more frequently because of their greater viability, down to mutation of genes whose effects were merely additive and would probably result only in intraspecific differences. The pattern of phylogeny and its timing would thus correspond to Valentine's palaeontological conception according to which major *Baupläne* are distinctive phenomena that arise relatively quickly. Wimsatt (1986; Schank and Wimsatt 1987) has developed ideas similar to Arthur's with his "developmental lock" metaphor.

Our final criticism of the Synthetic Theory and of Natural Selection itself requires little space for its discussion, but is in my view quite devastating in its implications. It has been stated most forcibly by Rosen (1984):

> Darwinism, neo-Darwinism, punctuated equilibrium, epigenetic neo-Darwinism . . . and even the various versions of the creationist theory of life all share another shortcoming of explanations of evolution. Each can equally well explain any evolutionary history with a minimum of empirical restraint. None of them uniquely determines one hierarchy. They prohibit no theory of relationship.

This should be compared with the views of two orthodox "Neo-Darwinists", Stebbins and Ayala (1981), quoted more extensively above in this section: "Macroevolution and microevolution are decoupled in the important sense that macroevolutionary patterns cannot be deduced from microevolutionary principles". Most orthodox "Neo-Darwinists" would probably accept Stebbins and Ayala's statement but probably not Rosen's more challenging version; so the question must be, Does it matter? It is a question I shall postpone until the end of the next, summary, section.

V. Conclusions: The Theory of Natural Selection

I shall now attempt to summarise some of my conclusions about the status of the Theory of Natural Selection; others must wait until we have considered the philosophy of science. Before doing so, however, I shall note two controversial and important aspects of theories of mechanism which I have not tackled. The first of these is the recent concern with directional internal factors in evolution, echoing the preoccupation of late-nineteenth-century palaeontologists with orthogenesis. New ideas have arisen from the parallel of irreversible evolutionary change with the universal increase in entropy postulated in the Second Law of Thermodynamics (Prigogine 1961). Brooks and Wiley (1986) have developed a new theory that relates the two and have been severely criticised (e.g., Morowitz 1986). As a result, they have recently produced a revision of their ideas (Brooks and Wiley 1988). Wicken (1987) has pursued a similar line, but there is disagreement about the term "entropy" and about whether what is being suggested is truly "evolution as entropy" (Brooks and Wiley's title) in the strict Second Law sense or merely a suggestive analogy (Depew and Weber 1988).

Another controversy concerns "levels of selection". In the previous section, we noted ideas on species selection and the selection of clades as a result of extinction events. At the other end of the spectrum, the idea that selection acts on genes or "replicators" rather than individual organisms was made notorious by Dawkins (1976, 1982); while at an intermediate level between these extremes is a controversy about group selection as proposed by Wynne-Edwards (1962, 1986) which has been rejected by most evolutionists (e.g., Williams 1966). The alternative proposal by Hamilton of kin selection suggested that apparent group selection could be explained in terms of individual selection. However, Hamilton's concepts have been widely misinterpreted (Grafen 1982). The whole subject of levels of selection is discussed by Sober (1984).

We saw in Chapter 7 that the Theory of Natural Selection was proposed by Darwin and Wallace more to explain the origin of adaptations than the pattern of phylogeny, but that both made gradualist assumptions about macroevolution so that selection could be extrapolated as the mechanism of evolution at every level. Darwin and Wallace differed, however, in that Wallace favoured a group selection hypothesis (as discussed by Ruse 1980b and Sober 1984) whereas usually Darwin did not. Wallace also accepted that selection was "hard", in Mayr's terminology, with no "inheritance of acquired characteristics". This view eventually prevailed amongst selectionists after Weismann had ushered in "Neo-Darwinism" with his theory of the germ-plasm.

In formulating the Theory of Natural Selection in Chapter 11 (Section II), I concluded that the theory is not tautologous, logically true, analytic, or *a priori*. However, if mathematical population genetics equates evolution with changes in gene frequency and substitutes "fitness" (measured as increase in gene frequency with time) for better adaptation, then it conflates my inferences 2 and 3 (Chapter 11, Section II) and thus makes the whole analytic.

It is generally agreed, however, that natural selection does not always result in change. A population at equilibrium can be one on which stabilising selection is acting. Thus the phenomenon of stabilising selection demonstrates that selection does not always result in anagenetic change; nor is selection a *necessary* condition for anagenetic change, as demonstrated by the phenomenon of genetic drift and, if true, the neutral theory of evolution. However, if drift or neutral evolution gives rise by chance to a new adaptive feature, then that feature will be "taken over" by selection. There is, therefore, an *a priori* case for saying that all the adaptations of organisms evolve as a result of natural selection. For this to happen, however, there must be a source of genotypic innovations, produced by mutations, recombination, and so forth, and those innovations, whatever their causality, must be "undirected with respect to the adaptive requirements of the organism" (Burian 1988); that is, inheritance must be "hard". If directed mutations of the type postulated by Cairns *et al.* (1988) (Chapter 11, Section III, this volume) do occur, then natural selection would, *in those cases,* merely have the negative culling action that ill-informed critics accuse it of.

In discussing the evolution of new adaptations, we have moved beyond selection as merely the condition for changes in gene frequency within populations to natural selection acting over periods of geological time and thus to macroevolution. In discussing macroevolution in the previous section, I said little about theories of speciation beyond Goldschmidt's ideas. The orthodox assumption is that selection is important in cladogenesis. Scenarios for the mechanism of speciation within the Synthetic Theory were discussed in Mayr's (1963) classic work on the subject. The controversy as to whether speciation could be other than allopatric was mentioned in Section I. The other current controversy is the role of selection in speciation. The question arises, Are those differences between two closely related species that lead to genetic isolation (the isolating mechanism) at least in part the result of selection in sympatry? Recent reviews of the problem include those of Templeton (1981), Carson and Templeton (1984), and Barton and Charlesworth (1984), along with discussions in Otte and Endler (1989).

Beyond the species level the controversy is between those who are content to assert that the patterns and reconstructed processes of macroevolution are compatible with the Synthetic Theory (e.g., Stebbins and Ayala 1981; Charlesworth, Lande, and Slatkin 1982) on the one hand, and those who emphasise emergent properties (Gould 1985), as well as those who regard the synthetic theory as false (Løvtrup 1987) or as the wrong sort of theory (Goodwin 1984) or merely irrelevant to its explanatory purpose (Rosen 1984), on the other hand.

In an excellent review, which could have stood alone as a summary of these two chapters, Burian (1988) quotes Depew and Weber (1988) as an epigraph. Their first sentence reads:

> Neo-Darwinism has been rightly construed more as a treaty than a theory because it laid down terms that allowed evolutionists and practitioners of the new science of genetics (and more generally molecular biology) to work together under common presuppositions.

Corresponding to the idea of the synthetic theory as a treaty is Popper's (1974b) characterisation of it as "metaphysical research programme", or Wassermann's (1981) as a hypertheory within which all evolutionary phenomena can be interpreted. But as Burian says:

> From just the synthetic theory plus suitable boundary conditions, one cannot derive claims about major patterns in the paleonto-

logical record, the causes or patterns of speciation, the importance of speciation (in contrast to other factors) in the dynamics of evolution, the evolutionary importance of sex, the evolutionary grounds for alternative life history strategies and so on.

I posed the question at the end of the previous section, Does it matter that the accepted theory of evolutionary mechanism fails to explain the pattern of evolution? To this question one might add another: Is a theory that would specify a pattern in any detail even possible?

I have emphasised repeatedly in previous chapters that the theory of evolution was proposed to explain the phenomenon of natural classification. As pattern, classification is taken to be hierarchical, divergent (despite evidence of reticulation in plants), and irregular. Its irregularity is explained by phylogeny as history, which is contingent and not explicable by universal statements, but the hierarchical and (usually?) divergent nature of phylogeny remain to be explained. Both features are dependent on speciation or cladogenesis; neither can be seen as a necessary outcome of natural selection as modelled by the Synthetic Theory, although the latter is entirely compatible with them.

As we saw in Chapter 8, the "transformed cladists" appeal to von Baer's Laws, both in general and in particular, to validate their methodology. Those laws are also congruent with divergence in phylogeny, and, given speciation, with a hierarchical branching pattern. The latter is explained by speciation. Thus an agreed mechanism, or series of mechanisms, for speciation or cladogenesis would explain both divergence and hierarchy. Those proposed mechanisms would probably incorporate selection at one or more levels from genes to "species selection", but should also give some account of the nature of the differences in genes, organisms, population groups, or clades that selection is enhancing. It seems to me that Arthur's "morphogenetic tree" concept, with its hierarchy of D-genes, might allow some increase in understanding here. I cannot, however, imagine what might be the nature of any theory of evolutionary mechanism that explained one pattern (the "true" one), or any aspect of it, and thus prohibited possible rival patterns. In that respect, I agree with Popper (1957) that there are no laws of history. With that in mind, we must now turn to the philosophical nature of theories in biology.

13

Scientific knowledge

In Chapter 6, I discussed the contribution that the writings of Plato and Aristotle made to the development of taxonomy. But we saw that the method of logical division was a method of gaining and systematising knowledge, not an exercise in systematics. We also saw that *idealism* derives from Plato. In its extreme form, developed in his later writings, "forms" or *Ideas* – concepts derived from the common feature of many phenomena (such as "beauty") – not only were real but were the only safe knowledge ἐπιστήμη (*episteme*). All data acquired directly by the senses were suspect, deceptive, or even illusory and thus could be expressed only as "opinions" (*doxa*). The parallel with modern taxonomy was to attribute some sort of reality to archetypes or *Baupläne* as originally developed in the morphology of Geoffroy Saint Hilaire, the *Naturphilosophen* and Richard Owen. I contrasted this with *essentialism,* derived from Aristotle's further development of logical division, which in taxonomy asserts that every taxon has an essence, a character, or series of characters that validate its reality independently of any explanatory theory such as phylogeny.

In order to lay a foundation for what I have to say about the nature of biology in the next and final chapter, I now have to draw on two other realms of discourse that can be traced back to the ancient Greeks. The first involves induction and deduction, the second causality. Plato believed that *Ideas* could be discovered by a purely rational process, the *dialectic.* Translated into the terminology of modern science, this would mean that by collecting sufficient observations within a particular discipline, it should be possible to arrive at general statements, phrased as hypotheses or theories – the process of *induction*. Such a general theory, if properly formulated, should then have the potentiality of yielding predictions of particular events under known (and preferably controlled) circumstances – the process of *deduction*. Thus the pattern of acquiring (particularly scientific) knowledge can be represented by an arch in which one ascends from the particular to the general and then descends again to the particular to test the general statements. Oldroyd's (1986) work, metaphorically entitled *The Arch of Knowledge,* is an account of the struggles of philosophers and scientists from Plato onwards to understand the nature of the world.

There is a feature of Aristotle's essentialism that I have not yet touched upon. He took the behaviour of an entity to be part of its essence just as much as its static characters. It was characteristic of fire that it rose into the air, but the reason was that it was attempting to reach its natural resting place in a hollow sphere within the orbit of the moon. This example can be related to Aristotle's doctrine of *causes,* a term he used in a much broader sense than that of modern philosophers and scientists (Oldroyd 1986, pp. 24ff.). The nature of objects and events was to be explained by four kinds of cause. The *material* cause explained the substance or "stuff" of which the entity was made; the *efficient* cause is close to the modern meaning of cause; the *formal* cause is related to essence in the sense of the example of fire, but can also be considered as blueprint, program, or essence in the sense of defining character; the *final* cause is a future event or configuration that in some way determines or is related to the present event or entity. The notion of final or *teleological* cause has dogged studies in natural history and biology ever since the time of Aristotle. A fertilized hen's egg has within it the potentiality to become a chicken; that chicken is a manifestation of the final cause of the egg. The idea of the future affecting the present is no longer acceptable in science, but final causes, as "God's plan", have always appealed to theologians and lurk in biology (apart from embryology) in concepts such as pre-adaptation (virtuously renamed "exaptation" by Gould and Vrba 1982) and ecological niche. We shall return to teleology in Chapter 14.

I don't wish to dwell further on the nature of "cause" in this chapter, except to note Hume's scepticism about the concept.

Hume (1748) could see no logical connection between a cause and its presumed effect, nor any way of apprehending "causality", and so was driven to talking of the "constant conjunction" of cause and effect, with cause *prior* and *contiguous* to effect (he did not deal with action at a distance). We shall, however, have to deal with a different aspect of Hume's scepticism, characterised as "the problem of induction".

I. The problem of induction

It is instructive to compare two well-known quotations from Darwin's writings, the first from his autobiography, the second from a letter written to Henry Fawcett in September 1861:

> I worked on true Baconian principles, and without any theory collected facts on a wholesale scale, more especially with respect to domesticated productions, by printed enquiries, by conversation with skillful breeders and gardeners, and by extensive reading.
>
> (*In F. Darwin, 1887, vol. 1, p. 83*)

> About thirty years ago there was much talk that geologists ought only to observe and not theorize; and I well remember someone saying that at this rate a man might as well go into a gravel-pit and count the pebbles and describe the colours. How odd it is that anyone should not see that all observation must be for or against some view if it is to be of any service!
>
> (*In F. Darwin and Seward, 1903, vol. 1, p. 195*)

The "Baconian principles" on which Darwin claimed he was working in assembling data for his theory of evolution, later to be published as the *Origin,* are those of Francis Bacon, Lord Verulam, in his *New Organon* (1620), consciously named as a replacement for Aristotle's *Organon,* the collected works on logic.

The simplest type of induction, supposedly leading to universal statements, is induction by simple enumeration. Hume (1739–40) showed that simple induction, consisting of repeated observations of the same phenomenon in similar circumstances, was not a way to arrive at the truth of universal statements: "[E]ven after the observation of the frequent constant conjunction of objects, we have no reason to draw any inference concerning any object beyond those of which we have had experience". There were no means in logic to proceed from a series of observational or experimental data to universal laws. This view, *the problem of induction,* is illustrated by Bertrand Russell's story of the philosophical turkey who observed, over nearly the whole period of his life, that every day he was fed at 9.00 a.m., whatever the weather or day of the week. Every day he added a further confirmation of his inductive theory that 9.00 a.m. signalled food – until Christmas Eve, when instead of being fed, his throat was cut.

Bacon's methodology for ascending the arch of knowledge was also dependent on induction but was somewhat more complicated (Oldroyd 1986, pp. 59–66). He was an empiricist; thus an understanding of nature must come through observation and experience, rather than there being concepts or ideas which the mind could arrive at directly. His method was to decide on a phenomenon for investigation (Oldroyd gives Bacon's example of "Heat"), and then compile tables of all bodies or whatever showing the phenomenon, and also of those not showing it, and then further, bodies showing variable degrees of it (bodies that showed different temperatures). From this the nature of heat ("motion" according to Bacon) was supposed to emerge. Importantly, however, Bacon believed that the hypothesis ("motion causes heat") could be tested in some way by new data.

In the early nineteenth century, Bacon was very much in favour, particularly after the publication of Herschel's *Preliminary Discourse on . . . Natural Philosophy* (1830) (incidentally, in the same self-education series, Lardner's *Cabinet Cyclopaedia,* as Swainson's quinarian works: see Chapter 2, Section III). Herschel's method was undoubtedly Baconian, with significant (causal) correlations said to emerge out of data, the latter collected without a pre-existing hypothesis (hence Darwin's "Baconian principles"). Thus the idea was that general statements could be arrived at by the unbiased tabulation of data; but what to observe? We see in Darwin's second quotation the realisation that "unbiased" data collection is an impossibility. The first quotation looks back to tradition, the second forward to the battles in the philosophy of science that were to extend forward into the twentieth century.

One aspect of the battle was engaged by two prominent nineteenth-century English philosophers, John Stuart Mill and William Whewell. The philosophical battle between these two has been represented by Medawar (1982) as that between naive inductivism and the championing of the hypothetico-deductive method, respectively. But there is more to it than that. Mill's great work on epistemology was *A System of Logic, Ratiocinative and Inductive*..., first published in 1843. But the word "logic" in the title did not denote the formal logic derived from Aristotle, the method of (deductive) inference of conclusions from premises, but rather methods of reasoning from one statement of fact to another. It is certainly the case that Mill was very much concerned with validating methods of induction. To do this, he produced a series of "Canons of Induction" very much in the tradition of Bacon as echoed by Herschel (Oldroyd 1986, pp. 149–54). As an example, the First Canon ("The Method of Agreement") is:

> If two or more instances of the phenomenon under investigation have only one circumstance, the circumstance in which alone all the instances agree, is the cause (or effect) of the given phenomenon.

Thus Mill suggested (wrongly) that *all* crystalline substances were deposited from a liquid, so that solidification from a liquid is at least a partial cause of the process of crystallisation. The Second Canon is "The Method of Difference", looking for a unique difference between the circumstances of two phenomena as the effect or the cause of the difference between them, and so on.

Whewell also acknowledged a debt to Bacon, actually titling one of his several works *Novum Organon Renovatum*. (1858). His major work was the *Philosophy of the Inductive Sciences* (1840), and here he attempted to introduce Kant's "transcendental idealism" into the tradition of British empiricism (Oldroyd 1986, pp. 156 – 64).

Insofar as it is possible briefly to summarise that part of Kant's (1781, 1783) epistemology for our theme, he states that there is no way that the human observer could have direct access to entities in the world "out there". They are in the realm of *noumena* – "things in themselves". All we can have is access to our sensory impressions, the realm of *phenomena*. But, furthermore, and importantly, our perception of phenomena requires the active participation of the human mind. It is part of the process of perceiving that sensory impressions are interpreted according to a series of "categories" (exactly twelve in number according to Kant) grouped as "Quantity", "Quality", "Relation", and "Modality" (see Oldroyd 1986, pp. 128–32).

For Whewell also there are "Fundamental Ideas" necessary to our knowledge of the world; these are not fixed categories but have evolved with the development of science. Perhaps it might be claimed by the cladists (although I am not aware that anyone has done so!) that the *a priori* taxonomic hierarchy is such a fundamental idea. Thus Whewell's (1849) assault on Mill's *Logic* was an attack on his empiricism, not only on his use of simple enumerative induction. Nevertheless, Mill's principal concern was to evolve a rational system of induction, whereas Whewell, as Medawar shows, elaborated on epistemology using approaches very close to the hypothetico-deductive methods now always associated with the name of Sir Karl Popper (next section). Medawar (1982, p. 130) quotes Whewell (1840) as follows:

> To form hypotheses, and then to employ much labour and skill in refuting, if they do not succeed in establishing them, is a part of the usual process of inventive minds. Such a proceeding belongs to the rule of the genius of discovery, rather than (as has often been taught in modern times) to the *exception*.

But Mill also accepted the validity of using the hypothetico-deductive method if only as a last resort, when direct experiment or observation were impossible. He characterised it as follows:

1. ascertaining a set of laws relevant to the phenomena under investigation, and which might together be deployed in an explanation of the phenomena – to be found with the help of Canons, or by formulating hypotheses;
2. deducing certain statements or conclusions from these laws;
3. testing or verifying the deduction by experiment and further observation.

However, laws arrived at by induction in (1) were preferable to hypotheses snatched out of the air.

It is worth noting in passing that both Whewell and Mill had something to say about taxonomy. Whewell, an Anglican clergyman and Master of Trinity College, Cambridge, rejected evolution (Ruse 1976). He saw adaptation in terms of *Final Causes* (just as in Aristotle) and was a "catastrophist" in geology (Chapter 5, Section I) (he also invented the term, as well as "uniformitarian"). He could not see any evolutionary process producing the adaptations of successive "creations". In taxonomy he was, following Richard Owen, an idealist, with archetypes giving an idea of the plan of Creation. Mill, as we have seen, proposed the criterion of naturalness of a classification which was ultimately taken up by the pheneticists via Gilmour (Chapter 7, Section II), emphasising their claim that phenetics is in the empiricist tradition. We might also want to say that phenetic classification is an inductive procedure, but the reader will recall that Whewell proposed criteria of naturalness similar to those of Mill.

Before moving to the philosophy of science of the twentieth century, we must note another movement in that of the nineteenth. This is *Positivism* named from a published series of free-lance lectures given by its founder Auguste Comte (*Course de Philosophie Positive,* published between 1830 and 1842). Positivism is an extreme form of empiricism in which any hypothetical entities are eschewed; thus "phlogiston" in the eighteenth century, the "subtle fluids" of Lamarck at the beginning of the nineteenth century, and the genes of the early twentieth century (before the elucidation of the structure of nucleic acids), as well as the concept of "atom" in all its changing manifestations, would have to be rejected as not available to observation. Comte drew inspiration from Cuvier's comparative anatomy, which could be regarded as an antidote to the excesses of *Naturphilosophie.* Positivism also rejected all metaphysics; only the direct objects of sensory perception constituted true knowledge, so that positivism embraced an extreme form of empiricism usually styled "phenomenalism". We shall see its twentieth-century development in the next section.

II. The hypothetico-deductive method

One of the odder claims of some cladists, as I noted in Chapter 8, is that cladistic analysis followed the hypothetico-deductive method of Karl Popper. I shall consider this claim further in Chapter 14, but meanwhile I want to give a brief critique of Popper's philosophy because it has held such a powerful influence over working scientists long after it has been largely rejected, at least in its original form, by fellow philosophers of science.

In the first half of the twentieth century, the interest of scientists in the philosophy of science was usually minimal, but that situation changed radically after the publication in English of Popper's *The*

Logic of Scientific Discovery (1959). Its reception by the scientific community, or at least by those scientists who thought about such problems, was rapid and largely positive. A review by Bondi and Kilmister (1959) is typical:

> a splendid book . . . it rings true . . . Popper speaks as a working scientist to working scientists. . . . The notion of falsifiability in particular is a concept of the most direct significance to science.

Bondi and Kilmister's reaction may be taken to represent the reaction of physical scientists, but Popper's most persuasive advocate was a biologist, the late Sir Peter Medawar, who in a series of essays advocated Popper's methods (Medawar 1967, 1969; reprinted in Medawar 1982).

As is so often the case, it helps the understanding of Popper's thesis to know the prevailing opinion against which his book was directed. *The Logic of Scientific Discovery* was first published in German in 1934 in Vienna as *Logik der Forschung* (Logic of [Scientific] Investigation) – a more accurate title. The spearhead of contemporary philosophy in Vienna was Logical Positivism, represented by a group calling themselves "the Vienna Circle", of which Schlick was the central figure with a changing membership including, *inter alia,* Carnap, Gödel, and Neurath. Their aim was to combine Comte's positivism with logical rigour. The precepts of logical positivism were introduced into Britain in Ayer's *Language, Truth and Logic* (1936), in combination with linguistic philosophy as in Wittgenstein's *Tractatus* (1922). One aim of the Vienna Circle was to urge a criterion of demarcation between propositions conveying information and those that were (to them) literally nonsense. The suggestion was that every meaningful proposition must be reducible to, or deducible from, "observation statements" based on sense data. Thus metaphysical propositions were meaningless. Unfortunately, however, as Popper pointed out, their criterion annihilates natural science as well. Scientific laws, whatever their source, are not reducible to observation statements.

So Popper sought a new criterion of demarcation, not between sense and nonsense, but between empirical science and metaphysics. That criterion was that empirical statements were potentially falsifiable. Popper claimed also to solved Hume's problem of induction, but this was done by simply ignoring it. No scientific theory can be established or proved by induction from singular data, and this led Popper virtually to disclaim any interest in the origin of hypotheses or theories. That was a problem for the psychologist rather than for the philosopher or logician. The business of the working scientist was to attempt to refute hypotheses by critical observations or experiments. Theories such as religious dogmas and those (according to Popper 1945, 1957) of Marxism and Freudian psychology are not framed in such a way as potentially to be tested; they are therefore metaphysical rather than scientific. But Popper is at pains to point out that metaphysical propositions are neither necessarily false nor necessarily nonsense. Future scientific theories may well be metaphysical when proposed, because the means do not then exist to test them.

Popper attached great importance to the nature and phrasing of scientific hypotheses to facilitate their testing. They were to be, in his terms, strictly universal statements ideally characterised by the formulation, "Of all points in space and time (or in all regions of space and time) it is true that . . . " Strictly universal statements must be contrasted with "numerically universal statements". Popper's example of the latter is "all human beings are less than 9 ft tall". It falls in this category because it concerns only a finite class of specific elements within a finite individual (or particular) spatio-temporal region. In principle, numerically universal statements could be replaced by a conjunction of singular statements. In his example, the height of all human beings could in principle be enumerated, an inductive procedure. He also distinguishes a related category of existential statements: "There are black ravens" is a given example. Such statements are equated with "there is" statements: in this example, "There exists at least one black raven". Universal statements include universal concepts. *Dictator, planet,* H_2O are Popper's examples. A corresponding list of individual concepts would be *Napoleon, Earth, Atlantic.*

If natural laws have the logical form of strictly universal statements, they can be expressed as negations of existential statements. The law of the conservation of energy implies that "there is no perpetual motion machine" (according to Popper); likewise the hypothesis of the electrical elementary charge can be expressed as "There is no electrical charge other than a multiple of the electrical elementary charge". Thus both these hypotheses could be falsified by a single instance: demonstration of a perpetual motion machine, or of an electrical charge that is not an elementary charge or a multiple thereof, respectively. This is the basis of Popper's method of falsification. An existential statement – "there is a perpetual motion machine in space–time region *K*" – can refute a strictly universal statement.

There is a complicating factor, however. The existential statement has a hypothetical component: "a perpetual motion machine has the following properties . . . " Thus in Popper's terms a theory or hypothesis is falsified by a reproducible effect that refutes it, so that a *low-level empirical hypothesis* describing the effect is proposed and corroborated and thus constituted as a falsifying hypothesis. However, the falsifying hypothesis can no more be proved than can the hypothesis under test. As we shall see below, this unfortunate fact, which Popper acknowledges in the *Logic of Scientific Discovery* as well as in *Conjectures and Refutations* (1963) proved to be the Achilles heel of his system. It is a matter of "decision" when a falsifying hypothesis is sufficiently corroborated.

As with a low-level falsifying hypothesis, a major hypothesis, which survives an attempt to falsify it, is not proved; it is *corroborated* and "lives to fight another day". However, a theory is not to be judged by the number of times it has been corroborated. That sounds too much like induction; other criteria must be used. According to Popper, these are the degree of falsifiability, corresponding to "empirical content", the level of universability, and the degree of precision. The empirical content of a statement is represented by the class of all its falsifiers. The corroboration of a seemingly improbable theory makes a greater contribution to science than does corroboration of an unsurprising or well-established one (contrariwise the falsification of a well-established theory is more significant than that of a new and/or improbable one). The level of universality and degree of precision of a theory are related to its empirical content. Popper (1959) gives an example of statements concerning the orbits of heavenly bodies. These are:

p: All orbits of heavenly bodies are circles.
q: All orbits of planets are circles.
r: All orbits of heavenly bodies are ellipses.
s: All orbits of planets are ellipses.

where circles are a sub-class of the class of ellipses. Whatever the truth or otherwise of these statements, the degree of universality decreases from p to q and the degree of precision from p to r, while s has the lowest level of both these attributes.

III. "Popper and after"

The title of this section is taken from that of a book by David Stove (1982) subtitled "Four Modern Irrationalists". In his preface, Stove says:

> This book is about a recent tendency in the philosophy of science: that tendency of which the leading representatives are Professor Sir Karl Popper, the late Professor Imre Lakatos, and Professors T. S. Kuhn and P. K. Feyerabend.
>
> These authors' philosophy of science is in substance irrationalist. They doubt, or deny outright, that there can be any reason to believe any scientific theory; and *a fortiori* they doubt or deny, for example, that there has been any accumulation of knowledge in recent centuries.

Popper's irrationalism can be seen as arising out of his whole system as set forth in *The Logic of Scientific Discovery:* if theories can only be corroborated, and never proved, one can never know whether truth has been reached. It might be thought that a well-corroborated theory was, in the technical sense, more probable than a less corroborated one, but Popper denies this (1959, chap. X). The number of possible corroborations of a strictly universal statement is infinite, so that the probability of any given number n of corroborations indicating that a hypothesis is true is n/∞, that is zero.

Meanwhile we have already noted that a logical flaw is apparent in Popper's system, at least if one assumes as (e.g.) Medawar did that the scientist "can perform with complete logical certainty . . . repudiation of what is false" (Medawar 1967; reprinted in Medawar 1982, p. 127). As we have seen, Popper himself accepted the unreliable nature of falsifying statements in both *The Logic of Scientific Discovery* and *Conjectures and Refutations*. Quoting from the latter:

> [T]he test statements which form the empirical basis . . . state observable simple facts about our physical environment. They

are, of course, facts interpreted in the light of theories; they are soaked in theory as it were.

As I pointed out in my *Logic of Scientific Discovery* (end of section 25), the statement 'Here is a glass of water', cannot be verified by any observational experience. The reason is that the universal terms which occur in this statement ('glass', 'water') are dispositional: they "denote physical bodies which exhibit a certain *law-like behaviour.*"

(*Popper 1974a, p. 387; italics as in the original*)

Philosophers were quick to point out the impossibility of falsification. Typical is the account by Achinstein (1968) in a review of *Conjectures and Refutations.* After pointing out that many scientific procedures that undoubtedly advance knowledge do not conform to Popper's protocol, Achinstein concluded as follows:

> My point is that if basic statements are to refute a theory they cannot merely be conjectures; they must be known to be true. Popper says that basic statements cannot be known to be true, they can only be conjectures. It follows that, for Popper, theories cannot be refuted.
>
> This yields an even more bizarre consequence. Popper says that if we subject a theory to severe tests and it is not refuted, then we can accept it, albeit tentatively (p. 51). But if we never can refute a theory, no matter how severe the tests, then we can accept any theory we like, and do so 'untentatively'.

Popper's reaction to this problem has been documented by Lakatos (1968), who set out to distinguish "three Poppers: *Popper$_0$, Popper$_1$, and Popper$_2$*" (also in Lakatos 1970; 1974, pp. 181–4). "Popper$_0$ is the dogmatic falsificationist who never published a word; he was invented – and 'criticised' – first by Ayer and then by many others" (and also virtually canonised by many scientists including Medawar, as Lakatos notes).

> Popper$_1$ is the naive falsificationist, Popper$_2$ the sophisticated falsificationist. The *real* Popper developed from dogmatic to a naive version of methodological falsification in the twenties; he arrived at the *'acceptance rules' of sophisticated falsificationism* in the fifties. The transition was marked by his adding to the original requirement of testability the 'second' requirement of 'independent testability', and then the 'third' requirement that some of these independent tests should result in corroborations. But the

> real Popper never abandoned his earlier (naive) *falsification rules.* He has demanded until this day, that '*criteria of refutation*' have to be laid down beforehand: it must be agreed, which observable situations, if actually observed, mean that the theory is 'refuted'. He still construes 'falsification' as the result of duel between theory and observation, without another, better theory *necessarily* being involved. The real Popper has never explained in detail the appeal procedure by which some 'accepted basic statements' may be eliminated. Thus the real Popper consists of Popper_1, together with some elements of Popper_2. [*italics and quotation marks as in the original*]

The meanings of *naive* and *sophisticated methodological falsificationism* are set out by Lakatos (1970). Against dogmatic falsification, Lakatos notes that "*no factual proposition can ever be proved from an experiment.* Propositions can be derived only from other propositions . . . " So *naive falsificationism* had to arise from the ashes of the dogmatic version. Falsifying basic statements had to be agreed by convention, but because they could not be proved, their use in the 'falsification' (to use quotation marks in the way that Stove notes is endemic in Lakatos) could result in rejection of a true theory. Thus, according to Lakatos, "the *methodological falsificationist separates rejection and disproof,* which the dogmatic falsificationist has conflated". As a result, he (the methodological falsificationist) proposes a new demarcation criterion: "only those theories . . . which forbid certain 'observable' [*sic*] states of affairs, and therefore may be 'falsified' [again *sic*] and rejected, are 'scientific' [!]." Popper (1974c, p. 1009) also distinguishes between falsification and rejection, but in the opposite sense, in asserting that, in the absence of a more successful rival theory, a theory may be *falsified* but *not necessarily rejected.*

This leads us to Lakatos's "*sophisticated methodological falsificationism,* which involves confrontation between theories:

> [A] scientific theory *T* is *falsified* if and only if another theory *T′* has been proposed with the following characteristics: (1) *T′* has excess empirical content over *T:* that is, it predicts *novel* facts, that is, facts improbable in the light of, or even forbidden, by *T;* (2) *T′* explains the previous success of *T,* that is, all the unrefuted content of *T* is included (within the limits of observational error) in the content of *T′*; and (3) some of the excess content of *T′* is corroborated.

Something very close to Lakatos's formulation is accepted by Popper (1972), in striking contrast to Popper's falsificationism in *The Logic of Scientific Discovery*. Lakatos further develops his scheme by the introduction of the concept of the "problem-shift". He goes from a consideration of the succession of theory T by T' to a series of theories, T_1, T_2, T_3 . . . in which each theory is derived from its predecessor by "adding auxiliary clauses to (or from semantical reinterpretations of) the previous theory in order to accommodate some anomaly, each theory having at least as much content as the unrefuted content of its predecessor". The progression from T_1 to T_2 and so on constitutes a "*theoretically progressive problem shift*" "if each new theory has some excess empirical content over its predecessor, that is if it predicts some novel, hitherto unexpected fact". The progression can also be "*empirically progressive*" if "some of this excess empirical content is also corroborated" so that a new fact is actually discovered. By fulfilling both these criteria, a problem shift is *progressive;* otherwise it is *degenerating*. Problem shifts are accepted as 'scientific' [*sic*] if they are at least theoretically progressive. (In the light of his own formulation, one might ask what the terms "unrefuted content", "corroborated", and "fact" mean in the above quotations from Lakatos). He then further distances himself from Popper's original stance by asserting, "Not an isolated *theory,* but only a series of theories can be said to be scientific or unscientific: to apply the term 'scientific' to one *single* theory is a category mistake". Thus the emphasis shifts from the confrontation of theories by data to the necessary confrontation between theories: "*no experiment, experimental report, observation statement or well-corroborated low-level falsifying hypothesis alone can lead to falsification*" (Lakatos 1970; 1974, p. 119).

Along with a number of other authors (e.g., Achinstein 1968; Kuhn 1962; Harris 1972; Brush 1974), Lakatos also notes that many of the most important case histories from the development of physics do not accord with 'dogmatic' or 'naive' falsificationisms, and thus sets the stage for his full-blown system. The logician's *modus tollens* (if p, then q; not q, therefore not p) is about the only feature that it has in common with Popper's original scheme. In the Lakatos system the problem shift ($T_1 \rightarrow T_2 \rightarrow T_3 \ldots$) is developed into the *research programme* characterised by a '*hard core*'.

> The negative heuristic of the programme forbids us to direct the *modus tollens* at this 'hard core'. Instead we must use our ingenuity to articulate or even invent 'auxiliary hypotheses', which form a *protective belt* around this core, and we must redirect the *modus tollens* to these. It is this protective belt of auxiliary hypotheses which has to bear the brunt of tests and get adjusted and re-adjusted, or even completely replaced, to defend the thus-hardened core. A research programme is successful if all this leads to a progressive problem shift; unsuccessful if it leads to a degenerating one.

Frankel (1979) develops an account of the development of the theory of continental drift as an example of a Lakatosian research programme. Frankel contrasts Continental Drift ("DRIFT") with two rival research programmes, Contractionism ("CON") and Permanentism ("PERM"). He characterises their 'hard cores' as follows:

> CON: The earth has been contracting periodically since its birth, with the result that the seafloor and continents have interchanged throughout the history of the earth.
> PERM: After an original contraction or settling out of continental and seafloor material in accordance with their densities, the oceans and continents have remained relatively the same . . .
> DRIFT: The continents have displaced themselves horizontally with respect to each other. Certain continents, now separated by vast oceans, were once combined.

He then goes on to give a history of DRIFT and its reception by the geological community from the publication of Wegener's *Die Entstehung der Kontinente und Ozeane* (1915). (For a recent review see Le Grand 1988.) The theory had little support until well after the Second World War despite palaeontological and geographical evidence from geological matching between Africa and South America, and later the interpretation of palaeomagnetic data. It was only the proposal of "seafloor spreading" in 1960 by Hess (1962) and in 1961 by Dietz, and its corroboration, principally by parallel patterns of polarity reversal (Vine and Matthews 1963), that produced the general acceptance of DRIFT. Wegener's original idea that the continents ploughed their way through the ocean floor was replaced by the more probable, and well-corroborated theory that the ocean floor is created at the mid-ocean ridges, and as it extends laterally, it pushes the continents apart.

Frankel accepts Lakatos's account of the growth of a scientific theory in his account except for two points. First, he discusses and refines the notion of *novel fact* in Lakatos's characterisation of sophisticated falsificationism. Second, Frankel suggests, with reference to DRIFT, that successive theories T_1, T_2, T_3... have a changing *hard core*. According to Lakatos the successive theories retain the same hard core, and it is the protective belt that changes, but according to Frankel there are four successive hard cores to DRIFT associated with the names of Wegener, Holmes, Du Toit, and Hess, respectively. They are:

Wegener: 1. The continents originally formed a single continent, broke up during the end of the Cretaceous, and migrated to their present positions.

2. The continents made of material lighter than the ocean floor ploughed their way through the ocean floor as they drifted to their present positions.

Du Toit: The continents originally formed two super continents, which both began to break apart during the Cretaceous. The separate landmasses continued to migrate to their present positions.

Holmes: 1. The continents originally formed two super continents, which both began to break apart during the Cretaceous.

2. The continents move passively on top of the uppermantle and lower layer of crust as seafloor is stretched due to convection currents.

Hess: 1. New seafloor is created at the ridges, spreads symmetrically ourwards from the ridges and eventually sinks back into the mantle forming trenches.

2. The drifting of the continents in a pattern similar to those proposed by DRIFTers is a consequence of seafloor spreading.

As a result, Frankel suggests that "theorists alter the hardcore of a programme whenever they believe that the alteration will add to the explanatory power of the programme or at least alleviate some of its major problems".

Frankel's account raises two important points. The first is my reason for choosing it as an illustration of Lakatos. Plate tectonics is a major theory in Natural History. The pattern and process of continental drift is, as far as we know, unique to the Earth's history; there is no suggestion that it embodies general statements about a definable class of planets or other heavenly bodies. There is no reason to believe that it is true "of all points in space and time". Thus, as well as being a theory in natural history, it is a historical theory in the most literal sense, a category whose existence was originally rejected by Popper (1945, 1957).

The second point is the nature of Frankel's account in particular, and Lakatos's in general, of the development of scientific research programmes. Is it normative or descriptive, or a mixture of both? In other words, does it consist of a general prescription, with exemplars, for scientists to follow, or of a simple historical (or sociological/psychological) characterisation of the work of successful scientists, or a combination of the two? Lakatos (1970; 1974, p. 177) is quite clear: "Kuhn's conceptual framework for dealing with continuity in science is socio-psychological: mine is normative. I look at continuity in science through 'Popperian spectacles'. Where Kuhn sees 'paradigms', I *also* see rational 'research programmes'". However, in the same volume, Kuhn (1970a; 1974a, p. 233) claims that, in agreement with Feyerabend, he (Kuhn) also makes normative claims. Furthermore, Kuhn claims that Lakatos "has yet to tell us how scientists are to select the particular statements [the hard core] that are to be unfalsifiable by their *fiat;* he must also still specify criteria which can be used at the time to distinguish a degenerative from a progressive research programme" (Kuhn 1974a, p. 239). If we again take DRIFT as an example, it would be difficult to point to a *methodological* lesson to be learned from the loyalty to their respective versions of the hard core by Wegener or du Toit.

In introducing reference to Thomas Kuhn above, I introduced Stove's third "irrationalist" and also, and perhaps more importantly, the philosopher of science, of whom, second only to Sir Karl Popper, the largest number of scientists seem to have heard. Kuhn's fame rests primarily on his book *The Structure of Scientific Revolutions* (1962; 2nd edn.: 1970) and to his usage of the word "paradigm". Kuhn's *paradigms* were in some sense equivalent to Lakatos's "research programmes"; they were systems of presuppositions within which scientists actually work. In a mature scientific discipline, scientists normally spend all their research time "puzzle-solving" within an accepted paradigm. They are not falsifying theories but contributing useful solutions to problems posed by an accepted system of theories. However, "paradigms" are not easily characterised. A number of writers, notably Masterman (1970), have pointed out that Kuhn uses paradigm in a number of senses. Masterman lists twenty-one, which, according to her, fall into three main groups. Of these three she first discusses *sociological paradigms* (i.e., the sociological use of the term).

Paradigms in this sense are sets of scientific habits that lead to successful problem solving by "normal scientists". They can in fact exist without explicit theories and are thus prior to theory. In this way they are distinguishable from Lakatos's research programmes. However, the second group of senses of the word yield *metaphysical paradigms,* which embrace such concepts as a shared myth or, more politely, "a successful metaphysical speculation", "a new way of seeing", or "an organising principle governing perception itself". The third usage Masterman refers to as *construct paradigms,* which, whereas metaphysical paradigms are something larger than research programmes, are themselves restricted. These are the artefacts that make puzzle solving possible; they could be a single piece of apparatus, a technique such as gel electrophoresis which enabled people to study genetic polymorphism at the protein level (Lewontin 1974), or a methodology such as cladistic analysis.

According to Kuhn any scientific discipline exists first in a pre-paradigm phase. The collection of data is unorganised; induction is Baconian in a derogatory sense. There can also be a state with multiple paradigms, in which rival research programmes are pursued with no useful communication among the different camps. Masterman cites the situation in "the psychological, social and information sciences". Finally, and in mature science, there is the single and dominant paradigm: Newtonian physics or the synthetic theory of evolution.

Subsequently, Kuhn attempted to correct the ambiguity inherent in his use of "paradigm" by introducing two new terms to characterise the unifying features of research programmes. These are introduced in the second edition, but more fully developed in Kuhn (1974b). The first is *exemplar.* Exemplars are the models, usually in the form of critical publications, that guide scientists in the application of new theories; Darwin's *Origin* and Newton's *Principia* are exemplars at the highest level. The second is *discipli-*

nary matrix, the whole technical and sociological background that conditions the work of normal scientists.

However, the most striking feature of Kuhn's picture is that which he used as a title: "the structure of scientific revolutions". It is also that which made Stove stigmatise him as an irrationalist. A scientific revolution is the replacement of an established paradigm by a new and untried one. Old and new paradigms are not only incompatible, they are frequently incommensurable. The breakdown of the old paradigm comes when a group, perhaps a very small one, of scientists become dissatisfied with its constraints, and problems accumulate that cannot be solved within the paradigm. But exponents of the new paradigm cannot usually communicate with those still working in the old: most scientists are not converted; they retire and die. Paradigm change involves not only methodology (construct paradigm) and *Weltanschauung* (metaphysical paradigm); thus science, when revolutions form a significant part of its history, cannot be an accumulation of knowledge. Successive paradigms cannot be shown to be successive steps towards truth because there is no criterion for comparison, let alone any criterion of truth.

Stove's fourth irrationalist is Paul Feyerabend, who, unlike the others, glories in his irrationalism. Originally a follower of Popper, Feyerabend, like Lakatos and Kuhn, was increasingly concerned that the history of science bore little resemblance to a "Popperian paradigm". Eventually he moved to a position, represented by his book *Against Method* (1975), where "anything goes". This is an expression of extreme relativism decorated all too often with assertions designed to shock; he is characterised by Chalmers (1982, p. xvii) as "philosopher and entertainer". Nevertheless, Feyerabend, as well as making the point that the history of science seldom supports any prescription as to scientific method, and further that the incommensurability of paradigms undermines any search for method, questions the status of science as a source of knowledge. He asserts, rightly, that other philosophers of science take the superiority of that status for granted without a correspondingly close investigation of other systems – Marxism, astrology, even voodoo!

In characterising the work of Lakatos, Kuhn, and Feyerabend, we have moved steadily away from what scientists in the 1960s regarded as the reassuring normative pattern of *The Logic of Sci-*

entific Discovery (which is no more about the logic of *discovery* than *On the Origin of Species* is about speciation). The writings of four irrationalists – Popper, Lakatos, Kuhn, and Feyerabend – seem to form a morphocline, terminating at the latter pole in a epistemological dead-end in research programmes. Nevertheless, Kuhn does paint a sociological picture that any working scientist will recognise, and Feyerabend's anarchic proposals do suggest disturbing questions about personal freedom in relation to the scientific establishment.

IV. Normative, descriptive, sociological, or cognitive?

My impression is that scientific colleagues believe that the "philosophy of science" has "gone quiet" since the heyday of Popper and Kuhn in the 1960s and 1970s. Working scientists have gone back to their benches and continued their investigations as before, unconcerned about the epistemological respectability of what they are doing, while perhaps regarding "falsification" (still associated with the name of Popper) in much the same light as a lapsed Catholic might regard confession. The more sophisticated will also talk about "paradigms", perhaps (for example) to explain how some research school of which they disapprove manages to get research funding more easily than they. But generally the situation has returned to what one might cynically believe to be normal: scientists once more ignore the epistemological basis of their research activity. There is, however, a striking exception to this generalisation. In recent years, there has been an unprecedented invasion of the territory of systematics and evolutionary theory by able philosophers, some of whom are so well versed in the science and so well acquainted with the scientists whom they have made the objects of their study, that their published work is scarcely distinguishable from – and sometimes co-authored with – that of their scientific colleagues. We have met many of them already in this book. But outside this particular area, philosophers of science largely seem to have ceased to attract the attention of scientists. Needless to say, the philosophy of science has not come to an end as a scholarly discipline, and I want to say something of more recent developments in the final section of this chapter.

In talking about Popper, Kuhn, Lakatos, and Feyerabend, I made the distinction between normative and descriptive philosophies of science. Those parts of Popper's and Lakatos's work that I discussed are largely normative. Their aim was to elucidate the correct, or at least a valid, methodology of science. Feyerabend reacted against these normative aims, and Kuhn, at least in the *The Structure of Scientific Revolutions,* was aiming to make sociological generalisations. To the normative and sociological approaches we can add a third, referred to by a prominent advocate, Ronald Giere (1988), as a "cognitive approach", in which the methodology of science is considered in the light of the cognitive abilities of scientists. I shall touch on each of these approaches in turn.

In considering the normative approach, two questions are of importance to us: First, is there a single scientific method, a whole range of scientific methods, or no rational scientific method at all? The first choice we can associated with the views of Popper; the last with those of Feyerabend. At this stage, I shall plump without further justification for the second; I shall give my reasons in Chapter 14. The second question amounts to an examination of Popper's dictum that all valid scientific hypotheses are, in his terminology, strictly universal statements. In Chapter 14, I shall discuss whether statements of that nature can be made in biological science. But Popper's original model of scientific method was based on physics, and there are indeed "laws" in physics – Boyle's Law, the Law of Universal Gravitation, and so on. Giere (1988) asks whether advances in physics are normally made by the proposal and testing of universal statements phrased as laws:

> Popper's whole philosophy of science rests on the deductive falsifiability of universal statements by singular statements (This F is not G). Nevertheless, the importance of universal generalizations is not supported by an examination of contemporary scientific practice. Let us look at how the venerated law of universal gravitation is treated in standard textbook presentations. Surprisingly, some textbooks of classical mechanics . . . never explicitly state the law of universal gravitation. Nothing of the form "For all bodies . . . " ever appears.
>
> (*Giere 1988, pp. 102–3*)

Giere goes on to assert that what is used in scientific practice is not *universal* generalisations but generalisations of limited extent incorporated into models. These verbal or mathematical models must be adequate for their explanatory purpose, but are the tools rather than the end of scientific research.

During the past two decades, the trend in studies of scientific method has been to become increasingly sociological. Fighting this trend, Laudan, in *Progress and Its Problems* (1977), continued the attempt to develop "a potentially more adequate model of scientific rationality", concentrating on the problem-solving of Kuhn's "normal science". He distinguished *empirical* problems – data that could not be explained by current theories – from *conceptual* problems – those detected in the nature of theories themselves. But he also introduced an "arationality assumption": "The sociology of knowledge may step in to explain beliefs if and only if those beliefs cannot be explained in terms of their rational merits" (Laudan 1977, p. 202).

Thus to Laudan it was a methodological principle that "beliefs" (i.e., theories) should be explained in terms of their rational status. Sociological factors should be invoked only to explain the irrational components. At the other extreme is David Bloor, representing a group at Edinburgh University, who came out with the "Strong Programme in the sociology of scientific knowledge" in which knowledge is not "true belief" but "whatever men take to be knowledge":

> The sociology of scientific knowledge should adhere to the following four tenets:
>
> 1. It would be causal, that is, concerned with the conditions which bring about belief or states of knowledge. Naturally there will be other types of causes apart from social ones which will co-operate in bringing about belief.
> 2. It would be impartial with respect to truth and falsity, rationality or irrationality, success or failure. Both sides of these dichotomies will require explanation.
> 3. It would be symmetrical in its style of explanation. The same types of cause would explain, say, true or false beliefs.
> 4. It would be reflexive. In principle its patterns of explanation would have to be applicable to sociology itself. . .
>
> *(Bloor 1976, pp. 4–5)*

Bloor did not confine himself to easy targets: mathematics was to be explained in sociological terms. Other sociological investigations have concerned themselves with scientific method as revealed by laboratory practice, notably those of Latour and Woolgar (1979) and Knorr-Cetina (1981). The former authors report on activity emanating from the Salk Laboratory in California. Their conclusions are that scientific knowledge is not wrested from nature; it is constructed: "Scientific activity is not 'about nature', it is a fierce fight to *construct reality*". In that fight, the status of individuals, that of the journals in which they publish, and the attitude of their colleagues, referees, and so on, all play their part. The scientist's aim is to push tentative hypotheses up a scale of accepted certainty until they become "taken-for-granted facts". Knorr-Cetina has less of the sociologist's cynicism: results, methods, and hypotheses are not accepted as matters of scientific opinion but by use:

> What we have, then, is not a process of opinion formation, but one in which certain results are solidified through continued incorporation into ongoing research. This means that the locus of solidification is the process of *scientific investigation*.

Thus scientific results are not to be equated with beliefs, but must be useful to the advancement of science itself.

It is ironic from the point of view of our present study that a number of authors have used the analogy of evolution by natural selection to explain the progress of science (e.g., Wuketits 1986; Plotkin 1987). This was actually done by Popper in his *Objective Knowledge: An Evolutionary Approach* (1972). The "objective knowledge" of the title is embodied in "World 3". "World 1" consists of the material objects of the universe; "World 2" of people's thoughts – ideas currently present in human minds. But "World 3" exists independently of World 2. It consists of beliefs, scientific theories, ideas, literature, music, and so forth, as they exist in books, recordings, and so on, independently of their creators – not the books, etc, themselves but the "software" embodied in them. The "evolutionary approach" is an overt analogy between competition among theories and ideas in World 3 for survival and competition among organisms in the theory of Natural Selection. But Popper's "selection" sounds remarkably Lamarckian: scientific theories, whatever their origin, are not well compared to "random" mutations (even by Popper!).

One of the philosophers of biology I spoke of, David L. Hull, has elaborated a more complex and sophisticated analogy between the growth of scientific knowledge and evolution by natural selection in *Science as a Process* (Subtitle: "*An Evolutionary Account of the Social and Conceptual Development of Science*") (1988). Hull's thesis is not one of austere competition between disembodied theories, but is represented by a blood-and-guts account of competition between scientists, with all their characters, enmities, alliances, and manoeuvres displayed. His study organisms are taxonomists, and his case history is the battles between and within phenetics and cladistics – taxonomy red in tooth and claw, as it were. While Hull's book is fun to read, except perhaps to those described therein, there is danger in such analogies: the more elaborate and detailed they are, the greater is the temptation to read inappropriately from one analogue to the other. Hull is anxious to show that research groups are social phenomena that do not have Aristotelian essences, in the sense of necessary shared beliefs: playing the analogies game, one can say that they are polythetic groups united by a historical continuity of relationships. But in no way does the polythetic nature of research groups corroborate the polythetic nature of natural taxa or *vice versa*.

Giere's book *Explaining Science: A Cognitive Approach* (1988) also claims to be an evolutionary explanation of scientific progress. Giere's models in scientific research (noted above, this section) are linked by him to cognitive "schemata" – patterns based on the innate and acquired abilities of scientists. Models are then judged, not for their linguistic truth, but for the similarity to real systems based on the data of research. He characterises his stance as naturalistic realism. Naturalism (in contrast to "rationalism" as in Popper) "is the view that theories come to be accepted (or not) through a natural process involving both individual judgement and social interaction. No appeal to supposed rational principles of theory choice is involved". Realism is taking most elements of a theory or model as representing aspects of the real world rather than entities proposed (e.g.) for "problem-solving effectiveness" as in Laudan (1977). As an example, Giere cites acceptance of the reality of protons and neutrons in his own study of work at the Indiana University Cyclotron Facility. He also, like Frankel (see Section III), does a study of the "evolution" of plate tectonics to show the development of the best model in terms of the analogy with natural selection.

It is obvious that any theory of the philosophy, sociology, or psychology of science must, to command attention, be based on case studies of scientists at work and/or the development of scientific theories. The case studies may be of current or recent developments or of those of the more distant past. In either case the techniques of history must be used, and therefore, the philosophy of science is inseparable from the history of science. The scientific developments, methods, and theories described so far in this book comprise the data on which I shall base my conclusions in the final chapter. But to round off this chapter, I would like to draw attention to a type of historical research that I regard as very important for future developments in the study of metascience (the scientific study of science). To some extent, it is utilized in Hull's (1988) study of recent squabbles in taxonomy referred to above, but the exemplar is Rudwick's *The Great Devonian Controversy* (1985). The controversy was initiated by the report by Henry de la Beche in 1834 of fossil plants of Coal Measure type in rocks thought to be of much greater age in Devon. The battle continued, with Roderick Murchison as de la Beche's principal opponent, until the 1840s, with the establishment of the Devonian period. Rudwick looks in great detail at all the personalities involved and the sociology and politics (particularly the organisation and meetings of the Geological Society) as well as the scientific issues involved. He does this not to corroborate a particular hypothesis about scientific progress, but to elucidate a particular case with the complex network of factors involved.

14

Philosophy and biology

> A scientist who seeks to explain the actual universe, or any (non-human) part of it, in detail as it is, in terms of its concrete history, can properly be called a natural historian – he is literally concerned with the history of nature. The natural philosopher on the other hand is devoted to the search for fundamental laws of the highest possible generality – laws which he hopes will apply throughout space and time. For the natural historian, the laws of natural philosophy are not ends in themselves, but tools for the understanding of the actual universe. . . . The mode of thought typical of the natural historian . . . is much more akin to that of Sherlock Holmes than to those of Newton, Einstein, Bohr or Rutherford.
>
> (*Crowson 1970, pp. 3–4*)

I began Chapter 1 of this book with an epigraph from Crowson's book on taxonomy, and I again quote from him in the final chapter. The distinction between Natural History and Natural Philosophy is not a new one. It is (or was) embodied in the titles of professorial chairs in the ancient Scottish universities, meaning Zoology (or Biology) and Physics, respectively, and can be regarded as explicit at the beginning of the systematic study of the Philosophy of Science. Bacon, as Crowson points out, coined the terms in his *De Dignitate et Augmentis Scientiarum* (1623). *Historia naturalis* was that part of science in which *memoria* was the dominant faculty, while *ratio* (reason) was dominant in *philosophia naturalis*. Further, as Crowson notes, Herbert Spencer (1864) makes the same distinction in overt disagreement with Comte's positivist ordering of sciences in a morphocline from the most fundamental (Mathematics) to the most derived (Sociology) (see Oldroyd 1986, pp. 169ff.).

Thus Natural Philosophy is said to be concerned with the generation of what Popperians would refer to as "strictly universal statements", whereas Natural History uses generalisations (that may or may not be strictly universal statements) to interpret individual contingent phenomena. Natural History is confined neither to field and systematic biology, nor yet to a study of the diversity of organisms plus aspects of Geology. Crowson lists its major divisions as Astronomy, Geology, Meteorology, Virology, Bacteriology, Botany, and Zoology. But it is not so easy to partition the disciplines of natural (as distinct from social) science into Natural Philosophy and Natural History. Popperian physicists would claim that the glory of their subject is the enunciation (discovery?) of natural laws and the proposal of theories, both of which are strictly universal statements. They may further claim that the proper function of "scientific" biology, often contrasted with "natural history", is to explain biological phenomena in terms of laws of physical science, the reductionist programme.

In this last chapter, in an attempt to draw conclusions from the material brought together in the whole book, I want to suggest that there is something special and distinctive about biology, which separates it from all other natural science; that distinction is not between Natural History and Natural Philosophy, but between biology and the rest of science. That something is the *taxonomic statement*.

The reader may have noticed that above, I spoke of *natural laws* and *theories* in physical science as distinct entities. In talking about the philosophy of science in Chapter 13, no such distinction was made. Additionally, I talked about *universal statements* and *hypotheses*. In order to develop the present argument, it is necessary to distinguish these terms.

In the hypothetico-deductive method a hypothesis is proposed for testing. As we saw, it should be in the form of a strictly universal statement ("Of all points in space and time . . . it is true that . . . "). But strictly universal statements, at least when corroborated or otherwise accepted, come in two forms – laws and theories:

> The tentative hypothesis which connects known facts is in the form of an empirical generalisation. Examples of such hypotheses are that the pressure of a gas is inversely proportional to its volume, at constant temperature (Boyle's law); that the extension of a wire is proportional to the extending force (Hooke's law); that the distance travelled by a dense body in free fall is proportional to the square of the time of fall (Galileo's law).

Thus hypotheses that are eventually accepted as *laws* can result from inductive inference or perhaps (*fide* Popper) from a single experiment or observation.

> However, if the tentative hypothesis is in the form of an explanatory theory it will do more than suggest relations between known facts. It will suggest new entities which will be related to the facts to be explained. (The entity may be entirely new, for example, the molecule, or it may a known effect acting in a new way, for example, the gravitational force.) An example of a scientific theory is Newton's theory of gravitational attraction where the new entities are the gravitational forces and the resulting gravitational field. Another example is the kinetic theory of gases where the new entities are the gas molecules.
>
> (*Both quotations from Trusted 1979, pp. 70–1*)

Trusted goes on to say that the distinction between an empirical generalisation and an empirical theory is that the former "relates or classifies facts which are already established", whereas the latter "*educes* new facts which themselves become established as *facts* when the theory comes to be accepted". Furthermore, "if the theory is to be ranked as a *scientific theory* it must be possible to relate new entities such as atoms, molecules etc. to observation, so that their existence can be inferred". *Eduction* is the form of "ampliative inference" (induction, *s.l.*) from known to unknown particulars, whereas *induction* (*s.s.*) is that from particulars to generalisations.

One further dichotomy needs to be explained. Empirical generalisations come in two forms. If they can be related to a scientific theory, then they are laws – "*A law is an empirical generalisation explained by a scientific theory*" – but if not (i.e., if they are "*mere* or simple empirical generalisations"), they correspond, more or less, to Popper's "numerically universal statements". Trusted's examples are "The orbits of the planets are ellipses with the sun in one focus" and "All crows are black", respectively. But the correspondence between "simple empirical generalisations" and "numerically universal statements" is not a perfect one. Popper's numerically universal statements could, he claimed, be replaced in principle by a conjunction of singular statements because they concern a finite aggregation of entities (Chapter 13, Section II). But one of Trusted's examples of a "simple generalisation" is "All electric charges are multiples of the charge on the electron". This, it will be recalled (somewhat rephrased) was one of Popper's examples of a strictly universal statement. A classification may help:

	Scientific Propositions	
I. Strictly universal statements	1. Theories	
	2. Natural laws	Empirical generalisations
	3. Simple empirical generalisations	
II. Numerically universal statements	4. Simple empirical generalisations	
III. Singular statements	5. Observation statements	

Part of the purpose of this chapter will be to suggest that a *taxonomic statement* is another type of proposition unique to biology.

I. Biological generalisations

It is clear that if we wish to say anything useful about the nature of biology as a science we must ask, Are there any strictly universal statements in biology? We shall also want to know whether, if such statements exist, they are in the form of theories, laws, or simple empirical generalisations.

In making his distinction between Natural Philosophy and Natural History, Crowson was dividing the activity and epistemology of scientists in a different way. The geologist as naturalist, investigating the stratigraphy of a particular area for its own sake, uses general principles, such as superposition and correlation, together with his or her knowledge of tectonics, "structure", and lithology, to produce a map. The meteorologist uses more or less reliable regularities in forecasting the weather, and the animal behaviourist uses supposed general principles in interpreting a particular piece of behaviour. All are trained specialists in "the mode of thought . . . akin to that of Sherlock Holmes".

But there are grey areas here. Until the sixteenth century, the accepted astronomical view was a geocentric one, based on the second-century views of Ptolemy. But Copernicus claimed a heliocentric solar system, with the Earth moving round the sun in a circular orbit with an annual cycle. At the beginning of the seventeenth century (1609), Kepler concluded that the planets moved in elliptical rather than circular orbits. Subsequently, Galileo (1632) supported the Copernican system. He also sought, by astronomical observation and by developing the concepts of inertia and acceleration, to explain planetary motion. Finally, of course, planetary motion was given a more complete explanation by Newton's law of gravitation and laws of motion. The whole history is usually presented as one of the greatest cumulative triumphs of natural philosophy. Undoubtedly, Galileo's insights into the nature of motion and Newton's laws represent "strictly universal statements". But are they being used to explain empirical generalisations about planetary systems, the realm of natural philosophy, or are general principles being used to explain a particular case, the configuration of *our* solar system, the realm of natural history?

Or again, how is one to interpret the theory of Plate Tectonics? I suggested in Chapter 13 that plate tectonics was a theory in Natural History – evidence and theories explaining what we believe to be a unique event, the tectonic history of the Earth. I further suggested that it could be paralleled with the theory of evolution. Can one then make the distinction between *explanandum* and evidence that I have developed for evolution? Giere quotes from the translated fourth edition of Wegener's book, *The Origin of Continents and Oceans* (1929, trans. 1966):

> The first concept of continental drift first came to me as far back as 1910, when considering the map of the world, under the direct impression produced by the congruence of the coastlines on either side of the Atlantic. At first I did not pay attention to the idea because I regarded it as improbable. In the fall of 1911, I came quite accidentally upon a synoptic report in which I learned for the first time of paleontological evidence for a former land

> bridge between Brazil and Africa. As a result I undertook a cursory examination of relevant research in the fields of geology and paleontology, and this provided immediately such weighty corroboration that a conviction of the fundamental soundness of the idea took root in my mind.

The match between the continents surrounding the Atlantic had been noticed before. It was noted by Bacon in *The New Organon,* and in 1858 Snider-Pellegrini proposed a biblical and catastrophic origin for the Atlantic to explain it. It is tempting, therefore, to suggest that the match of the Atlantic margins is the *explanandum* of plate tectonics and to draw a parallel with evolutionary theory thus:

	Plate Tectonics	Evolution
Explanandum	Fit of Atlantic continents	Natural Classification
Evidence	Geology and Palaeontology	Vestigial organs
	Palaeoclimatology	Palaeontology
	Palaeomagnetism	Biogeography
Mechanism	Seafloor spreading	Natural Selection, etc.

The comparison brings out a number of obvious and important differences. The first is in the nature of the suggested *explananda.* The fit of the continents on either side of the Atlantic Ocean appears to have started Wegener speculating about continental drift. But the fit is an empirical observation, rejected as an imperfect coincidence by some of Wegener's opponents, but shown to be much more convincing by Bullard *et al.* (1965), who matched the continental shelves rather than the coastlines. It can be seen, therefore, merely as a contingent fact that the "fit" was the *explanandum.* It might be the most obvious observation, but historically it could have been the case that one of the types of observation here listed as "evidence" was the *explanandum,* such as the geological matching between Africa and South America emphasised by du Toit (1937). There is no distinction *in logic* between *explanandum* and evidence in the case of plate tectonics, whereas my aim throughout this book has been to show that the natural arrangement of organisms, whether that seen by Lamarck or the hierarchy of Darwin and Wallace, is logically prior to the theory that evolution has occurred. Furthermore, the hierarchy, whether in some sense "real" or not, is not a matter of empirical observation. There can be no equivalent of photographs taken of continental outlines by satellite that would allow one to perceive the form of the hierarchy. On the definitions given above, the existence of the taxonomic hierarchy is presumably to be seen as a natural law – an empirical generalisation from numerous small-scale classifications explained by a scientific theory. That "the natural order of organisms is an irregular, divergent, inclusive hierarchy" is to be explained by the theory that the apparent relationships of classification are real. But the apparent *explanandum* of Plate Tectonics is a singular statement.

Another important series of differences are those between the proposed mechanisms. "Continental drift" (renamed "plate tectonics" because of the nature of the mechanism) was not generally accepted until the proximate cause – seafloor spreading – was accepted, although the ultimate cause – usually taken to be convection currents in the Earth's mantle – is not fully understood. But the occurrence of evolution was accepted by most naturalists soon after the publication of the *Origin,* while Natural Selection was not generally accepted until some eighty years later and still attracts minority opposition. Without a plausible mechanism, the concept of "continental drift" was literally incredible: geologists could not believe that continental blocks could plough through denser ocean floor. In contrast, it is a matter of historical fact that until the acceptance of the synthetic theory, most biologists found evolution plausible, whatever their views on mechanism. We have seen why this might have been the case in Chapters 11 and 12; here I want to emphasise two non-historical reasons. The first is the decoupling of *explanandum* and theory of mechanism (Chapter 12, Section IV): the latter in no way explains or constrains the former. The second is that natural selection is, even for the most ardent selectionist, neither a necessary "cause" nor an invariable "cause" of evolution. Evolution (i.e., anagenesis) can occur without it by drift or neutralism, and stabilising selection maintains the *status quo,* thus inhibiting evolution. Moreover the role of selection in producing genetic isolation and thus cladogenesis is under debate. The proposition that "natural selection is a necessary condition for *adaptive* anagenesis" (to phrase it more correctly) is, however, a better bet and would be a "strictly universal statement" (a testable theory?) if adaptive and non-adaptive characters could be distinguished. Hence the Panglossian strategy of asserting that all characters are adaptive. Nevertheless, I agree with Crowson (1970, p. 10) that if human beings land on Mars or Venus and

> find there anything analogous to living organisms, it is hardly to be expected that the understanding of these will be much helped by the most 'fundamental' recent discoveries in molecular biology; the terrestrial generalisations which are most likely to be applicable are those of Darwin.

II. The taxonomic statement

The taxonomic statement is a type of proposition found only in the construction of the *explanandum* of evolutionary theory. In order to demonstrate its nature, I want to discuss three possible candidates for the status of *universal* statements in biology.

The first is in the realm of molecular biology. Following the elucidation of the structure of DNA by Watson and Crick (1953a, 1953b, 1954), it was established that DNA embodied the primary genetic code in organisms other than some viruses, but RNA was involved in protein synthesis as "messenger RNA" (mRNA). The *operon model* of gene function in protein synthesis was proposed by Jacob and Monod (1961) for the lactose (*lac*) operon in the gut bacterium *Escherichia coli*. The whole system consists of a consecutive series of genes: three structural genes (each coding for a different enzyme) followed by (in order) the *operator*, the *promoter*, and the *regulator*. The regulator codes for a small molecule that binds to the operator, thus inhibiting its function of initiating transcription from DNA to mRNA in the three structural genes. Transcription, leading to enzyme synthesis, can therefore occur only if the regulator itself is inhibited, which happens when an extrinsic substance called the *inducer* binds to the regulator. In the *lac* operon, a number of substances have been found to act as inducers (Watson *et al.* 1987), but those having the highest "inductive value" are lactose or lactose derivatives, and lactose forms the substrate on which the three enzymes act. The function of the promoter is to retain the enzyme RNA polymerase so that mRNA synthesis at the three structural genes can take place. Thus the presence of the substrate (lactose) switches on mRNA and the resulting synthesis by immobilising the regulator.

Other operon systems, for other enzymes and in other bacteria, have been discovered, some with somewhat different methods of induction and repression. The essential features of an operon are that transcription from DNA to mRNA takes place in a polycistronic sequence (a *cistron* is a gene as a unit of function) that includes a series of structural genes (three in the *lac* operon) which produce a polycistronic mRNA which in its turn codes for the (three) enzymes. The whole is then controlled by a single operator–regulator system.

Discovery of the operon was understandably hailed as a triumph of fundamental biology. Here, surely, was a generalisation that would hold true with minor variation for all living organisms, and it was certainly taught as such to many undergraduates taking degrees in genetics. But when the situation in eukaryote organisms was investigated, this proved not to be the case. Eukaryotes (Protista, Fungi, Plants, and Animals) characteristically have a cell nucleus, and protein synthesis takes place outside the nucleus in the ribosomes (Chapter 9, Section III), whence the controlling program is taken by mRNA. Something like operon systems occur in the fungi *Saccharomyces* (yeast), *Aspergillus,* and *Neurospora* (both moulds), but the linked structural genes are not under unitary control (Arst 1981, 1983). Beyond that no eukaryote operons are known. Thus characterisation of the operon is not a universal statement in fundamental biology, but it could well be investigated as a taxonomic character: "the operon is an autapomorph character of bacteria"; or possibly, "the operon is plesiomorph for living things/bacteria" (having been lost or modified in eukaryotes).

Even my second possible universal statement, that nucleic acids (DNA or RNA) *always* embody the primary genetic code, may be false. The code-bearing causal agent in the mysterious group of brain diseases, "scrapie" in sheep, the recently newsworthy bovine spongiform encephalopathy (BSE) in cattle, and Creutzfeldt–Jacob disease in humans *may* be a protein molecule (a "prion") (Bock and Marsh 1988). As Hull (1987) says: "With very minor exceptions, all organisms here on Earth use the same genetic code. No one thinks that this regularity is a law of nature. To the contrary, it is a contingent fact of history". Hull's comment can be amplified by saying that the nucleic acid code is a potential taxonomic character, in which case the "regularity" can be rephrased as a taxonomic statement.

My third candidate for a universal statement was suggested to me by two departmental colleagues, both neurophysiologists, independently of each other.* It is what is known as "Dale's principle", so named by Eccles (1957, 1964) after a pronouncement by Dale (1935) that the same transmitter is liberated at all the synaptic terminals of a neurone. Thus for any neurone (nerve cell), the chemical messenger transmitted across any one of its synapses with other neurones is the same. The nature of the message, excitatory or inhibitory, is determined by the receiving neurone. Golding and Whittle (1977), reviewing the situation in annelid worms, noted that the principle applies even when the transmitter from one neurone acts as a neurotransmitter at one termination, and as a hormone passed into the blood system at another. The principle has had to be modified to specify a "cocktail" of chemicals in various proportions (Schmitt 1984; Whittaker 1984), but it appears still to hold. Most of the work has been carried out in vertebrates, but Eccles (1964) noted its corroboration for a mollusc, the "sea-hare" *Aplysia*. Once again the "principle" must be a contingent phenomenon, and with further taxonomically guided research it might be expressible in the form of a taxonomic statement.

Thus generalising from these three examples, I suggest that any apparent empirical generalisation in biology should be considered as a candidate taxonomic statement. These statements are in fact taxonomic hypotheses and come in two forms. As examples of the first, our three cases specify a potential taxonomic character – the operon system, the nucleic acid code, a single characteristic "cocktail" as a neurotransmitter – and ask the question, Of what taxon and at what rank are they characteristic? If systematic attempts were always made to answer that question, then biology could be placed on a proper comparative footing, and notably, the discipline known as "comparative physiology" would in fact become comparative physiology, analogous to comparative anatomy. It is worth noting what happens if such a *statement* is refuted. In the case of the first two, the original hypothesis was of the form "All organisms have operon systems/a nucleic acid code". The former is false, the latter subject to doubt. But in both cases one can ask whether the predicate is true of a lower-ranking taxon –

*I am grateful to Dr David Golding and Dr Henk Littlewood.

a purely cladistic procedure. In the case of Dale's principle, the taxon is not specified, but Dr. Golding suggests to me that it should be Sub-kingdom Metazoa (assuming that to be a natural taxon; see Chapter 9, Section III).

In this first type of taxonomic hypothesis, the character is specified, and the taxon is unknown or to be tested. The second type specifies the taxon; the hypothesis is that a particular feature characterises that taxon. This is of course the normal type of taxonomic hypothesis. Two old war-horses of philosophical discourse – "all crows are black", "all swans are white" – are vernacular examples, both of which are false. A bird-watching trip to the Highlands of Scotland to see hooded crows would refute the first, a trip to Australia (or some London parks) the second, but both are imprecise formulations of the same type of hypothesis as "dactyly is an autapomorphy of the taxon Tetrapoda".

Thus whereas the second type of taxonomic statement is easily recognised as such, the first type, where the rank and taxon are unspecified, is frequently taken by non-taxonomic biologists as an empirical generalisation of a lawlike type. By now it hardly needs stating that both types depend for their validity on the reality of the taxonomic hierarchy. So we must return to the hierarchy by way of questions about the nature of species.

III. What is being classified?

Most taxonomists would agree that a complete classification of the Animal (or Plant) Kingdom would have species as its terminal taxa (or as its OTUs or even EUs). The species category is usually regarded as occupying the lowest rank in the hierarchy and also as being "real" in a way that higher categories are not. Provision is made in the *International Code of Zoological Nomenclature* (Ride *et al.* 1985) for the category *sub-species,* and a sub-species taxon is indicated by a sub-specific name after the species binomen to give a trinomen such as *Homo sapiens sapiens.* But there is no refined "sub-species concept". The sub-species is usually a convenient sub-division where the boundaries of geography (or geology as *H. s. sapiens*) and morphology coincide. The *Code* suggests that this is the case (Article 45):

> [T]he original rank of an infraspecific name is deemed to be . . . subspecific, if the author . . . stated that the taxon was characteristic of a particular geographical area, or environmental or ecological context, or host species, or geological horizon.

But the species category seems to provide intractable problems of definition, not only to taxonomists but also to ecologists and evolutionists. Its unique nature has provoked controversies concerning the definition of species and the ontological nature of species. Both have generated an enormous literature, so I shall confine myself to brief summaries.

I noted in Chapter 12 (Introduction) that definitions of species in terms of interbreeding populations can be traced back to John Ray (1686). Ray's criterion, for both plants and animals, was that a member of one species never arises from the "seed" derived from a member of another (see Mayr 1982, pp. 256–7). Buffon (1749) more closely approached the modern biological species concept which identifies interbreeding explicitly as the principal criterion (Mayr 1982, pp. 260–3), but his views fluctuated during the writing of the many volumes of his *Histoire Naturelle*. Furthermore, both Ray and Buffon regarded their concepts as stipulating the separate creation and immutability of species.

Modern "biological" species definitions owe much to Dobzhansky and Mayr. Mayr (e.g., 1987) has emphasised frequently the replacement of the "typological" (i.e., idealist and essentialist) species concept by the "biological species concept", by which he means definitions in terms of interbreeding populations. Thus Dobzhansky (1951) says, "A Mendelian population is a reproductive community of sexual and cross-fertilizing individuals which share a common gene pool . . . [T]he biological species is the largest and most inclusive Mendelian population". One possible objection is that this definition denies species status to any group of non-sexual or non–cross-fertilizing individuals. Such organisms will form clones, and cladogenesis in the normal sense cannot occur. It is significant that the classification of microorganisms, mostly clonal, is a stronghold of phenetics (Austin and Priest 1986). Thus "species" in clonal organisms are typological, and they and their higher taxa are frequently polythethic. I agree with those who believe that the term "species" should not be used without qualification in the systematics of such organisms.

Mayr (1963) distinguished two ways of viewing species within his biological species concept. The "non-dimensional concept" would be that used in looking at a series of related sympatric species. The concern of the taxonomist, then, would be to decide whether two or more co-existing populations are or are not of the same species. The emphasis would be on discovering significant differences between them and, particularly, evidence of genetic isolation. The "multi-dimensional concept" looks at the world-wide range of populations forming a putative species in an attempt to judge whether they are "potentially interbreeding". Definition of species in the light of Mayr's "biological" concept is, however, more difficult, and he has made a number of attempts. The earliest (Mayr 1940, 1942) was that species are

> groups of actually or potentially interbreeding natural populations which are reproductively isolated from other such groups.

Later he was to remove the phrase "actually or potentially" (Mayr 1969, 1970). The latest version (Mayr 1982, p. 273) is:

> A species is a reproductive community of populations (reproductively isolated from others) that occupies a specific niche in nature.

As Ghiselin (1987) says, the phrase "occupies a specific niche in nature" is a truism. The only way it could fail to be so would be if "niche" were regarded as a series of ecological characteristics that defined a particular species and were thus necessary and sufficient for membership. But that would make the definition an essentialist one.

There are two other criticisms of Mayr's definition as a biological concept. Paterson (1985) regards Mayr's concept, particularly in its non-dimensional aspect, as an "*isolation concept*", whereas the emphasis should actually be on species recognition leading to mating within species, the "*recognition species concept*". But as Templeton (1989) points out, these are two sides of the same coin. Templeton goes on to develop a "*cohesion concept*", in which a "*species is the most inclusive population of individuals having the potential for phenotypic cohesion through intrinsic cohesion mechanisms*" (my italics). In doing so, he wishes to include "species" in which there is little or no inter-breeding between individuals and to exclude the situation where the inter-breeding criterion results in the inclusion together of a number of morphologically and ecologically distinct "species" that he would regard as separate – for example, the "syngameons" of botanists. Van Valen (1976) had also proposed a somewhat similar ecological species concept. Templeton also recognises another criticism of Mayr's concept – that it does not give any recognition of the species as a historical entity.

This was the concern of Simpson (1961a), whose definition has been somewhat modified by Wiley (1981, p. 25):

> An ***evolutionary species*** *is a single lineage of ancestor-descendant populations which maintains its identity from other such lineages and which has its own evolutionary tendencies and historical fate* [his italics]

This definition certainly takes account of the historical nature of species but, particularly in its last clause, is even less "operational" and more vague than Mayr's. Cracraft (1987b, 1989) produces a more overtly taxonomic effort: "*a species can be defined as an irreducible cluster or organism, within which there is a parental pattern of ancestry and descent, and which is diagnosably distinct from other such clusters*" (my italics). Mishler and Brandon (1987) also discuss a "phylogenetic species concept" and make the point that two operations are necessary to characterise the species category – grouping and ranking. Grouping is necessary to individuate species taxa; ranking, to define the level of the species category. Their "concept" is derived from Mishler and Donaghue (1982), who are botanists and, like Templeton, are concerned with the ability to characterise non-sexual "species" and also "species" as definable entities within syngameons.

Ridley (1989) advocates a specifically cladistic species concept based on Hennig's original views on the delimitation of species. Hennig's (1966) view was that a species was an inter-breeding lineage that came into existence and could be terminated as a result of speciation events (Chapter 8, Section I, and Fig. 8.2). His whole original system of "Phylogenetic Systematics" was an attempt to establish rules for delimiting and clustering species so defined. Ridley accepts that both the system of species recognition/genetic isolation and the species niche may be changed by anagenetic evolution during the history of a species. But Ridley's concept suffers from the same practical objections as that of Hennig on which it is based. If speciation results from the genetic differentiation of a peripheral isolate, while the parent stock remains unchanged, it seems intuitively ridiculous to claim that the latter is a different species before and after the isolate has become specifically distinct. In the limiting case of the isolate having a single gravid female as its origin, then if she failed to reach the (e.g.) new oceanic island, or was killed on arrival, the parent stock remains the same species; if she makes it, the parent species becomes a new species, even if it had world-wide continental distribution and had undergone no systematic genetical change. It was, of course, dissatisfaction with this type of anomaly that led to the development of transformed cladistics. There are, nevertheless, two things to be said for Ridley's revival of Hennig's concept. It has a (theoretically) unambiguous historical component, and it is non-arbitary in the sense of not cutting a continuum. It "carves nature at the joints", as Rosenberg (1985) puts it.

We can now recognise the following species concepts:

1. *Platonic:* All individuals comprising the species are more or less imperfect representations of the species archetype. This idealistic concept is rejected by most biologists, but powerfully reinforced, particularly in palaeontology, by the taxonomists' "type" system, where any newly described species must have a reference type specimen (or specimens) deposited in a museum as an exemplar.
2. *Essentialist* (Aristotle): Membership of a species depends on a series of characters necessary and sufficient for membership, although not all would be present at every stage in ontogeny of an individual, or manifest at one (e.g., the adult) stage. A fertilized egg has few species-specific characters, and closely related species, notably among insects, may have distinctive larvae and virtually indistinguishable adults (as in the European butterflies *Colias hyale* and *C. australis*). Both (1) and (2) are conflated by Mayr as "typological" concepts.
3. *Isolationist* (Mayr): This concept is "biological" in accepting "population thinking" (*sensu* Mayr), but it concentrates on genetic isolation between species in the "non-dimensional" context. It excludes non-sexual "species" and may conflate morphologically and ecologically distinct species without clear isolating mechanisms.
4. *Recognitionist* (Paterson): This concept is also "biological", but it concentrates on mutual recognition among members of a species. It still excludes non-sexual "species".

5. *Cohesional (or Ecological)* (Van Valen/Templeton): This concept is "biological" but can include non-sexual "species"; it distinguishes species separated on morphological and/or ecological grounds (corresponding to the niche element of Mayr 1982).
6. *Evolutionary* (Simpson/Wiley): This concept is based on the distinctness and coherence of a population or populations in time; it is highly non-operational.
7. *Phylogenetic* (Cracraft/Mishler and Brandon, and, more overtly cladistic, Hennig/Ridley): This concept emphasises cladistic monophyly and coherence in time but also the status of the species category as taxonomically irreducible.

Comparison of these species concepts leads one to question ontological status of species taxa. The species *category* is a class of all species *taxa* – past, present, and future. But is a species *taxon* a class, an individual, or something else? To Plato and Aristotle, species, as they would have used the term, were classes; they were also to Linnaeus, whose biological classification was closely Aristotelian (Chapter 6). But with the general acceptance of phylogeny, species taxa could no longer be treated as classes, whether membership was to be established by reference to an archetype as Plato (and Richard Owen) thought, or by reference to a series of defining characters as Aristotle (and Linnaeus) thought. Biological species consist of individual organisms that interact in competition, reproduction, social organisation, and so forth; and species have ecological niches, which may change with time, but at any given time have components that can be used as taxonomic characters. All this is implicit in the various "biological" species concepts. Furthermore (accepting evolution), species have unique histories, including "birth" at the speciation event which gives rise to them, and termination by extinction or a further speciation event. So the species category is a class of all species taxa and thus available as a subject for empirical generalisations, but any one species taxon is an ontological individual, a contingent entity. Species have been treated this way by evolutionists, notably by Mayr in the study of speciation, and by taxonomists aware of their (the species') histories notably Simpson and Hennig. But the credit for making their status explicit must go to Ghiselin (1966, 1974). Hull (1976) and Williams (1985) have supported the thesis as professional philosophers. Lively discussion has ensued with, as a recent example, the journal *Biology and Philosophy*, devoting nearly a whole number – vol. 2, no. 2 (1987) – to the subject with further contributions in vol. 3, no. 4 (1988).

Many authors, while rejecting at least Platonic classes, have found the notion of a whole species as an individual, with the individual organisms as "parts" rather than members, counter-intuitive. So various other collective nouns for species taxa have been suggested: "systems" (Stebbins 1987), "sets" (Kitcher 1987) with or without defining features, "extensionally defined classes" (a misunderstanding of Hennig; Dupuis 1984), disjunctively defined classes (Stent 1986), and "(bio)populations" (Mayr 1987). Mayr essentially agrees with Ghiselin and Hull, but cannot quite accept the nature of species as individuals. His usage of "population" is criticised by Hull (1987) as ontologically too vague, a criticism with which I agree.

If biological species are individuals, what then of higher taxa? The individuality of species taxa depends on two things: their reality and the fact that each taxon is unique and spatio-temporally bounded. They really are entities rather than useful operational abstractions. They are entities, whether defined by the exclusive interbreeding criterion (isolation or recognition concepts) or by their ecological role (ecological or cohesion concepts). Species also *do* things:

> They speciate, they evolve, they provide their component organisms with genetical resources, and they become extinct. They compete. . . . Above the level of the species, genera and higher taxa never do anything. Clusters of related clones . . . don't do anything either.
>
> (*Ghiselin 1987*)

But species are not universal entities like the class of all the hydrogen atoms in the universe. They are bounded in space and time, and those boundaries are potentially discoverable. Thus species are individuals, and their component organisms parts of the whole. There are of course difficulties. Some have compared, in effect, the culling of a species with chopping off (e.g.) the head of a vertebrate, to illustrate the lack of cohesion within a species. But this is a zoocentric prejudice, as anyone who has contemplated the individuality of a mass of algal "blanket weed" or pruned a rose-bush should realise. More cogently, species are difficult to delimit, isolation is not perfect, and incipient species exist. But more worrying is the fact that hybrid species and polyploid species could arise more than once, if only experimentally, and it seems arbitrary to insist that two entities, even if indistinguishable, must be different species. Finally, clones cannot be species, but here again the division is not absolute: there are insect and even vertebrate species that are always parthenogenetic within genera whose members are otherwise sexually reproducing (White 1978).

Claims about the individuality of higher taxa depend on the taxonomic methods used to create (discover!) them. Only one method of classification – that is, phylogenetic (or Hennigian) cladistics – can claim that its supra-specific taxa are real and spatio-temporally bounded. Every correctly identified monophyletic taxon (*sensu* cladistics) has an origin identical to the origin of its single ancestral species and is bounded in space and time by the aggregate histories of its component species. It is, in Ghiselin's oft-quoted phrase, "a particular chunk of the genealogical nexus". It is perhaps, surprising, therefore, that Wiley (1981), a phylogeneticist, shows the same reluctance to see higher taxa as individuals, dubbing them "historical groups", as Mayr does with species.

Phenetic taxa may be "real" on the criterion of "Gilmour naturalness" (Chapter 7), but the criterion incorporates no concept of being spatio-temporally bounded. They are classes, not individuals, but they differ from Platonic or Aristotelian classes in that they can be characterised only *post hoc;* the clusters are *created* (I use the word advisedly!), and then their characters can be investigated. The supra-specific taxa of "evolutionary" systematics may be spatio-temporally bounded in that, once delimited, their distribution in space and time could potentially be investigated and will probably be coherent. But Mayr's taxa, if different from Hennigian ones, are paraphyletic and so are "chunks of the genealogical nexus" with bits missing, as in "Sauropsida with Aves missing", or "Amniota with Aves plus Mammalia missing", equals "Reptilia" [depending on what you do with the mammal (theropsid) clade; see Chapter 7, Section IV]. Thus they may be bounded but they are not individuals. They are classes defined by presence of some characters and absence of others. Simpson's taxa, on the other hand, were avowedly arbitrary (Chapter 6, Section IV).

The taxa of transformed cladistics are a more difficult case. They are claimed to be real, a claim I investigate in the next section, but they can be shown to be spatio-temporally bounded only if phylogeny is accepted; and transformed cladists have set their faces against validating taxa by the *a priori* acceptance of phylogeny. The taxa, if real, must therefore be classes, whose diagnoses (in the medical sense – i.e., symptoms of membership; Ghiselin 1987) are also definitions. Reality inheres in the definition; this is Aristotelian essentialism, which leads us in to a final discussion of the natural hierarchy.

IV. The hierarchy again

The Natural Hierarchy I have talked about throughout this book is the taxonomic (or "Linnean") hierarchy in which the taxa at any rank are all members of the same category. The hierarchy is of the taxa themselves, so that (e.g.) each genus taxon contains one or (usually) more than one species taxon. Thus the hierarchy is inclusive. It is divergent by convention and irregular as a contingent fact and can be represented by a dendrogram. Evolutionary theory can be regarded as an empirical generalisation stating that the taxa at any rank, if correctly delimited, are real and spatio-temporally bounded individuals.

But the taxonomic hierarchy is not the only one employed by biologists in organising their data. There is, of course, a "metahierarchy" of taxonomic categories representing the ranks of the taxonomic hierarchy. This is a simple exclusive hierarchy like the ranks in an army, with the species category at the bottom and kingdom or some more inclusive category at the top. Other hierarchies, however, are also organisations of biological phenomena. An *ecological* hierarchy is generally recognised, and this is paralleled by a *genealogical* hierarchy by Eldredge and Salthe (1984). Both are reviewed and discussed in the context of evolutionary theory by Eldredge (1985; see also Vrba and Eldredge 1984; and Salthe 1985). They are tabulated thus by Eldredge and Salthe, with the lowest rank at the top of the table, but both can have extra categories interpolated between those in the basic list, as in taxonomy:

Genealogical hierarchy	Ecological hierarchy
Codons	Enzymes
Genes	Cells
Organisms	Organisms
Demes	Populations
Species	Local ecosystems
Monophyletic taxa	Biotic regions
(Special case: all life)	Entire biosphere

Eldredge (1985) inserts "chromosomes" between "genes" and "organisms" in the genealogical hierarchy.

It will be seen that the only category shared by both is "organisms", which in the table has been manipulated to appear at the same rank. Organisms are obviously both real and individuals, but as with the taxonomic hierarchy, it is claimed that the "taxa" at every rank are spatio-temporally bounded individuals. The genealogical hierarchy is one of reproduction and the transmission of information; the ecological hierarchy of "energy–matter transfer" (Eldredge). In the genealogical hierarchy, one codon is the lowest category and consists of three consecutive nucleotide bases, that together specify one amino acid. "Gene", as Eldredge uses the term, can be taken to be a cistron specifying a single polypeptide chain. The "glue", as Eldredge puts it, that associates the individuals at one rank so that they in their turn form an individual at the next higher rank, is replication ("more-making"). Above the organism level, demes are genetically defined local populations of a species that can subdivide and thus multiply. A species comprises the sum of all its demes. Monophyletic taxa (*sensu* Hennig) do not replicate, however, but are composed of species that do. All monophyletic taxa are alike in this hierarchy in consisting of all the descendant species of a single (included) ancestor. The genealogical hierarchy should not, therefore, be topped above "species" by all the ranks of the Linnean hierarchy. The two hierarchies are different entities characterised by different criteria.

The ecological hierarchy is different again. At the organism level, it is based on how individual organisms earn their living rather than how they replicate. Individuals of the same species interact with one another by association, competition, and so on, to form a population, the "taxon" at the next rank. Populations of a number of species in the same area interact to form an ecosystem and so on. In fact, the ecological hierarchy is more complex than that and can be split into two – one for interaction between biological entities, the other taking into account interactions with the abiotic environment (Eldredge 1985). It will be obvious that the theory of natural selection draws on components of both the genealogical and the ecological hierarchies, competitive reproduction and survival, to explain the phenomenon of phylogeny. Phylogeny is then the explanation of the taxonomic hierarchy.

But when we turn to the taxonomic hierarchy, a number of difficulties present themselves. We have seen that, following Ghiselin's logic, all the taxa at every rank of a (correct) Hennigian classification are individuals. As a result, any taxon has to be defined "ostensively" – by "pointing": individuals do not have defining properties (Ghiselin 1987). Thus the apomorph characters of a Hennigian classification are symptoms, literally providing a diagnosis as in medical practice, rather than a definition. Given anagenesis within species and higher taxa, they may also, of course, change with time. As the taxa are real, contingent, and spatio-temporally bounded, so are the taxic homologies that characterise them. Furthermore, if Hennigian cladistics reconstructs phylogeny in the form of a hierarchy (cladogram or tree), then the parallel hierarchy of homologies has individual groups of apomorph characters as its taxa; but it also consists of a coincident series of hierarchies, each a character state tree; and each character state tree reconstructs a "phylogeny" of transformational homologies. Such a reconstruction must be a legitimate pursuit in Hennigian or phylogenetic cladistics, but it must also be clear that with all the entities involved being individual and contingent, there is no way in which the hypothetico-deductive method of Popper could be used in phylogenetic practice. There are many other objections to such use which I dealt with in detail some years ago (Panchen 1982; see also below).

Turning to the taxonomic hierarchy as seen by "transformed" cladists, there seems to be some ontological confusion. The taxa are regarded as "real"; otherwise undetected homoplasies could not be regarded as "mistakes" (Chapter 8, Section III). But the taxa cannot be spatio-temporally bounded, because *a priori* phylogeny is rejected; therefore, taxa must be timeless entities – in fact, classes – with apomorph characters (taxic homologies) as defining properties: hence essentialism, that is, unless full-blown

Platonic idealism is accepted. There is some ambiguity in cladist writing about this. Patterson (1978) (admittedly before he "came out" as a transformed cladist) saw species and all monophyletic higher taxa as individuals, but subsequently (in Rosen *et al.* 1981, and in his personal communication cited in Panchen and Smithson 1987), he still seemed to favour defining taxa ostensively. Platnick, as we have seen, assigns a different ontological status to species from that of higher taxa, which is consistent with the former being individuals and the latter classes – a view at odds with the attitude expressed by Nelson and Platnick (1981, see our Chapter 8, Section II) that all taxonomy should be typological. If species are individuals but higher taxa are classes, they should not appear in the same hierarchy. That is to repeat Porphyry's and Linnaeus's mistake (Chapter 6, Section II). Furthermore, the synapomorphies uniting two sister-species would be mere diagnostic symptoms characterising both but, as autapomorphies of the taxon uniting both, would become defining properties of a class, which is irrational.

I sketched a solution to these paradoxes in Chapter 8 (Section I). Cladists should accept essentialism as a *methodological principle* despite the fact that it cannot be justified ontologically. Its epistemological justification is the *explanans–explanandum* principle. They should *then* regard any resulting cladogram not as a hypothesis of grouping, because of the individuality of taxa, but as a hypothesis of phylogeny, paralleling the grand hypothesis that the natural order represents the phylogeny of all life. Their cladogram then becomes a tree, and in agreement with the phylogeneticists I consider the two to be isomorphic (Chapter 8, Section I). The only realistic attack on this latter view involves reticulation, which is also an attack on the *a priori* hierarchy. The latter again can be a methodological assumption only, so that it is part of the cladogram-to-tree hypothesis that the tree represents a divergent hierarchy as well. It then becomes a particular hypothesis of phylogeny that may be tested by scenarios (Chapter 7, Section IV) or by evidence of reticulation, and the tree corrected if necessary. The substantive conclusion therefore is that the proper procedure is to follow the sequence cladogram → tree → classification.

One further test must be of hierarchical structure in the data, which can be applied in cladistic algorithms by way of the "consistency index" (Kluge and Farris 1969). Failure to reject the null hypothesis of no hierarchical structure in the whole natural order would of course render the theory of evolution unnecessary!

V. Propositions in biology

I suggested previously that *taxonomic statements* were a unique feature of biology, and I now wish to justify that statement and to suggest that such statements constitute a powerful method of inference. We saw above that the hypothetico-deductive method is inapplicable to test hypotheses of phylogeny that deal only with ontological individuals. I discussed this in Panchen (1982) for contingent entities, but most of the argument also applies to the methodological essentialism in which taxa are treated as classes. So hypothetico-deductive inference does not apply, but taxonomic inference does.

To justify this I shall again summarise two reasons why Popperian falsification, even if accepted otherwise, fails in taxonomy. I suggested, as the first reason, that its cladistic advocates were confusing the whole method with the *modus tollens* (if *p* then *q*; not *q*, therefore not *p*), which is merely its logical form. To demonstrate this I suggested the following argument, here modified to deal with classes, which is both valid and almost certainly true (Gardiner 1980; Panchen and Smithson 1987):

1. *If* the Porolepiformes [a fossil fish group] are the sister-group of the Tetrapoda or some group within the Tetrapoda *then* they will be choanate (if *p*, then *q*).
2. The Porolepiformes are *not* choanate [no internal nostrils] (not *q*).
3. *Therefore,* the Porolepiformes are *not* the sister-group of any tetrapod group (therefore not *p*).

(All tetrapods are choanate; only one fish group, the Osteolepiformes, is known to be.)

The inference, although correct, cannot be an example of Popperian falsification for one of two reasons, thus giving my conclusion the form of a classical logical dilemma: *Either* (1) the choanate or non-choanate status of the Porolepiformes is already known, so the consequent (*q*) does not constitute a *new* falsifiable prediction from the antecedent (*p*); *or* (2) if the choanate status of the Porolepiformes is unknown, *either* no choanate fishes are known, in which case choanae might be an autapomorphy of Tetrapoda, *or,* if any choanate fishes are known (other than the Porolepiformes), then the character could not be an exclusive synapomorphy uniting Tetrapoda and Porolepiformes as sister-groups (i.e., at that rank it would be a symplesiomorphy).

The second reason for rejecting Popperism in cladistics is embodied in a quotation from Gaffney (1979, pp. 96–7) explaining how it should be used! He refers to a three-taxon test:

> When multiple series of characters are used, it is often the case that all three possible hypotheses relating three taxa (1) (AB)C, (2)A(BC), (3)AC(B) will appear to be falsified by one or more character distributions. In this case, the least-rejected hypothesis, (the *one falsified the fewest number of times*) is the one incorporated into the internesting system of hypotheses. [my italics]

But falsification is an all-or-nothing affair, not a matter of parsimony.

So now we may ask whether there are any propositions in biology that are strictly universal statements (laws or theories) that are valid ontologically. Such statements would have to be about classes and not individuals – which immediately suggests that taxonomic categories, but not taxa, could be the subjects of laws. Thus Ghiselin (1987) points out the irony in the fact that Mayr, who "devoted much of a brilliant career in defending a law of allopatry in speciation", denies that there are any important laws in biology – its regularities being "either too obvious to be mentioned, or too trivial" (Mayr 1982, p. 37). The law in this case (whether true or false) can be phrased thus: *speciation in sexually reproducing organisms is always allopatric* (see Chapter 12, Section I). It is a law about the species category. An explanatory theory (a negative one) is that natural selection cannot produce genetic isolation, resulting in the separation of a population of a species into two or more species, in sympatry. Another example is Van Valen's (1973) hypothesis that the probability of extinction of a genus or family is independent of its prior duration. He proposed this as a law to explain his finding that the log of the number of taxa that survive for *t* years, plotted against *t* years, gives a straight line, with a negative slope. The result was explained by the *Red Queen Hypothesis,* that from the point of view of any species, the environment deteriorates continuously, because of the evolution of its competitors, predators, and parasites. Thus this is the law plus a theory about the categories of genus and family. They may well be false: Smith and Patterson (1988) suggest that the constant stochastic rate of extinction may be due to the data's being derived from non-monophyletic taxa, but the "law" and its explanation are far from being "obvious" or "trivial".

Thus there are laws and theories in biology including the theory of Natural Selection – *natural selection (as defined) is a necessary condition of adaptive anagenesis* – but why should I claim special status for taxonomic statements above other generalisations in natural history? Is there some important distinction in kind between (e.g.) "Dale's principle is characteristic of all Metazoa" or "all spiders have characteristic spinnerets" (Platnick's favourite example), on the one hand, and, on the other, "all undisturbed sediments retain residual palaeomagnetism" (if true)? I suggest that (1) only the taxa of organisms have autapomorphies, and that (2) only classifications of organisms, or their diagrammatic representation, constitute nested sets of congruent homologies. This explains the predictive power that resides in the diagnoses, indeed in the wider descriptions, of taxa at all levels, which John Stuart Mill saw as characteristic of natural classification. There is no necessary congruence in the classification of artefacts, such as library books: the subject matter of books, the usual character set used in such classifications, tells one nothing about size, shape, colour, binding, or price of an individual book. On the other hand, the features that characterise other natural entities, such as rocks, minerals, and so on, are not autapomorphies; neither is there one natural hierarchy of "homologies" into which they may be arranged. Any groupings of such entities are polythetic and any classifications phenetic.

VI. Postscript

At the time that I am completing the text of this book, the community of taxonomists throughout the world no doubt continues with what David Hull (1988) regards as its creative squabbles, but over one issue it has achieved an extraordinary and impressive consensus. The Natural History Museum in London, arguably the greatest institution for taxonomic research in the world, is threatened from within by a strategic plan that would destroy the broad sweep of its research. The plan was drawn up in response to the present British Government's policy of starving public institutions of funding. Its authors were the director (a former animal behaviourist) and chosen colleagues; its principal supporter, the chairman of the Museum's trustees (a medical geneticist). Undemanding entertainment is to be given precedence over scholarship. A list of research areas has been drawn up, selected for their appeal to sources of private funding rather than for their importance or cogency; and flourishing areas of taxonomic research – on extant "lower" plants, all fossil plants, recent and fossil mammals, and a number of invertebrate groups including many insect taxa – are to be closed down.

Protests have flooded in from around the world, so all may not be lost; there is, nevertheless, a sad lesson to be learned. Neither the director nor the chairman is a taxonomist (nor does either represent the only non-taxonomic department in the Museum, that of Mineralogy, which is also to be savaged). It would be optimistic to describe the status of taxonomy and its practitioners as at a low ebb; there is no evidence that the tide will turn, at least in Britain.

Few universities or polytechnics (at least in Britain) give courses in taxonomy, as distinct from the systematics of plants and animals; and yet, as I have tried to argue in this book, the results of taxonomic research must underlie not only evolutionary theory but any major principle in biology. Otherwise population geneticists will continue to believe that they are studying all that is worth studying in evolutionary theory; molecular biologists and physiologists will fail to investigate the taxonomic scope of their brilliant discoveries; and animal behaviourists will fail to understand that their generalisations are all contingent. Taxonomic statements will continue to appear in prestigious journals masquerading as laws of biology, but bad law is bad law, however disguised.

References

Abel, O. (1911). *Grundzüge der Palaeobiologie der Wirbeltiere* Stuttgart: Schweizerbart.

Abel, O. (1924) *Lehrbuch der Paläozoologie,* 2nd edn. Jena: G. Fischer.

Abel, O. (1929). *Paläobiologie und Stammesgeschichte.* Jena: G. Fischer.

Achinstein, P. (1968). [Review of] *Conjectures and Refutations* [by] Karl R. Popper. *British Journal for the Philosophy of Science, 19,* 159–80.

Adams, E.N. (1972). Consensus techniques and the comparison of taxonomic trees. *Systematic Zoology, 21,* 390–7.

Adanson, M. (1763). *Familles des Plantes.* Paris: Vincent.

Adey, M., Prentice, I.C., Bisby, F.A., and Harris, J.A. (1983). Instability and incongruence in the brooms and gorses (Leguminosae subtribe Genistinae). In *Numerical Taxonomy,* ed. J. Felsenstein, pp. 117–20. Berlin: Springer-Verlag.

Agassiz, J.L.R. (1833–44). *Recherches sur les Poissons Fossiles.* 5 vols. and suppl. Neuchâtel and Soleure.

Agassiz, J.L.R. (1857). *Essay on Classification. (Contributions to the Natural History of the United States, vol. 1, part 1).* Boston, Mass.: Little, Brown.

Ahlquist, J.E., Bledsoe, A.H., Sheldon, F.H., and Sibley, C.G. (1987). DNA hybridization and avian systematics: response to Houde. *Auk, 104,* 556–63.

Alberch, P., and Gale, E.A. (1983). Size dependence during development of the amphibian foot: colchicine-induced digital loss and reduction. *Journal of Embryology and Experimental Morphology, 76,* 177–97.

Alberch, P., and Gale, E.A. (1985). A developmental analysis of an evolutionary trend: digital reduction in amphibians. *Evolution, 39,* 8–23.

Allin, E.F. (1975). Evolution of the mammalian middle ear. *Journal of Morphology, 147,* 403–38.

Allin, E.F. (1986). The auditory apparatus of advanced mammal-like reptiles and early mammals. In *The Ecology and Biology of Mammal-like Reptiles,* ed. N. Hotton, P.D. MacLean, J.J. Roth, and C. Roth, pp. 283–94. Washington, D.C.: Smithsonian Institution.

Alvarez, L.W., Alvarez, W., Asaro, F., and Michel, H.V. (1980). Extraterrestrial cause for the Cretaceous/Tertiary extinction. *Science, 208,* 1095–108.

Andrews, P.J. (1985). Molecular evidence for catarrhine evolution. In *Major Topics in Primate and Human Evolution,* ed. B. Wood, L. Martin, and P. Andrews, pp. 107–29. Cambridge: Cambridge University Press.

Andrews, P.J. (1987). Aspects of hominoid phylogeny. In *Molecules and Morphology in Evolution: Conflict or Compromise?,* ed. C. Patterson, pp. 23–53. Cambridge: Cambridge University Press.

Andrews, P.J., and Cronin, J.E. (1982). The relationship of *Sivapithecus* and *Ramapithecus* and the evolution of the orang-utan. *Nature, 297,* 541–6.

Appel, T.A. (1987). *The Cuvier–Geoffroy Debate: French Biology in the Decades before Darwin.* New York: Oxford University Press.

Arnold, E.N. (1981). Estimating phylogenies at low taxonomic levels. *Zeitschrift für Zoologische Systematik und Evolutionsforschung, 19,* 1–35.

Arst, H.N. (1981). Aspects of the control of gene expression in fungi. In *Genetics as a Tool in Microbiology (Society for General Microbiology Symposium No. 31),* ed. S.W. Glover and D.A. Hopwood, pp. 131–60. Cambridge: Cambridge University Press.

Arst, H.N. (1983). Fungal systems. In *Eukaryotic Genes, Their Structure, Activity and Regulation,* ed. N. Maclean, S.P. Gregory, and R.A. Flavell, pp. 433–50. London: Butterworths.

Arthur, W. (1984). *Mechanisms of Morphological Evolution: A Combined Genetic, Developmental and Ecological Approach.* Chichester: John Wiley & Sons.

Arthur, W. (1988). *A Theory of the Evolution of Development.* Chichester: John Wiley & Sons.

Ashlock, P.D. (1971). Monophyly and associated terms. *Systematic Zoology, 20,* 63–9.

Ashlock, P.D. (1972). Monophyly again. *Systematic Zoology, 21,* 430–8.

Ashlock, P.D. (1980). An evolutionary systematist's view of classification. *Systematic Zoology, 28,* 441–50.

Austin, B., and Priest, F.G. (1986). *Modern Bacterial Taxonomy.* Wokingham: Van Nostrand Reinhold (UK) Ltd.

Ax, P. (1987). *The Phylogenetic System: The Systematization of Organisms on the Basis of Their Phylogenesis,* transl. R.P.S. Jefferies. Chichester: John Wiley & Sons.

Ayer, A.J. (1936). *Language, Truth and Logic.* London: Gollancz.

Bacon, F. (1620). *The New Organon: Or Directions Concerning the Interpretation of Nature (The Great Instauration, part 2),* ed. J.M. Robertson (1905). *New Organon To Be Found in the Philosophical Works of Francis Bacon. . . .* London: Routledge.

Bacon, F. (1623). *De Dignitate et Augmentis Scientiarum.* London.

Baer, K.E. von (1828). *Über Entwickelungsgeschichte der Thiere: Beobachtung und Reflexion.* Königsberg: Bornträger.

Bambach, R.K. (1986). Phanerozoic marine communities. In *Patterns and Processes in the History of Life (Dahlem Konferenzen, Life Sciences Research Report No. 36),* ed. D.M. Raup and D. Jablonski, pp. 407–28.

Barton, N.H., and Charlesworth, B. (1984). Genetic revolutions, founder effects, and speciation. *Annual Review of Ecology and Systematics, 15,* 133–64.

Bates, H.W. (1862). Contributions to an insect fauna of the Amazon Valley. Lepidoptera: Heliconidae. *Transactions of the Linnean Society, 23,* 495–566.

Bateson, W. (1894). *Materials for the Study of Variation: Treated with Especial Regard to Discontinuity in the Origin of Species.* London: Macmillan.

Bateson, W. (1913). *Problems of Genetics.* Oxford: Oxford University Press.

Beatty, J. (1982). Classes and cladists. *Systematic Zoology, 31,* 25–34.

Beckner, M. (1959). *The Biological Way of Thought.* New York: Columbia University Press.

Belon, P. (1555). *L'Histoire de la Nature des Oyseaux.* Paris.

Berg, L.S. (1958). *System der rezenten und fossilen Fischartigen und Fische.* Berlin: VEB deutscher Verlag der Wissenschaften.

Berry, R.J. (1990). Industrial melanism and peppered moths (*Biston betularia* (L.)). *Biological Journal of the Linnean Society, 39,* 301–22.

Bishop, M.J., and Friday, A.E. (1985). Evolutionary trees from nucleic acid and protein sequences. *Proceedings of the Royal Society [B], 226,* 271–302.

Bishop, M.J., and Friday, A.E. (1987). Tetrapod relationships: the molecular evidence. In *Molecules and Morphology in Evolution: Conflict or Compromise?,* ed. C. Patterson, pp. 123–39. Cambridge: Cambridge University Press.

Bishop, M.J., and Friday, A.E. (1988). Estimating the interrelationships of tetrapod groups on the basis of molecular sequence data. In *The Phylogeny and Classification of the Tetrapods, vol. 1: Amphibians, Reptiles and Birds,* ed. M.J. Benton, pp. 33–58. Oxford: Oxford University Press, for the Systematics Association.

Bjerring, H.C. (1967). Does a homology exist between the basicranial muscle and the polar cartilage? *Problèmes Actuels de Paléontologie (Colloques Internationaux du Centre National de la Recherche Scientifique No. 163),* pp. 223–67. Paris.

Bloor, D. (1976). *Knowledge and Social Imagery.* London: Routledge & Kegan Paul.

Bock, G., and Marsh, J. (eds.). (1988). *Novel Infectious Agents and the Central Nervous System. (CIBA Foundation Symposium no. 135).* New York: Wiley-Interscience.

Bock, W.J. (1969). Comparative morphology in systematics. In *Systematic Biology (National Academy of Sciences, publ. 1692),* pp. 411–48. Washington: National Academy of Sciences.

Bock, W.J. (1974). Philosophical foundations of classical evolutionary classification. *Systematic Zoology, 22,* 375–92.

Bock, W.J. (1977). Foundations and methods of evolutionary classification. In *Major Patterns in Vertebrate Evolution,* ed. M.K. Hecht, P.C. Goody, and B.M. Hecht, pp. 851–95. New York: Plenum Press.

Bode, H.R., and Steele, R.E. (1989). [In "Phylogeny and molecular data"]. *Science, 243,* 549.

Bohor, B.F., Foord, E.E., Modreski, P.J., and Triplehorn, D.M. (1984). Mineralogic evidence for an impact event at the Cretaceous/Tertiary boundary. *Science, 224,* 867–9.

Bohor, B.F., Modreski, P.J., and Foord, E.E. (1985). A search for shock-metamorphosed quartz at the KT boundary. *Abstracts of Papers: Sixteenth Lunar and Planetary Science Conference,* pp. 79–80.

Bondi, H., and Kilmister, C.W. (1959). The impact of *Logik der Forschung*. *British Journal for the Philosophy of Science*, *10*, 55–7.

Bonnet, C. (1764). *Contemplation de la Nature,* 2 vols, 1st edn. Amsterdam: Marc-Michel Rey.

Bowler, P.J. (1976). *Fossils and Progress: Paleontology and the Idea of Progressive Evolution in the Nineteenth Century*. New York: Science History Publications.

Bowler, P.J. (1983). *The Eclipse of Darwinism: Anti-Darwinian Evolution Theories in the Decades Around 1900*. Baltimore: John Hopkins University Press.

Brady, R.H. (1980). Natural selection and the criteria by which a theory is judged. *Systematic Zoology*, *28*, 600–21.

Brady, R.H. (1985). On the independence of systematics. *Cladistics*, *1*, 113–26.

Bremer, K., and Wanntorp, H-E. (1979). Hierarchy and reticulation in systematics. *Systematic Zoology*, *28*, 624–7.

Bronn, H.G. (1858). *Untersuchungen über die Entwickelungs-Gesetze der organischen Welt während der Bildungs-Zeit unsere Erd-Oberfläche*. Stuttgart. (Also as 1861, Essai d'une réponse à la question der prix proposée en 1850 . . . *Supplement aux Comptes-Rendus des Séances de l'Academie des Sciences*, *2*, 377–918).

Brooks, D.R., and Wiley, E.O. (1985). Theories and methods in different approaches to phylogenetic systematics. *Cladistics*, *1*, 1–11.

Brooks, D.R., and Wiley, E.O. (1986). *Evolution as Entropy: Toward a Unified Theory of Biology*, 1st edn. Chicago: University of Chicago Press.

Brooks, D.R., and Wiley, E.O. (1988). *Evolution as Entropy: Toward a Unified Theory of Biology*, 2nd edn. Chicago: University of Chicago Press.

Brooks, W.K. (1883). *The Law of Heredity: A Study of the Cause of Variation and the Origin of Living Organisms*. Baltimore: John Murphy.

Bruce, E.J., and Ayala, F.J. (1979). Phylogenetic relationships between man and the apes: electrophoretic evidence. *Evolution*, *33*,, 1040–56.

Brundin, L. (1966). Transantarctic relationships and their significance as evidenced by chironomid midges. *Kungliga Svenska Vetenskakademiens Handlingar* (4) *11*, no. 1, 1–472.

Brundin, L. (1968). Application of phylogenetic principles in systematics and evolutionary theory. In *Nobel Symposium no. 4: Current Problems of Lower Vertebrate Phylogeny*, ed. T. Orvig, pp. 473–95. Stockholm: Almqvist & Wiksell.

Brush, S.G. (1974). Should the history of science be rated X? *Science*, *183*, 1164–72.

Buch, C.L. von (1825). *Physicalische Beschreibung der Canarischen Inseln*. Berlin: Königliche Akademie der Wissenschaften.

Buckland, W. (1823). *Reliquiae Diluvianae; Or, Observations on the Organic Remains Contained in Caves, Fissures, and Diluvial Gravel, and on the Geological Phenomena, Attesting the Action of a Universal Deluge*, 1st edn. London: John Murray.

Buckland, W. (1836). *Geology and Mineralogy Considered with Reference to Natural Theology (Bridgewater Treatise IV)*, 2 vols., 1st edn. London: William Pickering.

Buffon, G.L.L. de (1749–1804). *Histoire Naturelle, Générale et Particulière, avec la Déscription du Cabinet du Roi*, 44 vols. plus Atlas. Paris: Imprimérie Royale, puis Plassan.

Bullard, E.C., Everett, J.E., and Gilbert Smith, A. (1965). The fit of the continents around the Atlantic. In *A Symposium on Continental Drift*, ed. P.M.S.

Blackett, E.C. Bullard, and S.K. Runcorn, pp. 41–51. London: The Royal Society.

Burian, R.M. (1983). "Adaptation". In *Dimensions of Darwinism: Themes and Counterthemes in Twentieth-Century Evolutionary Theory*, ed. M. Grene, pp. 287–314. Cambridge: Cambridge University Press.

Burian, R.M. (1988). Challenges to the evolutionary synthesis. *Evolutionary Biology, 23,* 247–69.

Burkhardt, R.W. (1977). *The Spirit of System: Lamarck and Evolutionary Biology*. Cambridge, Mass.: Harvard University Press.

Butler, P.M. (1982). Directions of evolution in the mammalian dentition. In *Problems of Phylogenetic Reconstruction,* ed. K.A. Joysey and A.E. Friday, pp. 235–44. London: Academic Press for the Systematics Association.

Butlin, R. (1989). Reinforcement of premating isolation. In *Speciation and Its Consequences,* ed. D. Otte and J.A. Endler, pp. 158–79. Sunderland, Mass.: Sinauer Associates.

Caccone, A., and Powell, J.R. (1989). DNA divergence among hominoids. *Evolution, 43,* 925–42.

Cain, A.J. (1958). Logic and memory in Linnaeus's system of taxonomy. *Proceedings of the Linnean Society of London, 169,* 144–63.

Cain, A.J., and Harrison, G.A. (1958). An analysis of the taxonomist's judgement of affinity. *Proceedings of the Zoological Society of London, 131,* 85–98.

Cain, A.J., and Harrison, G.A. (1960). Phyletic weighting. *Proceedings of the Zoological Society of London, 135,* 1–31.

Cain, S.A. (1944). *Foundations of Plant Geography*. New York: Harper & Row.

Cairns, J., Overbaugh, J., and Miller, S. (1988). The origin of mutants. *Nature, 335,* 142–5.

Camin, J.H., and Sokal, R.R. (1965). A method for deducing branching sequences in phylogeny. *Evolution, 19,* 311–26.

Camp, C.L., Allison, H.J., and Nichols, R.H. (1964). Bibliography of fossil vertebrates, 1954–1958. *The Geological Society of America, Memoir, 92,* 1–647.

Campbell, K.S.W., and Barwick, R.E. (1987). Paleozoic lungfishes – a review. In *The Biology and Evolution of Lungfishes,* ed. W.E. Bemis, W.W. Burggren, and N. E. Kemp, pp. 93–131. New York: Alan R. Liss.

Candolle, A.P. de (1820). Essai élémentaire de géographie botanique. In *Dictionnaire de sciences naturelles,* vol. 18, pp. 359–422. Strasbourg & Paris: Flevrault.

Candolle, A.P. de (1838). *Statistique de la Famille de Composées*. Paris & Strasbourg: Treuttel & Wurtz.

Cannon, H.G. (1958). *The Evolution of Living Things*. Manchester: Manchester University Press.

Cannon, H.G. (1959). *Lamarck and Modern Genetics*. Manchester: Manchester University Press.

Carpenter, W.B. (1839). *Principles of General and Comparative Physiology, Intended as an Introduction to the Study of Human Physiology, and as a Guide to the Philosophical Pursuit of Natural History,* 1st edn. London: John Churchill.

Carroll, R.L. (1988). *Vertebrate Paleontology and Evolution*. New York: W.H. Freeman.

Carson, H.L., and Templeton, A.R. (1984). Genetic revolutions in relation to speciation phenomena: the founding of new populations. *Annual Review of Ecology and Systematics, 15,* 97–131.

Cartmill, M. (1981). Hypothesis testing and phylogenetic reconstruction. *Zeitschrift für Zoologische Systematik und Evolutionsforschung, 19,* 73–96.

Castle, W.E. (1919). Piebald rats and selection, a correction. *American Naturalist, 53,* 370–6.

Castle, W.E., and Phillips, J.C. (1914). *Piebald Rats and Selection (Carnegie Institution of Washington Publication no. 195).* Washington, D.C.

Catzeflis, F.M., Sheldon, F.H., Ahlquist, J.E., and Sibley, C.G. (1987). DNA–DNA hybridization evidence of the rapid rate of rodent DNA evolution. *Molecular Biology and Evolution, 4,* 242–53.

Chalmers, A.F. (1982). *What Is This Thing Called Science? An Assessment of the Nature and Status of Science and Its Methods,* 2nd edn. Milton Keynes: Open University Press.

[Chambers, R.] (1844). *Vestiges of the Natural History of Creation,* 1st edn. London: Churchill.

[Chambers, R.] (1851). *Vestiges of the Natural History of Creation,* 9th edn. London: Churchill.

Charig, A.J. (1981). Cladistics: a different point of view. *Biologist, London, 28,* 19–20.

Charig, A.J. (1982). Systematics in biology: a fundamental comparison of some major schools of thought. In *Problems of Phylogenetic Reconstruction,* ed. K.A. Joysey and A.E. Friday, pp. 363–440. London: Academic Press for the Systematics Association.

Charlesworth, B., Lande, R., and Slatkin, M. (1982). A neo-Darwinian commentary on macroevolution. *Evolution, 36,* 474–98.

Chomsky, N. (1965). *Aspects of the Theory of Syntax.* Cambridge, Mass.: M.I.T. Press.

Churchill, S.P., Wiley, E.O., and Hauser, L.A. (1984). A critique of Wagner groundplan-divergence studies and a comparison with other methods of phylogenetic analysis. *Taxon, 33,* 212–32.

Clark, R.B., and Panchen, A.L. (1971). *Synopsis of Animal Classification.* London: Chapman & Hall.

Clarke, C.A., Clarke, F.M.M., and Dawkins, H.C. (1990). *Biston betularia* (the peppered moth) in West Kirby, Wirral, 1959–1989: updating the decline of *f. carbonaria. Biological Journal of the Linnean Society, 39,* 323–6.

Clarke, C.A., Mani, G.S., and Wynne, G. (1985). Evolution in reverse: clean air and the peppered moth. *Biological Journal of the Linnean Society, 26,* 189–99.

Cloud, P. (1960). Gas as a sedimentary and diagenetic agent. *American Journal of Science, 258A,* 35–45.

Comte, I.A. (1830–42). *Cours de Philosophie Positive,* 6 vols. Paris: Bachelier.

Cook, L.M., Mani, G.S., and Varley, M.E. (1986). Postindustrial melanism in the peppered moth. *Science, 231,* 611–13.

Cope, E.D. (1870). *On the Hypothesis of Evolution: Physical and Metaphysical.* New Haven, Conn.: C.C. Chatfield. [Reprinted in Cope, E.D. (1887). *The Origin of the Fittest,* pp. 128–72. New York: Macmillan].

Cope, E.D. (1896). *The Primary Factors of Organic Evolution*. Chicago: Open Court Publishing.

Corruccini, R., Baba, M., Goodman, M., Ciochon, R., and Cronin, J. (1980). Non-linear macromolecular evolution and the molecular clock. *Evolution, 34,* 1216–9.

Coyne, J.A., and Orr, H.A. (1989). Two roles of speciation. In *Speciation and Its Consequences,* ed. D. Otte and J.A. Endler, pp. 180–207.

Cracraft, J. (1978). Science, philosophy, and systematics. *Systematic Zoology, 27,* 213–6.

Cracraft, J. (1987a). DNA hybridization and avian phylogenetics. *Evolutionary Biology, 21,* 47–96.

Cracraft, J. (1987b). Species concepts and the ontology of evolution. *Biology and Philosophy, 2,* 329–46.

Cracraft, J. (1988). The major clades of birds. In *The Phylogeny and Classification of the Tetrapods, vol. 1: Amphibians, Reptiles and Birds,* ed. M.J. Benton, pp. 339–61. Oxford: Oxford University Press, for the Systematics Association.

Cracraft, J. (1989). Speciation and its ontology: the empirical consequences of alternative species concepts for understanding patterns and processes of differentiation. In *Speciation and Its Consequences,* ed. D. Otte and J.A. Endler, pp. 28–59. Sunderland, Mass.: Sinauer Associates.

Crowson, R.A. (1970). *Classification and Biology*. London: Heinemann.

Cuvier, G. (1795). Mémoire sur la structure interne et externe et sur les affinités des animaux auxquels on a donné le nom de vers. *La Décade Philosophique, 5,* 385–96.

Cuvier, G. (1827–35). *The Animal Kingdom Arranged in Conformity with Its Organisation,* 15 vols. London: William S. Orr.

Cuvier, G. (1830). Considérations sur les mollusques et en particulier sur les céphalopodes. *Annales des Sciences Naturelles, 19,* 241–59.

Cuvier, G., and Brongniart, A. (1811). Essai sur la géographie minéralogique des environs de Paris avec une carte géographique et des coupes de terrain. *Mémoires de la Classe des Sciences Mathématiques et Physiques de l'Institut Imperial de France, an 1810,* (part I), 1–278.

Cuvier, G., and Valenciennes, A. (1828–48). *Histoire Naturelle des Poissons,* 22 vols. Paris: F.G. Levrault.

Dale, H. (1935). Pharmacology and nerve endings. *Proceedings of the Royal Society of Medicine, 28,* 319–32.

Darwin, C.R. (1842). *The Structure and Distribution of Coral Reefs. Being the First Part of the Geology of the Voyage of the Beagle, Under Command of Capt. Fitzroy, R.N. during the Years 1832 to 1836,* 1st edn. London: Smith, Elder.

Darwin, C.R. (1851a, 1854a). *A Monograph of the Sub-Class Cirripedia, with Figures of All the Species,* 2 vols. London: Ray Society.

Darwin, C.R. (1851b). *A Monograph of the Fossil Lepadidae, or Pedunculated Cirripedes, of Great Britain*. London: Palaeontographical Society.

Darwin, C.R. (1854b). *A Monograph of the Fossil Balanidae and Verrucidae of Great Britain*. London: Palaeontographical Society.

Darwin, C.R. (1859). *On the Origin of Species by Means of Natural Selection, or the*

Preservation of Favoured Races in the Struggle for Life, 1st edn. London: John Murray.

Darwin, C.R. (1868). *The Variation of Animals and Plants Under Domestication,* 2 vols., 1st edn. London: John Murray.

Darwin, C.R. (1869). *On the Origin of Species by Means of Natural Selection . . . ,* 5th edn. London: John Murray.

Darwin, C.R. (1871). *The Descent of Man, and Selection in Relation to Sex,* 2 vols., 1st edn. London: John Murray.

Darwin, C.R. (1987). *Charles Darwin's Notebooks, 1836–1844: Geology, Transmutation of Species, Metaphysical Enquiries,* ed. P.H. Barrett, P.J. Gautrey, S. Herbert, D. Kohn, and S. Smith. Cambridge: British Museum (Natural History), and Cambridge University Press.

Darwin, C.R., and Wallace, A.R. (1858). On the tendency of species to form varieties; and on the perpetuation of varieties and species by natural means of selection. *Journal of the Proceedings of the Linnean Society [Zoology] 3,* (1859), 46–62. [Reprinted as Darwin and Wallace 1958].

Darwin, C.R., and Wallace, A.R. (1958). *Evolution by Natural Selection,* ed. G.R. de Beer. Cambridge: Cambridge University Press.

Darwin, F. (ed.). (1887). *The Life and Letters of Charles Darwin,* 3 vols. London: John Murray.

Darwin, F., and Seward, A.C. (eds.). (1903). *More Letters of Charles Darwin,* 2 vols. London: John Murray.

Davis, M., Hut, P., and Muller, R.A. (1984). Extinction of species by periodic comet showers. *Nature, 308,* 715–17.

Dawkins, R. (1976). *The Selfish Gene.* Oxford: Oxford University Press.

Dawkins, R. (1982). *The Extended Phenotype.* Oxford: Oxford University Press.

Dawkins, R. (1986). *The Blind Watchmaker.* London: Longman.

Dayhoff, M.O. (1969). Computer analysis of protein evolution. *Scientific American, 221,* no. 1, 86–95.

Depew, D.J., and Weber, B.H. (1988). Consequences of nonequilibrium thermodynamics for the Darwinian tradition. In *Entropy, Information and Evolution: Perspectives on Physical and Biological Evolution,* ed. B.H. Weber, D.J. Depew, and J.D. Smith, pp. 317–42. Cambridge, Mass.: M.I.T. Press.

Desmond, A. (1982). *Archetypes and Ancestors: Palaeontology in Victorian London, 1850–1875.* London: Blond & Briggs.

Desmond, A. (1989). *The Politics of Evolution: Morphology, Medicine and Reform in Radical London.* Chicago: University of Chicago Press.

Diehl, S.R., and Bush, G.L. (1989). The role of habitat preference in adaptation and speciation. In *Speciation and its Consequences,* ed. D. Otte and J.A. Endler, pp. 345–65. Sunderland, Mass.: Sinauer Associates.

Dietz, R.S. (1961). Continental ocean basin evolution by spreading of the sea floor. *Nature, 190,* 854–7.

Dobzhansky, T. (1937). *Genetics and the Origin of Species.* New York: Columbia University Press.

Dobzhansky, T. (1951). Mendelian populations and their evolution. In *Genetics in the 20th Century,* ed. L.C. Dunn, pp. 573–89. New York: Macmillan.

Dobzhansky, T., Ayala, F.J., Stebbins, G.L., and Valentine, J.W. (1977). *Evolution.* San Francisco: W.H. Freeman.

Dubois, E. (1894). Pithecanthropus erectus, *eine menschenaenliche Ubergangsform aus Java.* Batavia: Landesdruckerei.

Dupuis, C. (1979). Permanence et actualité de la systématique: 2. La "Systematique phylogénétique" de W. Hennig. *Cahiers des Naturalistes (N.S.), 34,* 1–69.

Dupuis, C. (1984). Willi Hennig's impact on taxonomic thought. *Annual Review of Ecology and Systematics, 15,* 1–24.

Eccles, J.C. (1957). *The Physiology of Nerve Cells.* Baltimore: Johns Hopkins Press.

Eccles, J.C. (1964). *The Physiology of Synapses.* Berlin: Springer-Verlag.

Eck, R.V., and Dayhoff, M.O. (1966). *Atlas of Protein Sequence and Structure, 1966.* Silver Spring, Md.: National Biomedical Research Foundation.

Edwards, A.W.F., and Cavalli-Sforza, L.O. (1963). The reconstruction of evolution [abstract]. *Heredity, 18,* 553.

Edwards, A.W.F., and Cavalli-Sforza, L.O. (1964). Reconstruction of evolutionary trees. In *Phenetic and Phylogenetic Classification,* ed. V.H. Heywood and J. McNeill, pp. 67–76. London: Systematics Association.

Eldredge, N. (1979a). Alternative approaches to evolutionary theory. *Bulletin of the Carnegie Museum of Natural History, 13,* 7–19.

Eldredge, N. (1979b). Cladism and common sense. In *Phylogenetic Analysis and Paleontology,* ed. J. Cracraft and N. Eldredge, pp. 165–98. New York: Columbia University Press.

Eldredge, N. (1985). *Unfinished Synthesis: Biological Hierarchies and Modern Evolutionary Thought.* New York: Oxford University Press.

Eldredge, N., and Cracraft, J. (1980). *Phylogenetic Patterns and the Evolutionary Process.* New York: Columbia University Press.

Eldredge, N., and Gould, S.J. (1972). Punctuated equilibria: an alternative to phyletic gradualism. In *Models in Paleobiology,* ed. T.J.M. Schopf, pp. 82–115. San Francisco: Freeman, Cooper.

Eldredge, N., and Gould, S.J. (1988). Punctuated equilibrium prevails. *Nature, 332,* 211–2.

Eldredge, N., and Salthe, S.N. (1984). Hierarchy and evolution. In *Oxford Surveys in Evolutionary Biology,* vol. 1, ed. R. Dawkins and M. Ridley, pp. 184–208. Oxford: Oxford University Press.

Eldredge, N., and Tattersall, I. (1975). Evolutionary models, phylogenetic reconstruction and another look at hominid phylogeny. *Contributions in Primatology, 5,* 218–42.

Engelmann, G.F., and Wiley, E.O. (1977). The place of ancestor–descendent relationships in phylogeny reconstruction. *Systematic Zoology, 26,* 1–11.

Estabrook, G.F., Strauch, J.G., and Fiala, K.L. (1977). An application of compatibility analysis to the Blackith's data on orthopteroid insects. *Systematic Zoology, 26,* 269–76.

Falconer, D.S. (1960). *Introduction to Quantitative Genetics.* Edinburgh: Oliver & Boyd.

Farris, J.S. (1969). A successive approximation approach to character weighting. *Systematic Zoology, 18,* 374–85.
Farris, J.S. (1970). Methods for computing Wagner trees. *Systematic Zoology, 19,* 83–92.
Farris, J.S. (1972). Estimating phylogenetic trees from distance matrices. *American Naturalist, 106,* 645–68.
Farris, J.S. (1973). On the use of the parsimony criterion for inferring evolutionary trees. *Systematic Zoology, 22,* 250–6.
Farris, J.S. (1974). Formal definitions of paraphyly and polyphyly. *Systematic Zoology, 23,* 548–54.
Farris, J.S. (1976). Phylogenetic classification of fossils with Recent species. *Systematic Zoology, 25,* 271–82.
Farris, J.S. (1977). On the phenetic approach to vertebrate classification. In *Major Patterns of Vertebrate Evolution,* ed. M.K. Hecht, P.C. Goody, and B.M. Hecht, pp. 823–50. New York: Plenum Press.
Farris, J.S. (1978). Inferring phylogenetic trees from chromosome inversion data. *Systematic Zoology, 27,* 275–84.
Farris, J.S. (1981). Distance data in phylogenetic analysis. In *Advances in Cladistics,* vol. 1, ed. V.A. Funk and D.R. Brooks, pp. 3–23. New York: New York Botanical Garden.
Farris, J.S. (1983). The logical basis of phylogenetic analysis. In *Advances in Cladistics,* vol. 2, ed. N.I. Platnick and V.A. Funk, pp. 7–36. New York: Columbia University Press.
Farris, J.S. (1985). Distance data revisited. *Cladistics, 1,* 67–85.
Farris, J.S. (1986). Distance and statistics. *Cladistics, 2,* 144–57.
Farris, J.S., Kluge, A.G., and Eckardt, M.J. (1970). A numerical approach to phylogenetic systematics. *Systematic Zoology, 19,* 172–91.
Felsenstein, J. (1973). Maximum likelihood and minimum-steps methods for estimating evolutionary trees from data on discrete characters. *Systematic Zoology, 22,* 240–9.
Felsenstein, J. (1978). Cases in which parsimony and compatibility methods will be positively misleading. *Systematic Zoology, 27,* 401–10.
Felsenstein, J. (1979). Alternative methods of phylogenetic inference and their interrelationship. *Systematic Zoology, 28,* 49–62.
Felsenstein, J. (1981). Evolutionary trees from DNA sequences: a maximum likelihood approach. *Journal of Molecular Evolution, 17,* 368–76.
Felsenstein, J. (1982). Numerical methods for inferring evolutionary trees. *Quarterly Review of Biology, 57,* 379–404.
Felsenstein, J. (1984). Distance methods for inferring phylogenies: a justification. *Evolution, 38,* 16–24.
Felsenstein, J. (1985). Confidence limits in phylogenies: an approach using the bootstrap. *Evolution, 38,* 783–91.
Felsenstein, J. (1986). Distance methods: a reply to Farris. *Cladistics, 2,* 130–43.
Felsenstein, J. (1988a). Phylogenies from molecular sequences: inference and reliability. *Annual Review of Genetics, 22,* 521–65.
Felsenstein, J. (1988b). Perils of molecular introspection. *Nature, 335,* 118.

Feyerabend, P.K. (1975). *Against Method: Outlines of an Anarchistic Theory of Knowledge*. London: New Left Books; Atlantic Highlands: Humanities Press.

Field, K.G., Olsen, G.J., Giovannoni, S.J., Raff, E.C., Pace, N.R., and Raff, R.A. (1989). [In "Phylogeny and molecular data"] – Response. *Science, 243,* 550–1.

Field, K.G., Olsen, G.J., Lane, D.J., Giovannoni, S.J., Ghiselin, M.T., Raff, E.C., Pace, N.R., and Raff, R.A. (1988). Molecular phylogeny of the Animal Kingdom. *Science, 239,* 748–53.

Fisher, D.C. (1981). The role of functional analysis in phylogenetic inference: examples from the history of the Xiphosura. *American Zoologist, 21,* 47–62.

Fisher, R.A. (1918). The correlation between relatives on the supposition of Mendelian inheritance. *Transactions of the Royal Society of Edinburgh, 52,* 399–433.

Fisher, R.A. (1930). *The Genetical Theory of Natural Selection*. Oxford: Oxford University Press.

Fitch, W.M. (1970). Distinguishing homologous from analogous proteins. *Systematic Zoology, 19,* 99–113.

Fitch, W.M. (1971). Toward defining the course of evolution: minimum change for a given tree topology. *Systematic Zoology, 20,* 406–16.

Fitch, W.M. (1979). Numerical taxonomy: a special project (II). *Systematic Zoology, 28,* 254–5.

Fitch, W.M., and Margoliash, E. (1967). Construction of phylogenetic trees. *Science, 155,* 279–84.

Fitzhugh, K. (1990). Invertebrates and ignorance: the book. [Review of] *Invertebrate Relationships: Patterns in Animal Evolution* [by] Pat Willmer. 1990. Cambridge University Press . . . *Cladistics, 6,* 403–9.

Ford, E.B. (1937). Problems of heredity in the Lepidoptera. *Biological Reviews, 12,* 461–503.

Ford, E.B. (1940). Genetic research in the Lepidoptera. *Annals of Eugenics, 10,* 227–52.

Ford, E.B. (1945). Polymorphism. *Biological Reviews, 20,* 73–88.

Ford, E.B. (1964). *Ecological Genetics,* 1st edn. London: Methuen.

Ford, E.B. (1975). *Ecological Genetics,* 4th edn. London: Chapman & Hall.

Forey, P.L. (1987). Relationships of lungfishes. In *The Biology and Evolution of Lungfishes,* ed. W.E. Bemis, W.W. Burggren, and N.E. Kemp, pp. 75–91. New York: Alan R. Liss.

Fothergill, P.G. (1952). *Historical Aspects of Organic Evolution*. London: Hollis & Carter.

Foxon, G.E.H. (1955). Problems of the double circulation in vertebrates. *Biological Reviews of the Cambridge Philosophical Society, 30,* 196–228.

Frankel, H. (1979). The career of continental drift theory: an application of Imre Lakatos' analysis of scientific growth to the rise of drift theory. *Studies in History and Philosophy of Science, 10,* 21–66.

Friday, A.E. (1980). The status of immunological distance data in the construction of phylogenetic classifications: a critique. In *Chemosystematics: Principles and Practice,* ed. F.A. Bisby, J.G. Vaughan, and C.A. Wright, pp. 289–304. London: Academic Press, for the Systematics Association.

Futuyma, D.J., and Mayer, G.C. (1980). Non-allopatric speciation in animals. *Systematic Zoology, 29,* 254–71.

Gadow, H. (1901). *Amphibia and Reptiles. (The Cambridge Natural History, vol. 8).* London: Macmillan.

Gaffney, E. (1979). An introduction to the logic of phylogeny reconstruction. In *Phylogenetic Analysis and Paleontology,* ed. J. Cracraft and N. Eldredge, pp. 79–111. New York: Columbia University Press.

Galilei, G. (1632). *Dialogo . . . dove nei Congressi di Quattro Gionate si Discorre sopra i Due Massimi Sistemi del Mondo Tolemaico e Copernicano . . .* Florence. [Reprinted as *Dialogue Concerning the Two Chief World Systems – Ptolemaic and Copernican,* transl. S. Drake. Berkeley: University of California Press, 1953].

Gardiner, B.G. (1980). Tetrapod ancestry: a reappraisal. In *The Terrestrial Environment and the Origin of Land Vertebrates,* ed. A.L. Panchen, pp. 177–85. London: Academic Press, for the Systematics Association.

Gardiner, B.G. (1982). Tetrapod classification. *Zoological Journal of the Linnean Society, 74,* 207–32.

Gass, I.G. *et al.* (1972). *Palaeontology and Geological Time (Science: A Second Level Course).* Bletchley: The Open University.

Gauld, I., and Underwood, G. (1986). Some applications of the Le Quesne compatibility test. *Biological Journal of the Linnean Society, 29,* 191–222.

Gauthier, J. (1986). Saurischian monophyly and the origin of birds. *Memoirs of the California Academy of Sciences, 8,* 1–56.

Gauthier, J., Kluge, A.G., and Rowe, T. (1988). Amniote phylogeny and the importance of fossils. *Cladistics, 4,* 105–209.

Gegenbaur, C. (1859). *Grundzüge der vergleichenden Anatomie,* 1st edn. Leipzig: W. Engelmann.

Gegenbaur, C. (1870). *Grundzüge der vergleichenden Anatomie,* 2nd edn. Leipzig: W. Engelmann.

Gegenbaur, C. (1874). *Grundriss der vergleichenden Anatomie,* 1st edn. Leipzig: W. Engelmann.

Geoffroy Saint-Hilaire, E. (1818). *Philosophie Anatomique (vol. 1): Des Organes Respiratoires sous le Rapport de la Détermination et de l'Identité de Leurs Pièces Osseuses.* Paris: for the author.

Ghiselin, M.T. (1966). An application of the theory of definitions to taxonomic principles. *Systematic Zoology, 15,* 127–30.

Ghiselin, M.T. (1969). *The Triumph of the Darwinian Method.* Berkeley & Los Angeles: University of California Press.

Ghiselin, M.T. (1974). A radical solution to the species problem. *Systematic Zoology, 23,* 536–44.

Ghiselin, M.T. (1987). Species concepts, individuality, and objectivity. *Biology and Philosophy, 2,* 127–43.

Giere, R.N. (1988). *Explaining Science, a Cognitive Approach.* Chicago: University of Chicago Press.

Gill, T. (1872). Arrangement of the families of fishes. *Smithsonian Miscellaneous Collections, 247,* 49pp.

Gilmour, J.S.L. (1937). A taxonomic problem. *Nature, 139,* 1040–2.

Gilmour, J.S.L. (1940). Taxonomy and Philosophy. In *The New Systematics*, ed. J.S. Huxley, pp. 461–74. Oxford: Clarendon Press.

Gilmour, J.S.L. (1961). Taxonomy. In *Contemporary Botanical Thought*, ed. A.M. MacLeod and L.S. Cobley, pp. 27–45. Edinburgh: Oliver & Boyd.

Gilmour, J.S.L., and Walters, S.M. (1963). Philosophy and classification. In *Vistas in Botany*, ed. W.B. Turrill, vol. 4, pp. 1–22. London: Pergamon Press.

Gingerich, P.D. (1979). The stratophenetic approach to phylogeny reconstruction in vertebrate paleontology. In *Phylogenetic Analysis and Paleontology*, ed. J. Cracraft and N. Eldredge, pp. 41–77. New York: Columbia University Press.

Gingerich, P.D. (1983). Rates of evolution: effects of time and temporal scaling. *Science, 222*, 159–61.

Gingerich, P.D. (1985). Species in the fossil record: concepts, trends and transitions. *Paleobiology, 11*, 27–41.

Gish, D.T. (1979). *Evolution? The Fossils Say No!* San Diego: Creation-Life Publishers.

Goethe, J.W. von (1807). Bildung und Umbildung organischer Naturen [new introduction], in *Versuch die Metamorphose der Pflanzen zu erklären* (reprint). Gotha: Carl Wilhelm Ettinger.

Goethe, J.W. von (1817–24). *Zue Naturwissenschaft überhaupt, besonders zur Morphologie*, 2 Bde. Stuttgart & Tübingen: J.G. Cotta.

Golding, D.W., and Whittle, A.C. (1977). Neurosecretion and related phenomena in annelids. In *Aspects of Cell Control Mechanisms. (International Review of Cytology, suppl. 5)*, ed. G.H. Bourne and J.F. Danielli, pp. 189–302.

Goldman, N. (1990). Maximum likelihood inference of phylogenetic trees, with special reference to a Poisson process model of DNA substitution and to parsimony analysis. *Systematic Zoology, 39*, 345–61.

Goldschmidt, R. (1940). *The Material Basis of Evolution*. New Haven, Conn.: Yale University Press.

Goodman, M., Braunitzer, G., Stangl, A., and Schrank, B. (1983). Evidence on human origins from haemoglobins in African apes. *Nature, 303*, 546–8.

Goodman, M., Miyamoto, M.M., and Czelusniak, J. (1987). Pattern and process in vertebrate phylogeny revealed by coevolution of molecules and morphologies. In *Molecules and Morphology in Evolution: Conflict or Compromise?*, ed. C. Patterson, pp. 141–46. Cambridge University Press.

Goodman, M., Moore, G.W., and Matsuda, G. (1975). Darwinian evolution in the genealogy of haemoglobin. *Nature, 253*, 603–8.

Goodrich, E.S. (1909). *Vertebrata Craniata, First Fascicle: Cyclostomes and Fishes. (A Treatise on Zoology, part 9*, ed. E.R. Lankaster). London: Adam & Black.

Goodrich, E.S. (1916). On the classification of the Reptilia. *Proceedings of the Royal Society of London [B], 89*, 261–76.

Goodrich, E.S. (1930). *Studies on the Structure and Development of Vertebrates*. London: Macmillan.

Goodwin, B.C. (1984). Changing from an evolutionary to a generative paradigm in biology. In *Evolutionary Theory: Paths into the Future*, ed. J.W. Pollard, pp. 99–120. Chichester: John Wiley & Sons.

Gould, S.J. (1972). Allometric fallacies and the evolution of *Gryphaea*. *Evolutionary Biology, 6*, 91–119.

Gould, S.J. (1974). The evolutionary significance of "bizarre" structures: antler size and skull size in the "Irish Elk", *Megaloceros giganteus*. *Evolution, 28,* 191–220.

Gould, S.J. (1977a). Eternal metaphors of palaeontology. In *Patterns of Evolution,* ed. A. Hallam, pp. 1–26. Amsterdam: Elsevier Scientific Publishing.

Gould, S.J. (1977b). *Ontogeny and Phylogeny*. Cambridge, Mass.: Harvard University Press.

Gould, S.J. (1977c). The return of the hopeful monster. *Natural History, 86,* no. 6, 22–30.

Gould, S.J. (1980a). *The Panda's Thumb*. New York: Norton.

Gould, S.J. (1980b). Hen's teeth and horse's toes. *Natural History, 89,* no. 7, 24–8.

Gould, S.J. (1980c). G. G. Simpson, paleontology, and the Modern Synthesis. In *The Evolutionary Synthesis: Perspectives on the Unification of Biology*, ed. E. Mayr and W.B. Provine, pp. 153–72. Cambridge, Mass.: Harvard University Press.

Gould, S.J. (1980d). Is a new and general theory of evolution emerging? *Paleobiology, 11,* 2–12.

Gould, S.J. (1983). *Hen's Teeth and Horse's Toes*. New York: Norton.

Gould, S.J. (1985). The paradox of the first tier: an agenda for paleobiology. *Paleobiology, 11,* 2–12.

Gould, S.J. (1987). *Time's Arrow, Time's Cycle: Myth and Metaphor in the Discovery of Geological Time*. Cambridge, Mass.: Harvard University Press.

Gould, S.J., and Eldredge, N. (1977). Punctuated equilibria: the tempo and mode of evolution reconsidered. *Paleobiology, 3,* 115–51.

Gould, S.J., and Eldredge, N. (1986). Punctuated equilibria at the third stage. *Systematic Zoology, 35,* 143–8.

Gould, S.J., and Lewontin, R.C. (1979). The spandrels of San Marco and the Panglossian paradigm: a critique of the adaptationist programme. *Proceedings of the Royal Society [B], 205,* 581–98.

Gould, S.J., and Vrba, E.S. (1982). Exaptation – a missing term in the science of form. *Paleobiology, 8,* 4–15.

Grafen, A. (1982). How not to measure inclusive fitness. *Nature, 298,* 425–6.

Le Grand, H.E. (1988). *Drifting Continents and Shifting Theories*. Cambridge: Cambridge University Press.

Grant, P.R. (1986). *Ecology and Evolution of Darwin's Finches*. Princeton, N.J.: Princeton University Press.

Grant, P.R., and Grant, B.R. (1989). Sympatric speciation and Darwin's finches. In *Speciation and Its Consequences,* ed. D. Otte and J.A. Endler, pp. 433–57. Sunderland, Mass.: Sinauer Associates.

Gray, G.S., and Fitch, W.M. (1983). Evolution of antibiotic resistance genes: the DNA sequence of a kanamycin resistance gene from *Staphylococcus aureus*. *Molecular Biology and Evolution, 1,* 57–66.

Greuter, W. *et al.* (eds.). (1988). *International Code of Botanical Nomenclature*. (Adopted by the Fourteenth International Botanical Congress, Berlin, 1987). Königstein: Koeltz Scientific Books.

Griffiths, G.C.D. (1974). On the foundations of biological systematics. *Acta Biotheoretica, 13,* 85–131.

Haeckel, E. (1866). *Generelle Morphologie der Organismen: Allgemeine Grundzüge der organischen Formen-Wissenschaft, mechanisch begründet durch die von Charles Darwin reformirte Descendenz-Theorie,* 2 vols. Berlin: Georg Reimer.

Haeckel, E. (1868). *Natürliche Schöpfungsgeschichte,* 1st edn. Berlin: Georg Reimer.

Haeckel, E. (1874). *Anthropogenie: Keimes- und Stammes-Geschichte des Menschen,* 1st edn. Leipzig: W. Engelmann.

Haeckel, E. (1889). *Natürliche Schöpfungsgeschichte,* 8th edn. Berlin: Georg Reimer.

Haeckel, E. (1910). *The Evolution of Man: A Popular Scientific Study,* 2 vols., transl. by J. McCabe from 5th edn. London: Watts & Co., for the Rationalist Press.

Haldane, J.B.S. (1935). Darwinism under revision. *Rationalist Annual,* 19–29.

Hallam, A. (1968). Morphology, palaeoecology and evolution of the genus *Gryphaea* in the British Lias. *Philosophical Transactions of the Royal Society [B], 254,* 91–128.

Hallam, A., and Gould, S.J. (1975). The evolution of British and American Middle and Upper Jurassic *Gryphaea:* a biometric study. *Proceedings of the Royal Society [B], 189,* 511–42.

Harris, E.E. (1972). Epicyclic Popperism. *British Journal for the Philosophy of Science, 23,* 55–67.

Harris, H. (1966). Enzyme polymorphism in man. *Proceedings of the Royal Society [B], 164,* 298–310.

Hasegawa, M., Kishino, H., and Yano, T-a. (1985). Dating of the human–ape splitting by a molecular clock of mitochondrial DNA. *Journal of Molecular Evolution, 22,* 160–74.

Hecht, M.K. (1976). Phylogenetic inference and methodology as applied to the vertebrate record. *Evolutionary Biology, 9,* 335–63.

Hecht, M.K., and Edwards, J.L. (1977). The methodology of phylogenetic inference above the species level. In *Major Patterns in Vertebrate Evolution,* ed. M.K. Hecht, P.C. Goody, and B.M. Hecht, pp. 3–51. New York: Plenum Press.

Hecht, M.K., and Hoffman, A. (1986). Why not Neodarwinism? A critique of paleobiological challenges. In *Oxford Surveys in Evolutinary Biology,* vol. 3, ed. R. Dawkins and M. Ridley, pp. 1–47. Oxford: Oxford University Press.

Hecht, M.K., Ostrom, J.H., Viohl, G., and Wellnhofer, P. (eds.). (1985). *The Beginnings of Birds.* Eichstätt: Freunde des Jura-Museums.

Hennig, W. (1950). *Grundzüge einer Theorie der phylogenetischen Systematik.* Berlin: Deutscher Zentralverlag.

Hennig, W. (1965). Phylogenetic Systematics. *Annual Review of Entomology, 10,* 97–116.

Hennig, W. (1966). *Phylogenetic Systematics,* transl. D.D. Davis and R. Zangerl. Urbana, Ill.: University of Illinois Press.

Hennig, W. (1968). *Elementos de Una Systematática Filogenética.* Buenos Aires: University Press.

Hennig, W. (1969). *Die Stammesgeschichte der Insekten.* Frankfurt: Kramer.

Hennig, W. (1981). *Insect Phylogeny,* transl. A.C. Pont, notes D. Schlee. Chichester: John Wiley & Sons.

Hennig, W. (1982). *Phylogenetische Systematik.* Berlin & Hamburg: Paul Parey.

Herschel, J.F.W. (1830). *A Preliminary Discourse on the Study of Natural Philosophy. (The Cabinet Cyclopaedia, vol. 56,* ed. D. Lardner). London: Longman, Rees, Orme, Brown & Green. . . .

Hess, H.H. (1962). History of ocean basins. In *Petrologic Studies: A Volume To Honor A.F. Buddington,* ed. A.E.J. Engel, H.L. James, and B.F. Leonard, pp. 599–620. New York: Geological Society of America.

Hill, L.R. (1975). Patterns arising from some tests of Le Quesne's concept of uniquely derived characters. In *Proceedings of the Eighth International Conference on Numerical Taxonomy,* ed. G.F. Estabrook, pp. 375–98. San Francisco: W.H. Freeman.

Hoffman, A. (1985). Patterns of family extinction depend on definition and geological timescale. *Nature, 315,* 659–62.

Holland, G.P. (1964). Evolution, classification and host relationships of Siphonaptera. *Annual Review of Entomology, 9,* 123–46.

Holmes, E.B. (1975). A reconsideration of the phylogeny of the tetrapod heart. *Journal of Morphology, 147,* 209–28.

Holmgren, N. (1933). On the origin of the tetrapod limb. *Acta Zoologica, 14,* 185–295.

Holmgren, N. (1939). Contribution to the question of the origin of the tetrapod limb. *Acta Zoologica, 20,* 89–124.

Holmgren, N. (1949). Contribution to the question of the origin of tetrapods. *Acta Zoologica, 30,* 459–84.

Holmquist, R., Miyamoto, M.M., and Goodman, M. (1988a). Higher primate phylogeny – why can't we decide? *Molecular Biology and Evolution, 5,* 201–16.

Holmquist, R., Miyamoto, M.M., and Goodman, M. (1988b). Analysis of higher-primate phylogeny from transversion differences in nuclear and mitochondrial DNA by Lake's methods of evolutionary parsimony and operator metrics. *Molecular Biology and Evolution, 5,* 217–236.

Hopson, J.A. (1969). The origin and adaptive radiation of mammal-like reptiles and nontherian mammals. *Annals of the New York Academy of Sciences, 167,* article 1, 199–216.

Hopson, J.A., and Crompton, A.W. (1969). Origin of mammals. *Evolutionary Biology, 3,* 15–72.

Hubby, J.L., and Lewontin, R.C. (1966). A molecular approach to the study of genic heterozygosity in natural populations. I. The number of alleles at different loci in *Drosophila pseudoobscura. Genetics, 54,* 577–94.

Hull, D.L. (1976). Are species really individuals? *Systematic Zoology, 25,* 174–91.

Hull, D.L. (1980). The limits of cladism. *Systematic Zoology, 28,* 416–40.

Hull, D.L. (1987). Genealogical actors in ecological roles. *Biology and Philosophy, 2,* 168–84.

Hull, D.L. (1988). *Science as a Process: An Evolutionary Account of the Social and Conceptual Development of Science.* Chicago: University of Chicago Press.

Hume, D. (1739–40). *A Treatise of Human Nature: Being an Attempt To Introduce the Experimental Method of Reasoning into Moral Subjects,* 2 vols. London: John

Noon. [Reprint ed. from L.A. Selby-Bigge. Rev. edn., ed. P.H. Nidditch. Oxford: Oxford University Press, 1978.]

Hume, D. (1748). *Philosophical Essays Concerning Human Understanding By the Author of the Essays Moral and Political,* 1st edn. [place and publisher not stated]. [Reprint edn. from L.A. Selby-Bigge. Nidditch, P.H. ed. (1975). *An Enquiry Concerning the Human Understanding, and an Enquiry Concerning the Principles of Morals . . . ,* 3rd edn., rev. Oxford: Oxford University Press].

Hume, D. (1779). *Dialogues concerning Natural Religion.* London: Robinson, for the author.

Humphries, C.J. (1983). Primary data in hybrid analysis. In *Advances in Cladistics,* vol. 2., ed. N.I. Platnick and V.A. Funk, pp. 89–103. New York: Columbia University Press.

Humphries, C.J., and Parenti, L.R. (1986). *Cladistic Biogeography.* Oxford: Oxford University Press.

Huxley, J.S. (1942). *Evolution: The Modern Synthesis.* London: Allen & Unwin.

Huxley, T.H. (1869). On *Hyperodapedon. Quarterly Journal of the Geological Society of London, 25,* 138–52, 157–8.

Huxley, T.H. (1870). Further evidence of the affinity between the dinosaurian reptiles and birds. *Quarterly Journal of the Geological Society of London, 26,* 12–31.

Huxley, T.H. (1880). On the application of the laws of evolution to the arrangement of the *Vertebrata,* and more particularly, of the *Mammalia. Proceedings of the Zoological Society of London, 1880,* 649–62.

Jablonski, D. (1986a). Background and mass extinctions: the alternation of macroevolutionary regimes. *Science, 231,* 129–33.

Jablonski, D. (1986b). Evolutionary consequences of mass extinctions. In *Patterns and Processes in the History of Life (Dahlem Konferenzen, Life Sciences Research Report No. 36),* ed. D.M. Raup and D. Jablonski, pp. 313–29. Berlin: Springer-Verlag.

Jacob, F., and Monod, J. (1961). Genetic regulatory mechanisms in the synthesis of proteins. *Journal of Molecular Biology, 3,* 318–56.

Jameson, R. (ed.). (1813). [Free translation of] G. Cuvier, *Essay on the Theory of the Earth. With Geological Illustrations by Professor Jameson.* 1st edn. Edinburgh: Blackwood.

Jardine, N., & Sibson, R. (1971). *Mathematical Taxonomy.* London: Wiley.

Jarvik, E. (1960). *Théories de l'Evolution des Vertébrés, Reconsidérées à la Lumière des Récentes Découvertes sur les Vertébrés Inférieurs.* Paris: Masson et Cie.

Jarvik, E. (1980). *Basic Structure and Evolution of Vertebrates,* 2 vols. London: Academic Press.

Jarvik, E. (1986). The origin of the Amphibia. In *Studies in Herpetology,* ed. Z. Roček, pp. 1–24. Prague: Charles University.

Jefferies, R.P.S. (1979). The origin of the chordates – a methodological essay. In *The Origin of the Major Invertebrate Groups,* ed. M.R. House, pp. 443–77. London: Academic Press for the Systematics Association.

Johannsen, W. (1903). *Ueber Erblichkeit in Populationen und in reinen Linien.* Jena: Gustav Fischer.

Johannsen, W. (1909). *Elemente der Exacten Erblichkeitslehre.* Jena: Gustav Fischer.

Jordon, D.S. (1905). The origin of species through isolation. *Science, 22,* 545–62.

Jukes, T.H., and Cantor, C.R. (1969). Evolution of protein molecules. In *Mammalian Protein Metabolism,* ed. H.N. Munro, pp. 21–132. New York: Academic Press.

Kant, I. (1781). *Kritik der reinen Vernunft.* Riga: J.F. Hartknoch. (1933). *Immanuel Kant's Critique of Pure Reason,* transl. N.K. Smith, 2nd edn. London: Macmillan.

Kant. I. (1783). *Prolegomena zu einer jeden kunftigen Metaphysic die als Wissenschaft wird auftreten können.* Riga. (1953). *Prolegomena to any Future Metaphysics that Will be Able to Present Itself as a Science,* transl. P.G. Lucas. Manchester: Manchester University Press.

Kellogg, D.E. (1988). "And then a miracle occurs" – weak links in the chain of argument from punctuation to hierarchy. *Biology and Philosophy, 3,* 3–28.

Kemp, T.S. (1982). *Mammal-like Reptiles and the Origin of Mammals.* London: Academic Press.

Kemp, T.S. (1985). Models of diversity and phylogenetic reconstruction. In *Oxford Surveys in Evolutionary Biology,* vol. 2, ed. R. Dawkins and M. Ridley, pp. 135–58. Oxford: Oxford University Press.

Kemp, T.S. (1988a). Haemothermia or Archosauria?: the interrelationships of mammals, birds and crocodiles. *Zoological Journal of the Linnean Society, 92,* 67–104.

Kemp, T.S. (1988b). Interrelationships of the Synapsida. In *The Phylogeny and Classification of the Tetrapods, vol. 2: Mammals,* ed. M.J. Benton, pp. 1–22. Oxford: Oxford University Press, for the Systematics Association.

Kepler, J. (1609). *Astronomia Nova Αιτιολογητος seu Physica Coelestis, Tradita Commentariis de Motibus Stellae Martis.* Pragae.

Kettlewell, H.B.D. (1955a). Selection experiments on industrial melanism in the Lepidoptera. *Heredity, 9,* 323–42.

Kettlewell, H.B.D. (1955b). Recognition of appropriate backgrounds by the pale and black phases of Lepidoptera. *Nature, 175,* 943–4.

Kettlewell, H.B.D. (1956). Further selection experiments on industrial melanism in the Lepidoptera. *Heredity, 10,* 287–301.

Kettlewell, H.B.D. (1965). Insect survival and selection for pattern. *Science, 148,* 1290–6.

Kettlewell, H.B.D. (1973). *The Evolution of Melanism: The Study of a Recurring Necessity.* Oxford: Oxford University Press.

Kimura, M. (1983). *The Neutral Theory of Molecular Evolution.* Cambridge: Cambridge University Press.

King, T.L., and Jukes, T.H. (1969). Non-Darwinian evolution. *Science, 164,* 788–98.

Kishino, H., and Hasegawa, M. (1989). Evaluation of the maximum likelihood estimate of the evolutionary tree topologies from DNA sequence data, and the branching order in Hominoidea. *Journal of Molecular Evolution, 29,* 170–9.

Kitcher, P. (1982). *Abusing Science – the Case Against Creationism.* Cambridge, Mass.: the M.I.T. Press.

Kitcher, P. (1987). Ghostly whispers: Mayr, Ghiselin, and the "philosophers" on the ontological status of species. *Biology and Philosophy, 2,* 184–92.

Kitts, D.B. (1974). Paleontology and evolutionary theory. *Evolution, 28,* 458–72.

Kluge, A.G. (1985). Ontogeny and phylogenetic systematics. *Cladistics, 1,* 13–27.

Kluge, A.G. (1988). The characterization of ontogeny. In *Ontogeny and Systematics,* ed. C.J. Humphries, pp. 57–81. London: British Museum (Natural History).

Kluge, A.G., and Farris, J.S. (1969). Quantitative phyletics and the evolution of anurans. *Systematic Zoology, 18,* 1–32.

Knorr-Cetina, K.D. (1981). *The Manufacture of Knowledge. Toward a Constructivist and Contextual Theory of Science.* Oxford: Pergamon Press.

Kojima, K., and Tobari, Y.N. (1969). The pattern of viability changes associated with genotype frequency at the alcohol dehydrogenase locus in a population of *Drosophila melanogaster. Genetics, 61,* 201–9.

Kojima, K., and Yarbrough, K.M. (1967). Frequency-dependent selection at the esterase 6 locus in *Drosophila melanogaster. Proceedings of the National Academy of Sciences, U.S.A., 57,* 645–9.

Kollar, J., and Fisher, C. (1980). Tooth induction in chick epithelium: expression of quiescent genes for enamel synthesis. *Science, 207,* 993–5.

Koop, B.F., Goodman, M., Xu, P., Chan, K., and Slightom, J.L. (1986). Primate η-globin sequences and man's place among the great apes. *Nature, 319,* 234–8.

Kuhn, T.S. (1962). *The Structure of Scientific Revolutions,* 1st edn. Chicago: University of Chicago Press.

Kuhn, T.S. (1970a, 1974a). Reflections on my critics. In *Criticism and the Growth of Knowledge (Proceedings of the International Colloquium in the Philosophy of Science, London 1965, vol. 4),* 3rd impression, 1974, ed. I. Lakatos and A. Musgrave, pp. 231–78. Cambridge: Cambridge University Press.

Kuhn, T.S. (1970b). *The Structure of Scientific Revolutions,* 2nd edn. Chicago: University of Chicago Press.

Kuhn, T.S. (1974b). Second thoughts on paradigms. In *The Structure of Scientific Theories,* ed. F. Suppe, pp. 459–82. Urbana, Ill.: University of Illinois Press.

Lack, D. (1947). *Darwin's Finches.* Cambridge: Cambridge University Press.

Lakatos, I. (1968). Criticism and the methodology of scientific research programmes. *Proceedings of the Aristotelian Society, 69,* 149–86.

Lakatos, I. (1970, 1974). Falsification and the methodology of scientific research programmes. In *Criticism and the Growth of Knowledge (Proceedings of the International Colloquium in the Philosophy of Science, London, 1965, vol. 4),* 3rd impression, 1974, ed. I. Lakatos and A. Musgrave, pp. 91–196. Cambridge: Cambridge University Press.

Lake, J.A. (1987). A rate-independent technique for analysis of nucleic acid sequences: evolutionary parsimony. *Molecular Biology and Evolution, 4,* 167–91.

Lake, J.A. (1990). Origin of the Metazoa. *Proceedings of the National Academy of Sciences, U.S.A., 87,* 763–6.

Lamarck, J.-B.-.P-A. de M. de (1801). *Système des Animaux sans Vertèbres . . . précédé du Discours d'Ouverture du Cours de Zoologie, donné dans le Muséum National d'Histoire Naturelle, l'An VIII de la Republique*. Paris: Deterville.

Lamarck, J.-B.-P.-A. de M. de (1802). *Recherches sur l'Organisation de Corps Vivans . . .* Paris: Maillard.

Lamarck, J.-B.-P.-A. de M. de (1809). *Philosophie Zoologique, ou Exposition des Considerations Relatives à l'Histoire Naturelle des Animaux . . .* Paris: Dentu. (English transl. 1914. *Zoological Philosophy.*, transl. H. Elliot. London: Macmillan).

Lamarck, J.-B.-P.-A. de M. de (1815–22). *Histoire Naturelle des Animaux sans Vertèbres . . .* (7 vols.), 1st edn. Paris: Verdiere.

Lamarck, J.-B.-P.-A. de M. de (1837–9). *Ibid.* [Belgian edn.], 3 vols. Bruxelles: Meline.

Lande, R. (1978). Evolutionary mechanisms of limb loss in tetrapods. *Evolution, 32,* 73–92.

Lang, W.D. (1923). Evolution: a resultant. *Proceedings of the Geologists Association, 34,* 7–20.

Langley, C.H., and Fitch, W.M. (1974). An examination of the constancy of the rate of molecular evolution. *Journal of Molecular Evolution, 3,* 161–77.

Lankaster, E.R. (1870). On the use of the term homology in modern zoology, and the distinction between homogenetic and homoplastic agreements. *The Annals and Magazine of Natural History* (4th ser.), *6,* 35–43.

Latour, B., and Woolgar, S. (1979). *Laboratory Life.* Beverley Hills, Calif.: Sage.

Laudan, L. (1977). *Progress and Its Problems.* Berkeley: University of California Press.

Lederberg, J., and Lederberg, E.M. (1952). Replica plating and indirect selection of bacterial mutants. *Journal of Bacteriology, 63,* 399–406.

Lewontin, R.C. (1974). *The Genetic Basis of Evolutionary Change.* New York: Columbia University Press.

Lewontin, R.C. (1979). Sociobiology as an adaptationist program. *Behavioural Science, 24,* 5–14.

Lewontin, R.C. (1980). Theoretical population genetics and the evolutionary synthesis. In *The Evolutionary Synthesis: Perspectives on the Unification of Biology,* ed. E. Mayr and W.B. Provine, pp. 58–68. Cambridge, Mass: Harvard University Press.

Lewontin, R.C., and Hubby, J.L. (1966). A molecular approach to the study of genic heterozygosity in natural populations. II. Amount of variation and degree of heterozygosity in natural populations of *Drosophila pseudoobscura. Genetics, 54,* 595–609.

Lewontin, R.C., Moore, J.C., Provine, W.B., and Wallace, B. (eds.). (1981). *Dobzhansky's Genetics of Natural Populations* I–XLIII. New York: Columbia University Press.

Liebert, T.G., and Brakefield, P.M. (1987). Behavioural studies on the peppered moth *Biston betularia* and a discussion of the role of pollution and epiphytes in industrial melanism. *Biological Journal of the Linnean Society, 31,* 129–50.

Linnaeus, C. (1735). *Systema Naturae sive Regna Tria Naturae.* 1st edn. Leiden: Haak.

Linnaeus, C. (1751). *Philosophia Botanica.* Stockholm: Kiesewetter.

Linnaeus, C. (1753). *Species Plantarum.* Helmiae: Laurentii Salvii.

Linnaeus, C. (1758). *Systema Naturae per Regna Tria Naturae, Secundum Classes, Ordines, Genera, Species cum Characteribus, Differentiis, Synonymis, Locis,* tomus 1, 10th edn. Halmiae: Laurentii Salvii.

Linnaeus, C. (1792). *Praelectiones in Ordines Naturales Plantarum . . .* (posth. ed. P.D. Giseke). Hamburg: B.G. Hoffmann.

Locke, J. (1690). *An Essay concerning Human Understanding,* 1st edn., book 3. London: Thomas Basset. [Reprinted (ed. A.C. Fraser) New York: Dover, 1959].

Lovejoy, A.O. (1936). *The Great Chain of Being.* Cambridge, Mass.: Harvard University Press.

Løvtrup, S. (1985). On the classification of the taxon Tetrapoda. *Systematic Zoology, 34,* 463–70.

Løvtrup, S. (1987). *Darwinism: The Refutation of a Myth.* London: Croom Helm.

Lundberg, J.G. (1973). More on primitiveness, higher level phylogenies and ontogenetic transformations. *Systematic Zoology, 22,* 327–9.

Luria, S.E., and Delbruck, M. (1943). Mutations of bacteria from virus sensitivity to virus resistance. *Genetics, 28,* 491–511.

Lyell, C. (1830–3). *Principles of Geology, Being an Attempt to Explain the Former Changes of the Earth's Surface, by Reference to Causes Now in Operation,* 1st edn. (3 vols.). London: John Murray.

MacFadden, B.J. (1985). Patterns of phylogeny and rates of evolution in fossil horses: hipparions from the Miocene and Pliocene of North America. *Paleobiology, 11,* 245–57.

McKenna, M.C. (1973). Sweepstakes, filters, corridors, Noah's Arks, and beached Viking funeral ships in palaeogeography. In *Implications of Continental Drift to the Earth Sciences,* ed. D.H. Tarling and S.K. Runcorn, vol. 1, pp. 295–308. London: Academic Press.

McKenna, M.C. (1975). Towards a phylogenetic classification of the Mammalia. *Contributions in Primatology, 5,* 21–46.

MacLeay, W.S. (1819, 1821). *Horae Entomologicae: Or, Essays on the Annulose Animals,* 1 vol. in 2 parts. London: S. Bagster.

McNeill, J. (1980). Purposeful phenetics. *Systematic Zoology, 28,* 465–82.

Maddison, W.P., Donaghue, M.J., and Maddison, D.R. (1984). Outgroup analysis and parsimony. *Systematic Zoology, 33,* 83–103.

Maeda, N., Wu, C-I., Bliska, J., and Reneke, J. (1988). Molecular evolution of intergenic DNA in higher primates: pattern of DNA changes, molecular clock, and the evolution of repetitive sequences. *Molecular Biology and Evolution, 5,* 1–20.

Maglio, V.J. (1973). Origin and evolution of the Elephantidae. *Transactions of the American Philosophical Society* (n.s.), *63,* 1–149.

Malmgren, B.A., Berggren, W.A., and Lohmann, G.P. (1983). Evidence for punctuated gradualism in the Late Neogene *Globorotalia tumida* lineage of planktonic foramnifera. *Paleobiology, 9,* 377–89.

Mani, G.S. (1990). Theoretical models of melanism in *Biston betularia* – a review. *Biological Journal of the Linnean Society, 39,* 355–371.

Marsh, O.C. (1874). Notice of new equine mammals from the Tertiary formation. *American Journal of Science* (3rd ser.), *7*, 247–58.

Marsh, O.C. (1879). Polydactyle horses, recent and extinct. *American Journal of Science* (3rd ser.), *17*, 499–505.

Marsh, O.C. (1892). Recent polydactyle horses. *American Journal of Science, 43*, 339–55.

Maslin, T.P. (1952). Morphological criteria of phyletic relationships. *Systematic Zoology, 1*, 49–70.

Masterman, M. (1970, 1974). The nature of a paradigm. In *Criticism and the Growth of Knowledge (Proceedings of the International Colloquium in the Philosophy of Science, London, 1965, vol. 4)*, 3rd impression, 1974, ed. I. Lakatos and A. Musgrave, pp. 59–89. Cambridge: Cambridge University Press.

Maynard Smith, J. (1966). Sympatric speciation. *American Naturalist, 100*, 637–50.

Maynard Smith, J., Burian, R., Kauffman, S., Alberch, P., Campbell, J., Goodwin, B., Lande, R., Raup, D., and Wolpert, L. (1985). Developmental constraints and evolution: a perspective from the Mountain Lake conference on development and evolution. *Quarterly Review of Biology, 60*, 265–87.

Maynard Smith, J., & Sondhi, K.C. (1960). The genetics of a pattern. *Genetics, 45*, 1039–50.

Mayr, E. (1940). Speciation phenomena in birds. *American Naturalist, 74*, 249–78.

Mayr, E. (1942). *Systematics and the Origin of Species*. New York: Columbia University Press.

Mayr, E. (1954). Change of genetic environment and evolution. In *Evolution as a Process*, ed. J.S. Huxley, A.C. Hardy, and E.B. Ford, pp. 157–80. London: Allen & Unwin.

Mayr, E. (1963). *Animal Species and Evolution*. Cambridge, Mass.: Harvard University Press.

Mayr, E. (1969). *Principles of Systematic Zoology*. New York: McGraw-Hill.

Mayr, E. (1970). *Populations, Species and Evolution*. Cambridge, Mass.: Harvard University Press.

Mayr, E. (1974). Cladistic analysis or cladistic classification? *Zeitschrift für Zoologische Systematik und Evolutionsforschung, 12*, 94–128.

Mayr, E. (1981). Biological clasification: toward a synthesis of opposing methodologies. *Science, 214*, 510–16.

Mayr, E. (1982). *The Growth of Biological Thought: Diversity, Evolution, and Inheritance*. Cambridge, Mass.: Harvard University Press.

Mayr, E. (1987). The ontological status of species: scientific progress and philosophical terminology. *Biology and Philosophy, 2*, 145–66.

Mayr, E. (1988). A response to David Kitts. *Biology and Philosophy, 3*, 97–8.

Mayr, E., Linsley, E.G., and Usinger, R.L. (1953). *Methods and Principles of Systematic Zoology*. New York: McGraw-Hill.

Mayr, E., and Provine, W.B. (eds.). (1980). *The Evolutionary Synthesis: Perspectives on the Unification of Biology*. Cambridge, Mass.: Harvard University Press.

Meacham, C.A. (1981). A probability measure for character compatibility. *Mathematical Bioscience, 57*, 1–18.

Meacham, C.A., and Estabrook, G.F. (1985). Compatibility methods in systematics. *Annual Review of Ecology and Systematics, 16,* 431–46.

Medawar, P.B. (1967). *The Art of the Soluble.* London: Methuen & Co.

Medawar, P.B. (1969). *Induction and Intuition in Scientific Thought.* London: Methuen.

Medawar, P.B. (1982). *Pluto's Republic (Incorporating the Art of the Soluble and Induction and Intuition in Scientific Thought).* Oxford: Oxford University Press.

Meyranx, P.-S., and Laurençet (1830). [Some considerations on the organisation of molluscs]. In Geoffroy Saint-Hilaire, E., and Latreille, P.-A. *Procès-Verbaux, Académie des Science, 9,* 403–6.

Michener, C.D., and Sokal, R.R. (1957). A quantitative approach to a problem in classification. *Evolution, 11,* 130–62.

Mikkola, K. (1979). Resting site selection of *Oligia* and *Biston* moths (Lepidoptera: Noctuidae and Geometridae). *Annales Entomologici Fennici, 45,* 81–7.

Mikkola, K. (1984a). On the selective forces acting in the industrial melanism of *Biston* and *Oliga* moths (Lepidoptera: Geometridae and Noctuidae). *Biological Journal of the Linnean Society, 21,* 409–21.

Mikkola, K. (1984b). Dominance relationships among the melanic forms of *Biston betularia* and *Odontopera bidentata* (Lepidoptera, Geometridae). *Heredity, 52,* 9–16.

Mill, J.S. (1843). *A System of Logic, Ratiocinative and Inductive: Being a Connected View of the Principles of Evidence and the Methods of Investigation,* 2 vols. London: John W. Parker & Sons. [J.M. Robson (ed.). (1974). *John Stuart Mill: Collected Works,* vol. 8. Toronto: Toronto University Press].

Mills, S., and Beatty, J. (1979). The propensity interpretation of fitness. *Philosophy of Science, 46,* 263–86.

Milne-Edwards, H. (1844). Considérations sur quelques principes relatifs à la classification naturelle des animaux. *Annales des Sciences Naturelles* (3rd ser.), *1,* 65–99.

Milner, A.R. (1988). The relationships and origin of living amphibians. In *The Phylogeny and Classification of the Tetrapods, vol. 1: Amphibians, Reptiles and Birds,* ed. M.J. Benton, pp. 59–102. Oxford: Oxford University Press for the Systematics Association.

Milner, A.R., Smithson, T.R., Milner, A.C., Coates, M.I., and Rolfe, W.D.I. (1986). The search for early tetrapods. *Modern Geology, 10,* 1–28.

Mishler, B.D., and Brandon, R.N. (1987). Individuality, pluralism, and the phylogenetic species concept. *Biology and Philosophy, 2,* 397–414.

Mishler, B.D., and Donaghue (1982). Species concepts: a case for pluralism. *Systematic Zoology, 31,* 491–503.

Miyamoto, M.M., Slightom, J.L., and Goodman, M. (1987). Phylogenetic relations of humans and African apes from DNA sequences in the ψη globin region. *Science, 238,* 369–73.

Moore, G.W., Barnabas, J., and Goodman, M. (1973). A method for constructing maximum parsimony ancestral amino acid sequences on a given network. *Journal of Theoretical Biology, 38,* 459–85.

Morgan, T.H. (1903). *Evolution and Adaptation.* New York: Macmillan.

Morgan, T.H. (1916). *A Critique of the Theory of Evolution*. Princeton: Princeton University Press.

Morowitz, H. (1986). Entropy and nonsense: a review of Daniel R. Brooks and E.O. Wiley, *Evolution as Entropy . . . Biology and Philosophy, 1,* 473–6.

Morris, H.M. (1974). *Scientific Creationism* (general edn.). San Diego: Creation-Life Publishers.

Muller, H.J. (1914). The bearing of the selection experiments of Castle and Phillips on the variability of genes. *American Naturalist, 48,* 567–76.

Müller, J. (1844). Über den Bau und die Grenzen der Ganoiden und über das natürliche System der Fische. *Abhandlungen der Akademie der Wissenschaften, Berlin, 1841,* 117–216.

Myers, A.A., & Giller, P.S. (eds.). (1988). *Analytical Biogeography: An Integrated Approach to the Study of Animal and Plant Distribution*. London: Chapman & Hall.

Nelson, G. (1971). Paraphyly and polyphyly: redefinitions. *Systematic Zoology, 20,* 471–2.

Nelson, G. (1972a). Phylogenetic relationships and classification. *Systematic Zoology, 21,* 227–31.

Nelson, G. (1972b). Comments on Hennig's "Phylogenetic Systematics" and its influence on ichthyology. *Systematic Zoology, 21,* 364–74.

Nelson, G. (1973). Notes on the structure and relationships of certain Cretaceous and Eocene teleostean fishes. *American Museum Novitates,* no. 2524, 1–31.

Nelson, G. (1974). Classification as an expression of phylogenetic relationship. *Systematic Zoology, 22,* 344–59.

Nelson, G. (1978). Ontogeny, phylogeny, paleontology, and the biogenetic law. *Systematic Zoology, 27,* 324–45.

Nelson, G. (1979). Cladistic analysis and synthesis: principles and definitions, with a historical note on Adanson's Familles des Plantes (1763–1764). *Systematic Zoology, 28,* 1–21.

Nelson, G. (1983). Reticulation in cladograms. In *Advances in Cladistics,* vol. 2, ed. N.I. Platnick and V.A. Funk, pp. 105–11. New York: Columbia University Press.

Nelson, G., and Platnick, N.I. (1980). Multiple branching in cladograms: two interpretations. *Systematic Zoology, 29,* 86–91.

Nelson, G., and Platnick, N.I. (1981). *Systematics and Biogeography: Cladistics and Vicariance*. New York: Columbia University Press.

Nelson, G., and Platnick, N.I. (1984). Systematics and evolution. In *Beyond Neo-Darwinism,* ed. M.-W. Ho and P.T. Saunders, pp. 143–58. London & New York: Academic Press.

Newell, N.D. (1982). *Creation and Evolution: Myth or Reality?* New York: Columbia University Press.

Nielsen, C. (1989). [In "Phylogeny and molecular data"]. *Science, 243,* 548.

Officer, C.B., Hallam, A., Drake, C.L., and Devine, J.D. (1987). Late Cretaceous and paroxysmal Cretaceous/Tertiary extinctions. *Nature, 326,* 143–9.

Ohno, S. (1970). *Evolution by Gene Duplication*. New York: Springer-Verlag.

Oken, L. (1809–11). *Lehrbuch der Naturphilosophie,* 3 vols. Jena: F. Frommand.

Oken, L. (1839–41). *Elements of Physiophilosophy,* transl. A. Tulk, 1847. London: Ray Society.

Oldroyd, D.R. (1980). *Darwinian Impacts: An Introduction to the Darwinian Revolution*. Milton Keynes: Open University Press.

Oldroyd, D. (1986). *The Arch of Knowledge: An Introductory Study of the History of the Philosophy and Methodology of Science*. New York & London: Methuen.

Olson, E.C. (1947). The family Diadectidae and its bearing on the classification of reptiles. *Fieldiana: Geology, 11,* 2–53.

Olson, E.C. (1971). *Vertebrate Paleozoology*. New York: John Wiley & Sons.

Oosterbroek, P. (1987). More appropriate definitions of paraphyly and polyphyly, with a comment on the Farris 1974 model. *Systematic Zoology, 36,* 103–8.

Osborn, H.F. (1934). Aristogenesis, the creative principle in the origin of species. *American Naturalist, 68,* 193–235.

Ospovat, D. (1976). The influence of Karl Ernst von Baer's embryology, 1828–1859: a reappraisal in light of Richard Owen's and William B. Carpenter's "Palaeontological application of von Baer's Law". *Journal of the History of Biology, 9,* 1–28.

Ospovat, D. (1981). *The Development of Darwin's Theory: Natural History, Natural Theology, and Natural Selection 1838–1859*. Cambridge: Cambridge University Press.

Oster, G.F., Shubin, N., Murray, J.D., and Alberch, P. (1988). Evolution and morphogenetic rules: the shape of the vertebrate limb in ontogeny and phylogeny. *Evolution, 42,* 862–84.

Otte, D., and Endler, J.A. (eds). (1989). *Speciation and Its Consequences*. Sunderland, Mass.: Sinauer Associates.

Owen, R. (1843). *Lectures on the Comparative Anatomy and Physiology of the Invertebrate Animals*. London: Longman, Brown, Green & Longmans.

Owen, R. (1840–5). *Odontography*, 2 vols. London: Hippolyte Baillière.

Owen, R. (1848). *On the Archetype and Homologies of the Vertebrate Skeleton*. London: J. van Voorst.

Owen, R. (1949). *One the Nature of Limbs*. London: J. van Voorst.

Owen, R. (1866). *On the Anatomy of Vertebrates,* vols. 1 and 2. London: Longmans, Green.

Page, R.D.M. (1987). Graphs and generalized tracks: quantifying Croizat's panbiogeography. *Systematic Zoology, 36,* 1–17.

Paley, W. (1802). *Natural Theology: Or Evidences of the Existence and Attributes of the Deity, Collected from the Appearances of Nature*. London: J. Faulder.

Panchen, A.L. (1970). *Handbuch der Paläoherpetologie, Part 5A: Anthracosauria,* ed. O. Kuhn. Stuttgart: Gustav Fischer Verlag.

Panchen, A.L. (1979). [In "the cladistic debate, continued"]. *Nature, 280,* 541.

Panchen, A.L. (1980). The origin and relationships of the anthracosaur amphibia from the Late Palaeozoic. In *The Terrestrial Environment and the Origin of Land Vertebrates,* ed. A.L. Panchen, pp. 319–50. London: Academic Press, for the Systematics Association.

Panchen, A.L. (1982). The use of parsimony in testing phylogenetic hypotheses. *Zoological Journal of the Linnean Society, 74,* 305–28.

Panchen, A.L. (1985). On the amphibian *Crassigyrinus scoticus* Watson from the Carboniferous of Scotland. *Philosophical Transactions of the Royal Society [B], 309,* 505–68.

Panchen, A.L. (1991). The early tetrapods: classification and the shapes of cladograms. *Origins of the Higher Groups of Tetrapods: Controversy and Consensus,* ed. H.-P. Schultze and L. Trueb, pp. 110–44. Ithaca, N.Y.: Cornell University Press.

Panchen, A.L., and Smithson, T.R. (1987). Character diagnosis, fossils and the origin of tetrapods. *Biological Reviews of the Cambridge Philsophical Society, 62,* 341–438.

Panchen, A.L., and Smithson, T.R. (1988). The relationships of the earliest tetrapods. In *The Phylogeny and Classification of the Tetrapods, vol. 1: Amphibians, Reptiles and Birds,* ed. M.J. Benton, pp. 1–32. Oxford: Oxford University Press, for the Systematics Association.

Panchen, A.L., and Smithson, T.R. (1990). The pelvic girdle and hind limb of *Crassigyrinus scoticus* (Lydekker) from the Scottish Carboniferous and the origin of the tetrapod pelvic skeleton. *Transactions of the Royal Society of Edinburgh: Earth Sciences, 81,* 31–44.

Parsons, T.S., and Williams, E.E. (1963). The relationships of the modern Amphibia: a reexamination. *Quarterly Review of Biology, 38,* 26–53.

Paterson, H.E.H. (1985). The recognition concept of species. In *Species and Speciation (Transvaal Museum Monograph No. 4),* ed. S. Vrba, pp. 21–9. Pretoria: Transvaal Museum.

Patterson, C. (1977). The contribution of paleontology to teleostean phylogeny. In *Major Patterns in Vertebrate Evolution,* ed. M.K. Hecht, P.C. Goody, and B.M. Hecht, pp. 579–643. New York & London: Plenum Press.

Patterson, C. (1978). Verifiability in systematics. *Systematic Zoology, 27,* 218–22.

Patterson, C. (1980a). Cladistics. *Biologist, London, 27,* 234–40.

Patterson, C. (1980b). Origin of tetrapods: historical introduction to the problem. In *The Terrestrial Environment and the Origin of Land Vertebrates,* ed. A.L. Panchen, pp. 159–75. London: Academic Press, for the Systematics Association.

Patterson, C. (1981a). Methods of paleobiogeography. In *Vicariance Biogeography: A Critique,* ed. G. Nelson and D.E. Rosen, pp. 446–500. New York: Columbia University Press.

Patterson, C. (1981b). Significance of fossils in determining evolutionary relationships. *Annual Review of Ecology and Systematics, 12,* 195–223.

Patterson, C. (1982). Morphological characters and homology. In *Problems of Phylogenetic Reconstruction,* ed. K.A. Joysey and A.E. Friday, pp. 21–74. London: Academic Press, for the Systematics Association.

Patterson, C. (1983). Aims and methods in biogeography. In *Evolution, Time and Space: The Emergence of the Biosphere,* ed. R.W. Sims, J.H. Price, and P.E.S. Whalley, pp. 1–28. London: Academic Press, for the Systematics Association.

Patterson, C. (ed.). (1987). *Molecules and Morphology in Evolution: Conflict or Compromise?* (Introduction by Patterson, pp. 1–22). Cambridge: Cambridge University Press.

Patterson, C. (1989). Phylogenetic relations of major groups: conclusions and prospects. In *The Hierarchy of Life,* ed. B. Fernholm, K. Bremer, and H. Jörnvall, pp. 471–88. Amsterdam: Elsevier Science Publishing.

Patterson, C. (1990). Metazoan phylogeny – reassessing relationships. *Nature, 344,* 199–200.

Patterson, C., and Rosen, D.E. (1977). Review of ichthyodectiform and other Mesozoic teleost fishes and the theory and practice of classifying fossils. *Bulletin of the American Museum of Natural History, 158,* 81–172.

Patterson, C., and Smith, A.B. (1987). Is the periodicity of extinctions a taxonomic artefact? *Nature, 330,* 248–51.

Pease, C.M. (1990). Is evolutionary taxonomy science? [Review of] *Invertebrate Relationships: Patterns in Animal Evolution* – Pat Willmer, 1990 . . . *Systematic Zoology, 39,* 301–3.

Pellegrin, P. (1986). *Aristotle's Classification of Animals: Biology and the Conceptual Unity of the Aristotelian Corpus,* transl. A. Preus. Berkeley, Calif.: University of California Press.

Pilbeam, D. (1982). New hominoid skull material from the Miocene of Pakistan. *Nature, 295,* 232–4.

Pilbeam, D., Rose, M.D., Barry, J.C., and Ibrahim Shah, S.M. (1990). New *Sivapithecus* humeri from Pakistan and the relationship of *Sivapithecus* and *Pongo. Nature, 348,* 237–9.

Platnick, N.I. (1977). Cladograms, phylogenetic trees, and hypothesis testing. *Systematic Zoology, 26,* 438–42.

Platnick, N.I. (1980). Philosophy and the transformation of cladistics. *Systematic Zoology, 28,* 537–46.

Platnick, N.I. (1985). Philosophy and the transformation of cladistics revisited. *Cladistics, 1,* 87–94.

Platnick, N.I. (1987). An empirical comparison of microcomputer parsimony programs. *Cladistics, 3,* 121–44.

Platnick, N.I. (1989). An empirical comparison of microcomputer parsimony programs, II. *Cladistics, 5,* 145–61.

Platnick, N.I., and Gaffney, E. (1977). Review – systematics: a Popperian perspective – *The Logic of Scientific Discovery* by Karl R. Popper, *Conjectures and Refutations* by Karl R. Popper. *Systematic Zoology, 26,* 360–65.

Platnick, N.I., and Gaffney, E. (1978a). Evolutionary biology: a Popperian perspective. *Systematic Zoology, 27,* 137–41.

Platnick, N.I., and Gaffney, E. (1978b). Systematics and the Popperian paradigm. *Systematic Zoology, 27,* 381–8.

Plotkin, H.C. (1987). Evolutionary epistemology as science. *Biology and Philosophy, 2,* 295–313.

Popper, K.R. (1934). *Logik der Forschung: Zur Erkenntnistheorie der modernen Naturwissenschaft.* Vienna: Julius Springer.

Popper, K.R. (1945). *The Open Society and Its Enemies,* 2 vols. London: Routledge & Kegan Paul.

Popper, K.R. (1957). *The Poverty of Historicism.* London: Routledge & Kegan Paul.

Popper, K.R. (1959). *The Logic of Scientific Discovery,* 1st edn. London: Hutchinson.

Popper, K.R. (1963). *Conjectures and Refutations: The Growth of Scientific Knowledge,* 1st edn. London: Routledge & Kegan Paul.

Popper, K.R. (1972). *Objective Knowledge: An Evolutionary Approach*. Oxford: Oxford University Press.

Popper, K.R. (1974a). *Conjectures and Refutations: The Growth of Scientific Knowledge*, 5th edn. London: Routledge & Kegan Paul.

Popper, K.R. (1974b). Darwinism as a metaphysical research programme. In *The Philosophy of Karl Popper*, ed. P.A. Schilpp, pp. 133–43. La Salle, Ill.: Open Court.

Popper K.R. (1974c). Replies to my critics. In *The Philosophy of Karl Popper*, ed. P.A. Schilpp, pp. 961–1197. La Salle, Ill.: Open Court.

Prakash, S., Lewontin, R.C., and Hubby, J.L. (1969). A molecular approach to the study of genic heterozygosity in natural populations. IV. Patterns of genic variation in central, marginal, and isolated populations of *Drosophila pseudoobscura*. *Genetics, 61*, 841–58.

Prichard, J.C. (1826). *Researches into the Physical History of Mankind*, 2nd edn., 2 vols. London: J. & A. Arch.

Prigogine, I. (1961). *Introduction to Thermodynamics of Irreversible Processes*. New York: Interscience Publishers.

Provine, W.B. (1971). *The Origins of Theoretical Population Genetics*. Chicago: University of Chicago Press.

Queiroz, K. de, (1985). The ontogenetic method for determining character polarity and its relevance to phylogenetic systematics. *Systematic Zoology, 34*, 280–99.

Queiroz, K. de, (1988). Systematics and the Darwinian revolution. *Philosophy of Science, 55*, 238–59.

Le Quesne, W.J. (1969). A method of selection of characters in numerical taxonomy. *Systematic Zoology, 18*, 201–5.

Le Quesne, W.J. (1972). Further studies on the uniquely derived character concept. *Systematic Zoology, 21*, 281–8.

Le Quesne, W.J. (1974). The uniquely evolved character concept and its cladistic application. *Systematic Zoology, 23*, 513–7.

Le Quesne, W.J. (1977). The uniquely evolved character concept. *Systematic Zoology, 26*, 218–23.

Quine, W.V.O. (1966). *Elementary Logic*, rev. edn. Cambridge, Mass.: Harvard University Press.

Rampino, M.R., and Stothers, R.B. (1984). Terrestrial mass extinctions, cometary impacts and the Sun's motion perpendicular to the galactic plane. *Nature, 308*, 709–12.

Raup, D.M., and Gould, S.J. (1974). Stochastic simulation and evolution of morphology – towards a nomothetic paleontology. *Systematic Zoology, 23*, 305–32.

Raup, D.M., and Sepkoski, J.J. (1982). Mass extinctions in the fossil record. *Science, 215*, 1501–3.

Raup, D.M., and Sepkoski, J.J. (1984). Periodicity of extinctions in the geologic past. *Proceedings of the National Academy of Sciences, U.S.A., 81*, 801–5.

Ray, J. (1686). *Historia Plantarum: Species Hactenus Editas Aliasque Insuper Multas Noviter Inventas & Descriptas Complectens*, Tomus Primus. London: Henry Faithorne.

Ray, J. (1691). *The Wisdom of God Manifested in the Works of the Creation: Being the Substance of Some Common Places Delivered in the Chappel of Trinity College in Cambridge*. London: Samuel Smith.

Read, D.W. (1975). Primate phylogeny, neutral mutations and 'molecular clocks'. *Systematic Zoology, 24,* 209–21.

Reader, J. (1981). *Missing Links: The Hunt for Earliest Man*. London: Collins.

Reed, C.A. (1960). Polyphyletic or monophyletic ancestry of mammals, or: what is a class? *Evolution, 14,* 314–22.

Reichert, C. (1837). Über die Visceralbogen der Wirbeltiere im allgemeinen und deren Metamorphosen bei den Vögeln und Säugetieren. *Archiv für Anatomie und Physiologie,* 1837, 120–222.

Rensch, B. (1947). *Neuere Probleme der Abstammungslehre,* 1st edn. Stuttgart: Enke.

Rensch, B. (1959). *Evolution Above the Species Level,* transl. Dr. Allevogt, from Rensch (1954) [2nd edn. of Rensch (1947)]. London: Methuen.

Ride, W.L.D. *et al.* (eds.). (1985). *International Code of Zoological Nomenclature* (3rd edn. adopted by the XX General Assembly of the International Union of Biological Sciences). Berkeley & Los Angeles: University of California Press for International Trust for Zoological Nomenclature.

Ridley, M. (1986). *Evolution and Classification: The Reformation of Cladism*. London: Longman.

Ridley, M. (1989). The cladistic solution to the species problem. *Biology and Philosophy, 4,* 1–16.

Riedl, R. (1979). *Order in Living Organisms: A Systems Analysis of Evolution,* transl. R.P.S. Jefferies. Chichester: John Wiley & Sons.

Rieppel, O. (1979). Ontogeny and the recognition of primitive character states. *Zeitschrift für Zoologische Systematik und Evolutionsforschung, 17,* 57–61.

Rieppel, O.C. (1984). Atomism, transformism and the fossil record. *Zoological Journal of the Linnean Society, 82,* 17–32.

Rieppel, O.C. (1988). *Fundamentals of Comparative Biology*. Basel: Birkhäuser Verlag.

Ritterbush, P.C. (1964). *Overtures to Biology: The Speculations of Eighteenth Century Naturalists*. New Haven: Yale University Press.

Roberts, M.B.V. (1987). *Biology: A Functional Approach,* 4th edn. Walton-on-Thames: Thomas Nelson & Sons.

Rogers, D.J., and Tanimoto, T.T. (1960). A computer program for classifying plants. *Science, 132,* 1115–18.

Rohlf, F.J. (1963). Classification of *Aedes* by numerical taxonomic methods (Diptera: Culicidae). *Annals of the Entomological Society of America, 56,* 798–804.

Rohlf, F.J., and Sokal, R.R. (1982). Comparing numerical taxonomic studies. *Systematic Zoology, 30,* 459–90.

Romanes, G.J. (1896). *Life and Letters*. London: Longmans, Green.

Romer, A.S. (1945). *Vertebrate Paleontology,* 2nd edn. Chicago: University of Chicago Press.

Rosen, D.E. (1978). Vicariant patterns and historical explanation in biogeography. *Systematic Zoology, 27,* 159–88.

Rosen, D.E. (1979). Fishes from the uplands and intermontane basins of Guatemala: revisionary studies and comparative biogeography. *Bulletin of the American Museum of Natural History, 162,* 267–376.

Rosen, D.E. (1984). Hierarchies and history. In *Evolutionary Theory: Paths into the Future,* ed. J.W. Pollard, pp. 77–97. Chichester: John Wiley & Sons.

Rosen, D.E., Forey, P.L., Gardiner, B.G., and Patterson, C. (1981). Lungfishes, tetrapods, paleontology and plesiomorphy. *Bulletin of the American Museum of Natural History, 167,* 159–276.

Rosenberg, A. (1985). *The Structure of Biological Science.* Cambridge: Cambridge University Press.

Roth, V.L. (1984). On homology. *Biological Journal of the Linnean Society, 22,* 13–29.

Roth, V.L. (1988). The biological basis of homology. In *Ontogeny and Systematics,* ed. C.J. Humphries, pp. 1–26. London: British Museum (Natural History).

Rudwick, M.J.S. (1976). *The Meaning of Fossils: Episodes in the History of Palaeontology,* 2nd edn. New York: Science History Publications.

Rudwick, M.J.S. (1985). *The Great Devonian Controversy: The Shaping of Scientific Knowledge Among Gentlemanly Specialists.* Chicago: University of Chicago Press.

Ruse, M. (1976). The scientific methodology of William Whewell. *Centaurus, 20,* 227–57.

Ruse, M. (1980a). Falsifiability, consilience, and systematics. *Systematic Zoology, 28,* 530–6.

Ruse, M. (1980b). Charles Darwin and group selection. *Annals of Science, 37,* 615–30.

Russell, E.S. (1916). *Form and Function: A Contribution to the History of Animal Morphology.* London: John Murray. [Reprinted 1982, University of Chicago Press].

Ruvolo, M., and Pilbeam, D.R. (1985). Hominoid evolution: molecular and palaeontological pattern. In *Major Topics in Primate and Human Evolution,* ed. B. Wood, L. Martin, and P. Andrews, pp. 157–60. Cambridge: Cambridge University Press.

Salthe, S.N. (1985). *Evolving Hierarchical Systems: Their Structure and Representation.* New York: Columbia University Press.

Sarich, V.M., Schmid, C.W., and Marks, J. (1989). DNA hybridization as a guide to phylogenies: a critical analysis. *Cladistics, 5,* 3–32.

Sarich, V.M., and Wilson, A.C. (1967). Immunological time scale for hominid evolution. *Science, 158,* 1200–3.

Säve-Söderbergh, G. (1934). Some points of view concerning the evolution of the vertebrates and the classification of this group. *Arkiv för Zoologi, 26A,* no. 17, 1–20.

Säve-Söderbergh, G. (1945). Notes on the trigeminal musculature in non-mammalian tetrapods. *Nova Acta Regiae Societatis Scientiarum Upsaliensis,* (4), *13,* 1–59.

Scadding, S.R. (1981). Do "vestigial organs" provide evidence of evolution? *Evolutionary Theory, 5,* 173–6.

Schaeffer, B., Hecht, M.K., and Eldredge, N. (1972). Phylogeny and paleontology. *Evolutionary Biology, 6,* 31–46.

Schank, J.C., and Wimsatt, W.C. (1987). Generative entrenchment and evolution. In *Proceedings of the 1986 Biennial Meeting of the Philosophy of Science Association,* vol. 2, ed. A. Fine and P. Machamer, pp. 33–60. East Lansing, Mich: P.S.A.

Schindewolf, O.H. (1950). *Grundfragen der Paläontologie.* Stuttgart: Schweizerbart.

Schmid, C.W., and Marks, J. (1990). DNA hybridization as a guide to phylogeny: chemical and physical limits. *Journal of Molecular Evolution, 30,* 237–46.

Schmitt, F.O. (1984). Molecular regulators of brain function: a new view. *Neuroscience, 13,* 991–1001.

Sclater. P.L. (1858). On the general geographical distribution of the members of the class Aves. *Journal of the Linnean Society of London, 2,* 130–45.

Scott-Ram, N.R. (1990). *Transformed Cladistics, Taxonomy and Evolution.* Cambridge: Cambridge University Press.

Scrutton, C.T. (1988). Patterns of extinction and survival in Palaeozoic corals. In *Extinction and Survival in the Fossil Record,* ed. G.P. Larwood, pp. 65–88. Oxford: Oxford University Press, for the Systematics Association.

Sepkoski, J.J. (1982). A compendium of fossil marine families. *Milwaukee Public Museum, Contributions in Biology and Geology, 51,* 1–125.

Serres, E.R.A. (1830). Anatomie transcendante – quatrième memoire: loi de symétrie et de conjugasion de système sanguin. *Annales des Sciences Naturelles,* (1st ser.), *21,* 5–49.

Sheldon, F.H. (1987). Rates of single-copy DNA evolution in herons. *Molecular Biology and Evolution, 4,* 56–69.

Sheldon, P.R. (1987). Parallel gradualistic evolution of Ordovician trilobites. *Nature, 330,* 561–3.

Sibley, C.G., and Ahlquist, J.E. (1984). The phylogeny of the hominoid primates, as indicated by DNA–DNA hybridization. *Journal of Molecular Evolution, 20,* 2–15.

Sibley, C.G., and Ahlquist, J.E. (1986). Reconstructing bird phylogeny by comparing DNAs. *Scientific American, 254,* no. 2, 68–78.

Sibley, C.G., and Ahlquist, J.E. (1987). Avian phylogeny reconstructed from comparisons of the genetic material, DNA. In *Molecules and Morphology in Evolution: Conflict or Compromise?,* ed. C. Patterson, pp. 95–122. Cambridge: Cambridge University Press.

Simkins, J., and Williams, J.I. (1989). *Advanced Biology,* 3rd edn. London: Unwin Hyman.

Simpson, G.G. (1936). Data on the relationships of local and continental mammalian faunas. *Journal of Paleontology, 10,* 410–4.

Simpson, G.G. (1940). Mammals and land bridges. *Journal of the Washington Academy of Sciences, 30,* no. 4, 137–63.

Simpson, G.G. (1944). *Tempo and Mode in Evolution.* New York: Columbia University Press.

Simpson, G.G. (1945). The principles of classification and a classification of mammals. *Bulletin of the American Museum of Natural History, 85,* 1–350.

Simpson, G.G. (1951). *Horses.* New York: Oxford University Press.

Simpson, G.G. (1953). *The Major Features of Evolution.* New York: Columbia University Press.

Simpson, G.G. (1961a). *Principles of Animal Taxonomy*. New York: Columbia University Press.

Simpson, G.G. (1961b). Lamarck, Darwin and Butler, three approaches to evolution. *American Scholar, 30,* 238–49.

Simpson, G.G. (1965). *The Geography of Evolution*. Philadelphia: Chilton Brooks.

Smith, A.B., and Patterson, C. (1988). The influence of taxonomic method on the perception of patterns of evolution. *Evolutionary Biology, 13,* 127–216.

Smith, S. (1965). The Darwin collection at Cambridge with one example of its use: Charles Darwin and *Cirripedes*. *Actes du XIe Congrès International d'Histoire des Sciences, 5,* 96–100.

Smith, W. (1816–19). *Strata Identified by Organised Fossils,* 4 vols. London: W. Aarding.

Sneath, P.H.A. (1957a). Some thoughts on bacterial classification. *Journal of General Microbiology, 17,* 184–200.

Sneath, P.H.A. (1957b). The application of computers to taxonomy. *Journal of General Microbiology, 17,* 201–26.

Sneath, P.H.A. (1961). Recent developments in theoretical and quantitative taxonomy. *Systematic Zoology, 10,* 118–39.

Sneath, P.H.A. (1962). The construction of taxonomic groups. In *Microbial Classification (Symposia of the Society for General Microbiology No. 12),* ed. G.C. Ainsworth and P.H.A. Sneath, pp. 289–332. Cambridge: Cambridge University Press.

Sneath, P.H.A., and Sokal, R.R. (1962). Numerical taxonomy. *Nature, 193,* 855–60.

Sneath, P.H.A., and Sokal, R.R. (1973). *Numerical Taxonomy: The Principles and Practice of Numerical Classification*. San Francisco: W.H. Freeman.

Snider-Pellegrini, A. (1858). *La Création et ses Mystères Dévoilés*. Paris: Franck et Dentu.

Sober, E. (1983). Parsimony in systematics: philosophical issues. *Annual Review of Ecology and Systematics, 14,* 335–58.

Sober, E. (1984). *The Nature of Selection: Evolutionary Theory in Philosophical Focus*. Cambridge, Mass.: M.I.T. Press.

Sober, E. (1985). A likelihood justification of parsimony. *Cladistics, 1,* 209–33.

Sober, E. (1988). *Reconstructing the Past: Parsimony, Evolution and Inference*. Cambridge, Mass.: M.I.T. Press.

Sokal, R.R. (1983). A phylogenetic analysis of the Caminalcules. I. The data base. *Systematic Zoology, 32,* 159–84.

Sokal, R.R. (1986). Phenetic taxonomy: theory and methods. *Annual Review of Ecology and Systematics, 17,* 423–42.

Sokal, R.R., and Michener, C.D. (1958). A statistical method for evaluating systematic relationships. *University of Kansas Science Bulletin, 38,* 1409–38.

Sokal, R.R., and Sneath, P.H.A. (1963). *Principles of Numerical Taxonomy*. San Francisco: W.H. Freeman.

Spencer, H. (1864). *The Classification of the Sciences: To Which Are Added Reasons for Dissenting from the Philosophy of M. Comte*. London: for the author.

Spinoza, B. (1677). *The Ethics,* 1st Latin edn. Amsterdam. [republished London: Dent 1910].

Springer, M., and Krajewski, C. (1989). DNA hybrization in animal taxonomy: a critique from first principles. *Quarterly Review of Biology, 64,* 291–318.

Stanley, S.M. (1975). A theory of evolution above the species level. *Proceedings of the National Academy of Sciences, U.S.A., 72,* 646–50.

Stanley, S.M. (1987). *Extinction.* San Francisco: W.H. Freeman (Scientific American Library).

Stebbins, G.L. (1987). Species concepts: semantics and actual situations. *Biology and Philosophy, 2,* 198–203.

Stebbins, G.L., and Ayala, F.J. (1981). Is a new evolutionary synthesis necessary? *Science, 21,* 967–71.

Steele, E.J. (1979). *Somatic Selection and Adaptive Evolution: On the Inheritance of Acquired Characters.* Toronto: Williams & Wallace.

Steno, N. (1667). *Elementorum myologiae specimen, seu musculi descriptio geometrica, cui accedunt canis carchariae dissectum caput, et dissectus piscus ex canum genere.* Florentiae.

Steno, N. (1669). *The prodromus to a dissertation concerning solids naturally contained within solids. Laying a foundation for the rendering a rational accompt both of the frame and the several changes of the masse of the earth, as also of the various productions in the same. "English'd by H.O.".* London, 1671.

Stent, G.S. (1986). Glass bead game: a review of Alexander Rosenberg, *The Structure of Biological Science. Biology and Philosophy, 1,* 227–47.

Stevens, P.F. (1980). Evolutionary polarity of character states. *Annual Review of Ecology and Systematics, 11,* 333–58.

Stigler, S.M., and Wagner, M.J. (1987). A substantial bias in non-parametric tests for periodicity in geophysical data. *Science, 283,* 940–5.

Stove, D.C. (1982). *Popper and After: Four Modern Irrationalists.* Oxford: Pergamon Press.

Strickland, H.E. (1841). On the true method of discovering the natural system in zoology and botany. *The Annals and Magazine of Natural History* (1st ser.), *6,* 184–94.

Stringer, C.B., and Andrews, P. (1988). Genetic and fossil evidence for the origin of modern humans. *Science, 239,* 1263–8.

Swainson, W. (1834). *A Preliminary Discourse on the Study of Natural History (The Cabinet Cyclopaedia, vol. 59,* ed. D. Lardner). London: Longman, Rees, Orme, Brown & Green . . .

Swainson, W. (1835). *A Treatise on the Geography and Classification of Animals (The Cabinet Cyclopaedia, vol. 66,* ed. D. Lardner). London: Longman, Rees, Orme, Brown & Green . . .

Syvanen, M. (1987). Molecular clocks and evolutionary relationships: possible distortions due to horizontal gene flow. *Journal of Molecular Evolution, 26,* 16–23.

Tattersall, I., and Eldredge, N. (1977). Fact, theory, and fantasy in human paleontology. *American Scientist, 65,* 204–11.

Tauber, C.A., and Tauber, M.J. (1989). Sympatric speciation in insects: perception and perspective. In *Speciation and Its Consequences,* ed. D. Otte and J.A. Endler, pp. 307–44. Sunderland, Mass.: Sinauer Associates.

Teilhard de Chardin, P. (1959). *The Phenomenon of Man,* transl. B. Wall (Introd. J.S. Huxley). London: Collins.
Templeton, A.R. (1981). Mechanisms of speciation – a population genetic approach. *Annual Review of Ecology and Systematics, 12,* 23–48.
Templeton, A.R. (1985). The phylogeny of the hominoid primates: a statistical analysis of the DNA–DNA hybridization data. *Molecular Biology and Evolution, 2,* 420–33.
Templeton, A.R. (1989). The meaning of species and speciation: a genetic perspective. In *Speciation and Its Consequences,* ed. D. Otte and J.A. Endler, pp. 3–27. Sunderland, Mass.: Sinauer Associates.
Thompson, D'A.W. (1917). *On Growth and Form.* Cambridge: Cambridge University Press.
Thompson, D'A.W. (1961). *On Growth and Form,* abridged edn., ed. J.T. Bonner. (citation from paperback edn., 1966). Cambridge: Cambridge University Press.
Thompson, E.A. (1975). *Human Evolutionary Trees.* Cambridge: Cambridge University Press.
Thorpe, R.S. (1976). Biometrical analysis of geographical variation. *Biological Reviews of the Cambridge Philosophical Society, 51,* 407–52.
du Toit, A.L. (1937). *Our Wandering Continents.* Edinburgh: Oliver & Boyd.
Trusted, J. (1979). *The Logic of Scientific Inference, an Introduction.* London: Macmillan.
Uzzell, T.M., Gunther, R., and Berger, L. (1977). *Rana ridibunda* and *Rana esculenta:* a leaky hybridogenetic system (Amphibia, Salientia). *Proceedings of the Academy of Natural Sciences, Philadelphia, 128,* 147–71.
Valentine, J.W. (1986). Fossil record of the origin of *Baupläne* and its implications. In *Patterns and Processes in the History of Life (Dahlem Konferenzen, Life Sciences Research Report 36),* ed. D.M. Raup and D. Jablonski, pp. 209–22. Berlin: Springer-Verlag.
Van Valen, L. (1973). A new evolutionary law. *Evolutionary Theory, 1,* 1–30.
Van Valen, L. (1976). Ecological species, multispecies and oaks. *Taxon, 25,* 233–9.
Vernon, K. (1988). The founding of numerical taxonomy. *British Journal for the History of Science, 21,* 143–59.
Vicq-d'Azyr, F. (1792). *Système Anatomique des Quadrupèdes: Encyclopédie Méthodique,* vol. 2. Paris: Baudouin.
Vine, F.J., and Matthews, D.H. (1963). Magnetic anomalies over oceanic ridges. *Nature, 199,* 947–9.
Vrba, E.S. (1984). What is species selection? *Systematic Zoology, 33,* 318–28.
Vrba, E.S., and Eldredge, N. (1984). Individuals, hierarchies and processes: towards a more complete evolutionary theory. *Paleobiology, 10,* 146–71.
de Vries, H. (1889). *Intracelluläre Pangenesis.* Jena: Gustav Fischer.
de Vries, H. (1901–3). *Die Mutationstheorie.* Leipzig: von Veit.
Waddington, C.H. (1957). *The Strategy of the Genes.* London: Allen & Unwin.
Waddington, C.H. (1960). Evolutionary adaptation. In *Evolution after Darwin, The University of Chicago Centennial, vol. 1: The Evolution of Life,* ed. S. Tax, pp. 381–402. Chicago: University of Chicago Press.

Wagner, W.H. (1961). Problems in the classification of ferns. *Recent Advances in Botany,* vol. 1, pp. 841–4. Toronto: University of Toronto Press.
Wagner, W.H. (1968). Hydridization, taxonomy, and evolution. In *Modern Methods in Plant Taxonomy,* ed. V.H. Heywood, pp. 113–38. New York: Academic Press.
Wagner, W.H. (1969). The role and taxonomic treatment of hybrids. *Bioscience, 19,* 785–9.
Wagner, W.H. (1983). Reticulistics: the recognition of hybrids and their role in cladistics and classification. In *Advances in Cladistics,* vol. 2, ed. N.I. Platnick and V.A. Funk, pp. 63–79. New York: Columbia University Press.
Walker, W.F. (1989). [In "Phylogeny and molecular data"]. *Science, 243,* 548–9.
Wallace, A.R. (1855). On the law which has regulated the introduction of new species. *Annals and Magazine of Natural History* (n.s.), *16,* 184–96.
Wallace, A.R. (1858). On the tendency of varieties to depart indefinitely from the original type. In Darwin and Wallace (1958), pp. 268–79.
Wallace, A.R. (1863). On the physical geography of the Malay Archipelago. *Journal of the Royal Geographical Society, 33,* 217–34.
Wallace, A.R. (1870). *Contributions to the Theory of Natural Selection.* London: Macmillan.
Wallace, A.R. (1876). *The Geographical Distribution of Animals, with a Study of the Relations of Living and Extinct Faunas as Elucidating the Past Changes of the Earth's Surface,* 2 vols. London: Macmillan.
Wanntorp, H.-E. (1983). Reticulated cladograms and the identification of hybrid taxa. In *Advances in Cladistics,* vol. 2, ed. N.I. Platnick and V.A. Funk, pp. 81–8. New York: Columbia University Press.
Wassermann, G.D. (1981). On the nature of the theory of evolution. *Philosophy of Science, 48,* 416–37.
Watrous, L.E., and Wheeler, Q.D. (1981). The outgroup comparison method of character analysis. *Systematic Zoology, 30,* 1–11.
Watson, D.M.S. (1951). *Paleontology and Modern Biology.* New Haven, Conn.: Yale University Press.
Watson, D.M.S. (1956). The brachyopid labyrinthodonts. *Bulletin of the British Museum (Natural History): Geology, 2,* 315–91.
Watson, J.D., and Crick, F.H.C. (1953a). A structure for deoxyribose nucleic acid. *Nature, 171,* 737–8.
Watson, J.D., and Crick, F.H.C. (1953b). Genetical implications of the structure of deoxyribonucleic acid. *Nature, 171,* 964–7.
Watson, J.D., and Crick, F.H.C. (1954). The structure of DNA. *Cold Spring Harbor Symposia on Quantitative Biology, 18,* 123–31.
Watson, J.D., Hopkins, N.H., Roberts, J.W., Steitz, J.A., and Weiner, A.M. (1987). *Molecular Biology of the Gene,* 2 vols., 4th edn. Menlo Park, Calif.: Benjamin, Cummings.
Wegener, A.L. (1915). *Die Entstehung der Kontinente und Ozeane.* Brunswick: Friederich Vieweg & Sohn.
Wegener, A.L. (1929). *The Origin of Continents and Oceans,* transl. J. Biram from 4th German edn. New York: Dover, 1966.

Weichert, C.K. (1965). *Anatomy of the Chordates,* 3rd edn. New York: McGraw-Hill.

Weinberg, R.A. (1985). The molecules of life. *Scientific American, 253,* no. 4, 34–43.

Weismann, A. (1892). *Das Keimplasma: Eine Theorie der Vererbung.* Jena: Gustav Fischer. (English: 1893. *The Germ Plasm: A Theory of Heredity,* transl. W.N. Parker and H. Ronfeldt. London: Walter Scott).

West, D.A. (1977). Melanism in *Biston* (Lepidoptera: Geometridae) in the rural Appalachians. *Heredity, 39,* 75–81.

Weston, P.H. (1988). Indirect and direct methods in systematics. In *Ontogeny and Systematics,* ed. C.J. Humphries, pp. 27–56. London: British Museum (Natural History).

Whewell, W. (1840). *The Philosophy of the Inductive Sciences, Founded upon Their History,* 2 vols. London: John W. Parker & Sons.

Whewell, W. (1849). *Of Induction, With Especial Reference to Mr. J. Stuart Mill's System of Logic.* London: John W. Parker & Sons.

Whewell, W. (1858). *Novum Organon Renovatum (Being the Second Part of the Philosophy of the Inductive Sciences,* 3rd edn). London: John W. Parker & Sons.

White, M.J.D. (1978). *Modes of Speciation.* San Francisco: W.H. Freeman.

Whitehead, A.N. (1929). *Process and Reality: An Essay in Cosmology (Gifford Lectures: Edinburgh, 1927–1928).* Cambridge: Cambridge University Press.

Whitmire, D.P., and Jackson, A.A. (1984). Are periodic mass extinctions driven by a distant solar companion? *Nature, 308,* 713–15.

Whitmire, D.P., and Matese, J.J. (1985). Periodic comet showers and planet X. *Nature, 313,* 36–8.

Whittaker, V.P. (1984). What is Dale's Principle? In *Coexistence of Neuroactive Substances in Neurons,* ed. V. Chan-Palay and S.L. Palay, pp. 137–40. New York: John Wiley & Sons.

Wicken, J.S. (1987). *Evolution, Thermodynamics and Information.* Oxford: Oxford University Press.

Wiley, E.O. (1979a). Cladograms and phylogenetic trees. *Systematic Zoology, 28,* 88–92.

Wiley, E.O. (1979b). Ancestors, species, and cladograms – remarks on the symposium. In *Phylogenetic Analysis and Paleontology,* ed. J. Cracraft and N. Eldredge, pp. 211–25. New York: Columbia University Press.

Wiley, E.O. (1979c). An annotated Linnean hierarchy, with comments on natural taxa and competing systems. *Systematic Zoology, 28,* 308–37.

Wiley, E.O. (1981). *Phylogenetics: The Theory and Practice of Phylogenetic Systematics.* New York: John Wiley & Sons.

Williams, G.C. (1966). *Adaptation and Natural Selection.* Princeton: Princeton University Press.

Williams, M.B. (1985). Species are individuals: theoretical foundations for the claim. *Philosophy of Science, 52,* 578–89.

Willmer, P. (1990). *Invertebrate Relationships: Patterns in Animal Evolution.* Cambridge: Cambridge University Press.

Wilson, A.C., Cann, R.L., Carr, S.M., George, M., Gyllensten, U.B., Helm-

Bychowski, K.M., Higuchi, R.G., Palumbi, S.R., Prager, E.M., Sage, R.D., and Stoneking, M. (1985). Mitochondrial DNA and two perspectives on evolutionary genetics. *Biological Journal of the Linnean Society, 26,* 375–400.

Wilson, E.O. (1965). A consistency test for phylogenies based on contemporaneous species. *Systematic Zoology, 14,* 214–20.

Wilson, E.O., and Bossert, W.H. (1971). *A Primer of Population Biology.* Stamford, Conn.: Sinauer Associates.

Wimsatt, W.C. (1986). Developmental constraints, generative entrenchement, and the innate-acquired distinction. In *Integrating Scientific Disciplines,* ed. W. Bechtel, pp. 185–208. Dordrecht: Martinus-Nijhoff.

Winsor, M.P. (1976). *Starfish, Jellyfish, and the Order of Life: Issues in Nineteenth-Century Science.* New Haven: Yale University Press.

Wittgenstein, L. (1922). *Tractatus Logico-Philosophicus* (Introd. B. Russell). London: Paul, Trench, Trubner.

Wolbach, W.S., Gilmour, I., Anders, E., Orth, C.J., and Brooks, R.R. (1988). Global fire at the Cretaceous–Tertiary boundary. *Nature, 334,* 665–9.

Wolbach, W.S., Lewis, R.S., and Anders, E. (1985). Cretaceous extinctions: evidence for wildfires and search for meteoric material. *Science, 230,* 167–70.

Wolpert, L. (1969). Positional information and the spatial pattern of cellular differentiation. *Journal of Theoretical Biology, 25,* 1–47.

Wolpert, L. (1971). Positional information and pattern formation. *Current Topics in Developmental Biology, 6,* 183–224.

Wright, S. (1931). Evolution in Mendelian populations. *Genetics, 16,* 97–159.

Wright, S. (1967). Comments on the preliminary working papers of Eden and Waddington. In *Mathematical Challenges to the Neo-Darwinian Theory of Evolution (Wistar Institute Symposia no. 5),* ed. P.S. Moorehead and M.M. Kaplan, pp. 117–20.

Wuketits, F.M. (1986). Evolution as a cognition process; towards an evolutionary epistemology. *Biology and Philosophy, 1,* 191–206.

Wynne-Edwards, V.C. (1962). *Animal Dispersion in Relation to Social Behaviour.* Edinburgh: Oliver & Boyd.

Wynne-Edwards, V.C. (1986). *Evolution by Group Selection.* Oxford: Blackwell Scientific.

Yule, G.U. (1902). Mendel's Laws and their probable relations to intra-racial heredity. *New Phytologist, 1,* 193–207, 222–38.

Zuckerhandl, E., and Pauling, L. (1965). Evolutionary divergence and convergence in proteins. In *Evolutionary Genes and Proteins,* ed. V. Bryson and H.J. Vogel, pp. 97–166. New York: Academic Press.

Author index

Subject index